Karl Schlagenhauf   Fanji Gu

The Brain and AI
Correspondence between a German Engineer and a Chinese Scientist

Volume III

The Third Spring of AI

SHANGHAI EDUCATIONAL PUBLISHING HOUSE

The Authors

Karl Schlagenhauf

Karl Schlagenhauf is a serial entrepreneur in the field of new technologies and has founded numerous start-ups in the high-tech sector in Europe and the USA. He was CEO of IFAO (Institute for applied organizational research) and ADI Software, a company with a focus on RDBMS, Multimedia and Internet-Applications in industry and banking, both of which he spun off from the University of Karlsruhe in the 1980s.

He stepped down as CEO in 2003, and is now chairman of the board of ADI Innovation and runs a Family Office. He also serves on the board of companies where he is invested and has served on the boards of several technology companies such as AP Automation + Productivity (now Asseco Solutions), Brandmaker, CAS Software, JPK Instruments and Web.de.

As an inventor he holds patents in the field of secure remote control via the internet and Nano-robots for protein analysis based on atomic force spectroscopy.

He is an advisor to private equity firms and governmental bodies and also a coach to young entrepreneurs.

While his focus has always been on leading edge technologies and their impact on social systems and the delicate orchestration of human, intellectual und financial resources, he has extended his interest from software, electronics and manufacturing to the life sciences and artificial intelligence.

He holds a Master's degree in Economics/Industrial Engineering, a PhD in Philosophy and a "Venia Legendi" for Sociology and Theory of Science from University of Karlsruhe (now KIT).

The Authors

Fanji Gu

Prof. Fanji Gu is an emeritus professor of computational neuroscience at the School of Life Sciences, Fudan University, Shanghai, China.

He graduated in mathematics from Fudan University in 1961, worked in Department of Biophysics, Science & Technology University of China from 1961 to 1979, then worked in School of Life Sciences, Fudan University since 1979. He worked as a visiting research associate in the Department of Physiology & Biophysics, University of Illinois at Urbana-Champaign from 1983 to 1985.

His major is computational neuroscience. He published 3 monographs and about 100 papers. After his retirement in 2004, he was the managing editor of the journal *Cognitive Neurodynamics* from 2006 to 2011. He edited three proceedings of international conferences.

After that he became a popular science writer on brain science. He wrote 6 popular science books on brain science and translated seven books, including Walter Freeman's *Neurodynamics*, Christof Koch's *Quest for Consciousness*, Gerald Edelman's *A Universe of Consciousness* and V. S. Ramachandran's *Phantoms in the Brain*. His popular science books won seven awards, including the 2017 best seller in China, the 2016 Shanghai Municipal Award of Science and Technology (science popular book, third grade), the 2015 Shanghai Science Popularization Education Innovation Award (science popular book, second grade) etc. He himself also won a Merit Award conferred by the 2013 Fourth International Conference on Cognitive Neurodynamics held in Sweden, the 2017 Shanghai Science Popularization Education Innovation Award (individual contribution, second grade), and the 2018 Shanghai Award for Science & Technology (third grade).

He is now the honorary president of Shanghai Chapter of University of Illinois Alumni Association and a member of the editorial board of WeChat's subscription channel "Fanpu".

Recommendation

This book is a special treat. It has something to offer to everyone who is curious about the nature of human insight and how we can make progress on its understanding.

——Matthias Bethge (Chair of the Bernstein Center for Computational Neuroscience, Professor, University Tübingen and cofounder of deepart.io)

An unusual book which clearly stands out from many other neuroscience publications. It should be recommended to all IT experts and neuroscientists who wish to escape well-trodden paths of Main Stream research.

——Hans Albert Braun (Professor of Philipps University of Marburg, Head of the Neurodynamics Group, Institute of Physiology and Pathophysiology)

A fascinating dialogue of a German engineer and a Chinese biophysicist. We should thank the authors for publishing their in-depth discussion on AI, brain activity and consciousness.

——Yizhang Chen (Academician of Chinese Academy of Sciences, Professor of The Second Military Medicine University)

I have read "The Brain & AI" completely spellbound and would be on fire to film it one day.

——Serdar Dogan (German Director and Filmmaker)

An intellectual delight like very few I have encountered. It is treasure trove of ideas and insights, sometimes playful, sometimes deeply serious, yet always optimistic.

——Dietmar Harhoff (Director at the Max Planck Institute for Innovation and Competition in Munich)

Chinese venerable wisdom and expertise in neurophysiology meets Western social science plus information technology. "Do we need such a book?" My answer is definitely yes.

——Gert Hauske (Retired Professor at Technische Universität München, former Editor-in-chief of *Biological Cybernetics*)

I really hope Fanji Gu's and Karl Schlagenhauf's book, so full of original thoughts and insights, sparks new ways of collaboration between science, politics and business, to create the next wave of "real" A(G)I to the benefit of all human beings.

——Rafael Laguna (CEO of Open-Xchange, Director of the German Agency for Disruptive Innovation)

With wisdom and vision the authors bring readers together to discuss the leap and myth behind the age of intelligence.

——Peiji Liang (Director of the Chinese Neuroscience Society, Professor of Shanghai Jiaotong University)

Perhaps you are just interested in artificial intelligence and the human brain, or you have delved into a specific related area and would like to see the bigger picture, or you just want to learn something fun in your spare time. No matter what the case is, I believe that this book will bring you unique and fun experiences.

——Amanda Song (PhD candidate, University of California, San Diego)

A book to arouse thinking. All readers interested in brain science and artificial intelligence deserve to read this.

——Xiaowei Tang (Academician of Chinese Academy of Sciences, Professor of Zhejiang University)

A model of how to conduct scientific discussions. All interested readers in brain science, cognitive science, artificial intelligence and the philosophy of science will love this collision of views and ideas. I am convinced that this dialog is one of the most profound books on human brain and artificial intelligence published in recent years.

——Pei Wang (Associate Professor of Temple University, Vice Chair of Artificial General Intelligence Society, Chief Executive Editor, *Journal of Artificial General Intelligence,* one of the Translators of *Gödel, Escher, Bach: An Eternal Golden Braid*)

A science book as thrilling as an Agatha Christie Crime Novel.
The style of the "old debate culture of scientists", i.e. respectfully exchanging arguments in letters and developing knowledge based on solid information of arguments, is very refreshing. We should not unlearn these skills, they are extremely important in an era of twitter and short messages.

——Ehrenfried Zschech (Department Head, Fraunhofer Institute for Ceramic Technologies and Systems IKTS, Professor, Member of the Senate, European Materials Research Society)

Introduction 1

"AI, Science and Everything" would also have been an appropriate title for this refreshing dialog between a Chinese brain scientist and a German "all-round" engineer. Fanji Gu and Karl Schlagenhauf cover a lot of ground and tease out new and original insights and thoughts relating to some of the key issues facing proponents and practitioners working in AI, science, policy making, business and politics, but also young, aspiring humans that want to understand where the puck is going, not where it is.

This fascinating read doesn't pull any punches when it comes to reviewing some of the more controversial discussion taking place and is a good antidote to the AI hype often disseminated by popular media and the associated spin doctors.

The authors take us on a far-ranging journey starting with a look at issues in scientific methodology, through problems and shifts in brain research, and the way AI discussions are framed, and after a quick grapple with the amorphous concept of consciousness finish on the thorny issue of "productivity", in this sense being how best to accelerate progress in academia by creating an environment where resources are not wasted on false starts and incremental progress on overly broad and philosophical diversions, but guiding activity so that breakthroughs in research in Germany and Europe can be applied effectively in a business context and improve the competitiveness of the region as both the US and China ramp up spending on the AI.

Although the authors do not discuss the issue in great detail, they make clear that attention is required along the value chain in order to bring AI solutions to market and that knowledge transfer and cooperation between business and academia is an essential part of this and helps cut through some of the VC- and media-driven hyperbole.

There are already a number of business approaches which could be adapted to meet these needs, the open source software and hardware approach springs to mind, and a more detailed chapter on this topic would be a great addition to the next edition, that I am very much looking forward to!

The observations that "a huge amount of research doesn't pile up to [create] a theory that helps us explain and understand the functioning of the mind ..." and the lack of major breakthroughs in the field over the last fifty years or so is an indication that alternative models/approaches are duly missed. This gap between expectations and practical application allows space for a lot of exaggerated claims and "terminator" type positioning of any progress in the press. However, as the authors point out, there is now a window of opportunity to correct this, considering for example the significant funding targeted at accelerating progress that has been made available through the Human Brain Project (HBP) that may be ill-spent, as we know too little of how the brain actually works to simulate it. But it proves that break-through amounts of money can be made available to create leaps in progress and knowledge.

In Europe, anyone proposing solutions based even loosely on a free market or neo-liberal approach runs the risk of being accused of cheerleading, or down playing the negative aspects, but in this case maybe we can learn something from the success of this model in terms of driving rapid technology dispersion and uptake while adopting a more balanced methodology which mitigates against some of the more excessive and exaggerated claims made for AI.

There are numerous examples coming out of Silicon Valley and other tech centers around the globe, and while there are positive and negative aspects to this (data security & abuse, fake news, echo chamber effects and unscrupulous marketing and advertising practices to name a few) the social & political impact of IT and the positive impact it has on so many people when looked at through the lens of ICT's for development is undeniable. As the genie is already out of the bottle in terms of the use and abuse of technology in the commercial and public spheres then maybe a new synthesis/coalition of academic and commercial interest and know-how can provide a "better" solution to some of these negative impacts. Particularly by accelerating the creation of, and demand for, "intelligent systems" across the machine learning, expert system, AI spectrum.

I really hope Fanji and Karl's book sparks new ways of collaboration between science, politics and business, to create the next wave of "real" A(G)I to the benefit of all human beings.

Rafael Laguna

CEO of Open-Xchange

Director of the German Agency for Disruptive Innovation

Introduction 2

It is my great honor to accept the invitation of Mr. Fanji Gu and Mr. Karl Schlagenhauf to comment on the collection of their correspondence.

Last year I read the emails between the two senior researchers with a refreshing feeling. The virtue that touched me the most was the scientific attitude and the gentleman's manner they showed in the discussion. Both are well-established figures, but they still have humble hearts with respect to science, and are sensitive and curious about the progress in the related fields. They are willing to learn new knowledge with an open mind, and at the same time remain skeptical about claims that have not been supported by enough evidence. They show full respect to each other in the discussion, while do not avoid confrontation in opinions. They do not easily give up their own beliefs, nor do they feel that being convinced by the other party is a shame. They desire to contribute to the enterprise of science, but are not aimed at recent fame and fortune. Isn't this the attitude people should keep in conducting scientific discussions?

The qualities mentioned above can be said to be both widely known and rarely found, especially in discussions related to artificial intelligence (AI). In recent years, this field has suddenly changed from deep winter to hot summer, and various types of character have joined the show. On this crowded stage, there are many monologues, and the occasional dialogues are often either mutual flattery or mutual belittlement, while it is unusual to see the two parties to carefully consider the other side's opinions and to respond frankly. As for the opinions expressed, they often follow either intuition or fashion, rather than result from independent thought after considering the previous works.

This situation is certainly caused by the overall atmosphere of the academia, but these bizarre phenomena are more commonplace in AI, as they are also due to the particularity of this field. On the one hand, it is because the puzzle of thinking is widely recognized as one of the most complicated problems in science, and in this field, there is still no well-tested research paradigm to become a consensus. On the other hand, out of direct perception of the phenomenon of thinking, everyone seems to have strong opinions. A particularly remarkable phenomenon is the "outsider" opinions from celebrities, as if smart people naturally know what "smartness" is about. Within the AI community, due to the multifaceted nature and historical changes of the concept of "AI", many "AI Experts" are only familiar with a certain subarea on which they earned their fame, and even a few years ago they still didn't consider themselves as "AI researchers", but now they are giving "authoritative opinions" on topics that they have not really studied. Given all these, it is no wonder that the academic circles here began to be indistinguishable from that of business, politics, and even entertainment.

It is exactly because that I am not optimistic about the current "prosperous scene" of AI, that I appreciate the role models provided by these two senior researchers. Specific views can be debated, but if scientific issues cannot be discussed in a scientific manner, the current AI fever can only lead to winter again. I hope that readers (especially youth who are interested in science) will learn not only scientific knowledge from this book, but also the correct attitude and manner when discussing scientific issues. I am convinced that this dialog is one of the most profound books on human brain and artificial intelligence published in recent years.

Pei Wang

Associate Professor

Temple University

Vice Chair

Artificial General Intelligence Society

Chief Executive Editor

Journal of Artificial General Intelligence

Introduction 3

During the holiday season 2018/2019, coincidentally, I read Agatha Christie's crime novel "*Death on the Nile*" and the manuscript of the correspondence between Karl Schlagenhauf and Fanji Gu "*The brain and AI*" simultaneously. Both were in the same way thrilling for me, because of the gripping contents and the style of writing.

The topic of science book "*The brain and AI*" is of timely societal relevance! The authors are addressing it very openly, supported by numerous facts from brain research and artificial intelligence, demonstrating their solid knowledge in these fields. I am not an expert in these fields, however, as solid-state physicist and materials scientist I touched these topics several times during my research activities in the previous years, when I was studying materials and processes for neuromorphic systems or when I applied machine learning algorithms in advanced data analysis in 3D imaging from high-resolution tomography and in spectroscopy data processing. I am convinced that AI approaches will be very useful to get better and faster information, however, the strategy should be to combine both, solid knowledge based on mathematics, natural science and engineering on the one hand and (big) data-driven informatics on the other hand.

The style of the "old debate culture of scientists", i.e. respectfully exchanging arguments in letters, in a logic approach (rational causally thinking, as we physicists were trained), step by step, and developing knowledge based on solid information, was very refreshing for me. We should not unlearn these skills, they are extremely important in an era of twitter and short messages.

I hope that a large number of people will enjoy reading this book and will be inspired to contribute to a discussion of the up-to-date topics addressed by the authors — in the same way as the book was written, i.e. passionately and respectful.

Ehrenfried Zschech

Prof. Dr.

Department Head

Fraunhofer Institute for Ceramic Technologies and Systems IKTS

Member of the Senate, European Materials Research Society

Preface

This book is special, if not unique. We, the two authors, were raised in different cultural environments and have never met face to face. Karl an engineer and entrepreneur, is German, and Fanji a scientist and popular science writer on brain science, is Chinese. A mutual friend, Prof. Hans Braun, a neuro-physiologist, introduced us nearly 6 years ago and we have been corresponding since on the subjects of brain research and artificial intelligence (AI).

We have become good friends and correspond frequently, in some cases even more frequently than with some of our old friends. Our common interest in the puzzles of the brain, the mind, consciousness and AI binds our friendship, and with great enthusiasm we pay close attention to the rapid progress in these fields.

We prefer the method of rational thinking, and are always eager to scrutinize the whys and wherefores of everything, as opposed to following mainstream or scholastic thinking. Our points of view are significantly different due to our experience — Karl worked in industry and Fanji in academia.

Fanji's focus was on creating knowledge and how to teach it while Karl was engaged in how to make use of knowledge via technological applications.

After several decades of working in interdisciplinary fields with various technologies we both have reached the official age of retirement. However, scientists and entrepreneurs never retire and now we can use our time and experience to indulge our passion, and enjoy the freedom that we don't have to nurture a career anymore or impress our peers. It is a wonderful privilege to follow your interests without such restrictions and we both enjoy this extraordinarily.

As Karl has often said "we are like two free and happy birds that can sit on any tree they like and discuss whatever they are interested in".

But that doesn't mean that we have no goals or high level objectives.

First of all, it is our own need and ambition to understand as well as possible what is happening in our complicated fields of interest and to assess how they are likely to develop.

Second, Fanji has a large audience of interested and discerning readers and Karl has a number of people (young scientists, engineers and entrepreneurs and also managers in industry and government) asking his advice on these topics. Both groups can legitimately expect that what we say is neither careless nor superficial.

Our discussion started with a question Fanji had asked Hans Braun about the relevance of a specific type of neuron, after studying the technical concept of the later famous European HBP (Human Brain Project). Hans had redirected this question to Karl and so our little journey began in January 2013. It resulted in a series of e-mails which led us from neurons to the latest developments in AI, and what some people have called the technology and trade-war between China and the US.

This book is an organized collection of our correspondence. It consists of letters written in the tradition of the old debate culture in which scientists exchanged their arguments in elaborate letters and controversial discussions. Of course we didn't invent this method. In fact, it was the normal way scientists communicated and refined their ideas a hundred years ago in the golden age of science. It just looks somewhat outdated in the time of Twitter, and short message services, where everything has to be expressed in headlines, available in seconds and is quickly said and forgotten.

The adversarial tone in our letters may sound a little strange to young scientists more accustomed to a culture of consensus. It should be mentioned however that adversarial methods are very promising techniques now being introduced in the most advanced applications of artificial neural networks. To communicate in the old-fashioned adversarial way can be very time consuming and demanding, but it can also be very productive and rewarding for those who like to look under the hood in order to thoroughly understand what is going on.

Nowadays people are less used to writing long letters, but letters have a big advantage over regular publications. They are less formal, give more room for creativity and speculation, and they allow the correspondents to change their positions more easily and thereby learn more from their counterparts. You can also ask more critical questions in an easier tone and can present an idea in a non-protected form. We found this method very helpful in our highly dynamic fields of interest where nothing is engraved in stone yet and lots of myths and buzz-words are swirling around.

Especially since China has declared its ambition to become a main world center of AI innovation by 2030, many people have become interested in the questions we have been dealing with for quite some time.

To achieve clarity, a sober analysis of both the state of the art and the prospects and limits of what can be achieved in the related sciences and technologies is needed.

This is what we were trying to achieve before the hype started and we want to share our insights with those that are trying to understand what is going on in these fields and what the difference between facts, popular belief, realistic hopes, dreams and marketing buzz is.

What you have in your hands now is not a science textbook or a typical popular science book either. It is also not a systematic or complete introduction to the two disciplines.

What we did was more like a random walk; wandering from field to field, stopping to study something in depth whenever we wished. We were guided only by our curiosity, and simply put in more effort when we had the desire to understand things a little more precisely or when we felt the need to fill gaps in our knowledge. And we often enjoyed following the streams of knowledge back to their sources, including some excursions into the history of our very different cultures. But as chaotic as our journey was, we feel that through our continuous, and sometimes controversial debate, we gained insights we would have not acquired had we chosen a more systematic approach.

We both like to take the perspective of a child who asks simple questions in order to grasp what is happening. Sometimes the child can see that the emperor's new clothes are not as brilliant as they are presented. But we don't want to overdo it with this perspective because it would be presumptuous to say that we are the child in the famous fairy tale "The Emperor's New Clothes" who can see or can't see what others see.

However in the case of HBP, Karl insists that very early on, while others were still praising it, Fanji was able to recognize that there were flaws in this impressively presented project.

We spent a great deal of effort in our early letters demonstrating and assuring ourselves that more than one thing was wrong in the concept of the HBP, and that we shouldn't expect too much from it. It initiated our correspondence, and it became a good model to explore many of the basic elements of the brain and the mind and a possible link with artificial intelligence and computer-technology.

Today, after the facade of this project has been seriously damaged in public such criticism is common and our verve in the old days may seem to some as an attempt to kick a dead dog. Maybe the meanwhile common critique is even too much because in our eyes there are interesting parts in the HBP-concept that have deserved a second try.

In addition to discussing a variety of great myths about the brain and artificial intelligence, we deliberated over rational thinking and consciousness. During this period, several important events occurred both in brain science and artificial intelligence, such as the launching of the US BRAIN (Brain Research through Advancing Innovative Neuro technologies) Initiative, AlphaGo defeating the world Go game champion and a self-driving car running on the highway among many others. We followed these impressive events and discussed how to evaluate their significance. Some of the new advances supported our speculations, and this encouraged our further discussions. The progress in some areas was even beyond our best expectations, so we had to reconsider our views and learn lessons from our mistakes. All this inspired our discussion, redirected our focus, ignited new debates and sometimes made us change our mind. On some points we reached consensus, on some we still differ, and on some matters we just never reached any conclusions.

We don't expect all readers to read all letters. Some may be experts in neuro-physiology who just want to know what can be expected from "neuromorphic" chips while others may be skilled programmers of artificial neural networks and want to better understand what the problems of a link between biological and artificial neurons are. Others may be less interested in the physiological details of the brain or computer architectures but in aspects that have also occupied us in detail, like the emergence of consciousness, free will, relevance of memes, self-organization, circular causality or the organization of research in the life sciences and in engineering disciplines. Some of our friends who are concerned with science organization and that we have shown the draft of the book to, found our discussion about the misery of bureaucratic big science the most relevant to them. Others may ignore the technical details and prefer, as in a travelogue, to follow the story, as the two authors try to orient themselves in the difficult borderlands of biology and computer technology.

For those readers who are only interested in specific questions we have provided two different tables of content where they can identify letters containing such subjects.

The idea to publish our conversation occurred only after a while and the reader in the focus was the ambitious lay person with some interest in science and technology who wants to better understand what the facts and the myths around these two popular fields are. And while our discussions goes into the great detail here and there, which is hard to understand without the right background knowledge, the Chinese content of the book comes with explanations in textboxes, footnotes and drawings to give readers convenient reference material to better understand what the basics of the matters discussed are.

Adding this material takes a lot of time and also significantly expands the size of the book which makes a publication in three volumes necessary.

For those who are interested in a summary of our reasoning, and also in the consequences we have drawn, we have added a longer than usual epilogue where we summarize our most relevant findings. The curious readers who, as with a crime novel, can't wait and want to know in advance who the murderer is may read this epilogue first.

Of course, we do not claim to have answered all questions. In fact, our survey isn't complete and many questions have remained unanswered and new ones have been raised. We also don't claim that we understand the matter better than others. The self-confident and determined tone in some parts should not be misinterpreted as absolute certainty about one's own position. It is only due to the method of this old-fashioned debate culture, where a position is worked out as clearly as possible, not to protect it, but to invite others to refute it.

Many of our assumptions, conclusions and comments are probably erroneous or incomplete and will need to be corrected soon. The trouble is we don't know which of them are wrong.

In any event after we have often criticized others so generously we are prepared for this to happen to us.

We may not be happy to be proven wrong, but this is unavoidable if we want to make progress. We both believe that rational reasoning, as a continuous process of challenging theories by empirical research, is the best way to increase our knowledge. Rational reasoning alone can't replace intense wrestling with the matter in the real world and what also can't be replaced in this process are the curiosity, boldness and ambition of researchers and engineers, especially in the young generation.

For some it may be disappointing to see how little we know about the riddles of our own minds and how slowly our knowledge of them is increasing. It catches the eye that progress on the technical side of the field is developing much faster.

Others may see this as an opportunity to enter a very promising field of work in which many unbelievably rewarding discoveries await those who are ready to leave the beaten track.

We're both sure we haven't made any relevant discoveries. But we have the desire and hope that we can inspire some talents to try their luck and find new entrances to the magic castle.

Not all of them may succeed, but we hope that many readers will find our insights helpful and will enjoy this walk around the magic castle as much as we did in the last 6 years.

Karl Schlagenhauf and Fanji Gu

Table of Contents by Subjects

This series of books contains 31 pairs of letters, discussing a series of open problems about brain and artificial intelligence. These discussions and debates run through the series, but there are differences in the emphasis of each volume. Among them, the first volume *Brain Research, a New Continent to be Explored* contains 11 pairs of letters (No. Ⅰ 001 – 011), focusing on the open problems of brain research; the second volume *Mystery of Consciousness and Myth of Mind Uploading* contains 10 pairs of letters (No. Ⅱ 001 – 010), focusing on discussing different views on consciousness research and the possibility of mind uploading; and the third volume *The Third Spring of AI* contains 10 pairs of letters (No. Ⅲ 001 – 010), focusing on the potential and prospect of AI. Related scientific methodology and scientific organization are also discussed in all three volumes. The letters in the three books are sorted by time. In order to facilitate readers to understand the contents of the whole series, we provide the following list:

1. Preface (Ⅰ - 001 Fanji)

2. Scientific Methodology

"*System-*" *and* "*Interest-performer*" (Ⅰ – 002 Karl, Ⅰ – 003 Fanji, Ⅰ – 003 Karl, Ⅰ – 008 Karl, Ⅱ – 003 Fanji)

Nature and engineering adopt different methods (Ⅰ – 002 Fanji, Ⅰ – 003 Fanji, Ⅰ – 005 Fanji, Ⅱ – 004 Karl, Ⅱ – 009 Karl)

Different Thinking Habits in Different Disciplines (Ⅰ – 003 Fanji)

Competition and Cooperation between Scientists (Ⅰ – 003 Karl)

Academic Controversy (Ⅰ – 004 Fanji, Ⅰ – 004 Karl, Ⅰ – 005 Fanji, Ⅰ – 006 Fanji, Ⅱ – 002 Karl)

Theory, Models, and Simulations (Ⅰ - 001 Karl, Ⅰ - 002 Fanji, Ⅰ - 004 Karl, Ⅰ - 006 Karl, Ⅰ - 010 Fanji, Ⅰ - 011 Fanji, Ⅱ - 003 Karl, Ⅱ - 008 Karl)
Circular Causality and Linear Causality (Ⅲ - 002 Karl, Ⅲ - 003 Fanji, Ⅲ - 005 Karl, Ⅲ - 009 Karl)

3. Some Open Problems in Brain Research

The Brain Study is a Long March (Ⅰ - 001 Fanji, Ⅰ - 005 Fanji, Ⅰ - 005 Karl, Ⅰ - 006 Fanji)
Neurons (Ⅰ - 002 Fanji, Ⅰ - 009 Karl, Ⅰ - 010 Fanji)
Functional Column May Not be the Standard Module of the Cortex (Ⅰ - 002 Fanji)
Memories (Ⅰ - 007 Fanji, Ⅰ - 008 Karl, Ⅰ - 009 Fanji)
Emotion (Ⅰ - 007 Fanji, Ⅲ - 001 Karl)
Connectome (Ⅱ - 009 Karl)
Is the Brain an Information Processing System or a Machine Creating Meaning? (Ⅱ - 009 Fanji)
Social Brain (Ⅲ - 001 Karl, Ⅲ - 002 Fanji, Ⅲ - 002 Karl)

4. Disputing Problems in Artificial Intelligence

Intelligence (Ⅰ - 001 Karl, Ⅰ - 002 Karl, Ⅱ - 005 Fanji, Ⅲ - 003 Fanji, Ⅲ - 003 Karl, Ⅲ - 004 Fanji, Ⅲ - 005 Fanji, Ⅲ - 005 Karl, Ⅲ - 006 Fanji, Ⅲ - 006 Karl, Ⅲ - 007 Fanji)
"Turing Test" and "Chinese Room" Thought Experiment (Ⅱ - 005 Fanji, Ⅱ - 010 Fanji, Ⅱ - 010 Karl)
Neural Networks and Deep Learning (Ⅱ - 001 Karl, Ⅲ - 004 Karl, Ⅲ - 008 Karl)
Machine Translation (Ⅲ - 001 Fanji, Ⅲ - 001 Karl, Ⅲ - 002 Fanji, Ⅲ - 003 Karl, Ⅲ - 008 Fanji, Ⅲ - 008 Karl, Ⅲ - 009 Fanji)
AlphaGo (Ⅲ - 003 Karl, Ⅲ - 010 Fanji)
Paradigm Shifts in the Field of Artificial Intelligence (Ⅲ - 004 Karl)
Singularity (Ⅱ - 002 Karl, Ⅱ - 003 Fanji, Ⅱ - 004 Fanji)
Moore's Law (Ⅱ - 004 Fanji, Ⅱ - 004 Karl, Ⅱ - 005 Karl)
Brain-inspired Computing and Brain-like Computing (Ⅰ - 010 Fanji)
Neuromorphic Systems (Ⅰ - 010 Karl, Ⅰ - 011 Fanji, Ⅲ - 007 Fanji, and Ⅲ - 007 Karl)
Reverse Engineering (Ⅱ - 003 Karl, Ⅱ - 004 Karl)
Weak Artificial Intelligence and Strong Artificial Intelligence (Ⅲ - 004 Fanji)

Computer and the Brain (Ⅱ - 010 Karl, Ⅲ - 001 Fanji, Ⅲ - 001 Karl, Ⅲ - 002 Fanji)
The Next Step in Artificial Intelligence (Ⅲ - 003 Karl, Ⅲ - 004 Fanji, Ⅲ - 008 Fanji)
New Industrial Revolution (Ⅲ - 007 Karl, Ⅲ - 010 Karl)
China's "New Generation of Artificial Intelligence Development Planning" (Ⅲ - 009 Karl, Ⅲ - 010 Fanji, Ⅲ - 010 Karl)

5. The Mystery of Consciousness

Difficult Problems and Easy Problems (Ⅱ - 001 Fanji)
Neural Correlates of Consciousness (Ⅱ - 007 Karl)
Neural Global Workspace Hypothesis of the Consciousness (Ⅱ - 008 Fanji)
Subjectivity (Ⅱ - 001 Fanji, Ⅱ - 005 Fanji, Ⅱ - 006 Fanji, Ⅱ - 007 Fanji, Ⅱ - 007 Karl, Ⅱ - 008 Fanji, Ⅱ - 008 Karl, Ⅲ - 001 Fanji)
Privacy (Ⅱ - 001 Fanji, Ⅱ - 002 Fanji, Ⅱ - 005 Fanji, Ⅱ - 007 Fanji, Ⅱ - 008 Fanji, Ⅱ - 009 Fanji, Ⅱ - 009 Karl)
Mind Uploading (Ⅱ - 005 Fanji, Ⅱ - 005 Karl, Ⅱ - 006 Fanji, Ⅱ - 006 Karl, Ⅱ - 007 Fanji, Ⅱ - 007 Karl, Ⅱ - 008 Fanji, Ⅱ - 009 Karl)
The Perspective of Consciousness Research (Ⅱ - 005 Fanji, Ⅱ - 005 Karl, Ⅱ - 006 Fanji, Ⅱ - 006 Karl, Ⅱ - 007 Karl, Ⅱ - 008 Fanji)
Free Will (Ⅲ - 003 Fanji, Ⅲ - 009 Fanji, Ⅲ - 009 Karl)

6. Big Science Plan

Big Science (Ⅰ - 005 Karl, Ⅰ - 006 Fanji, Ⅲ - 005 Fanji, Ⅲ - 006 Fanji, Ⅲ - 006 Karl)
EU Human Brain Project (Ⅰ - 001 Fanji, Ⅰ - 001 Karl, Ⅰ - 005 Karl, Ⅰ - 007 Karl, Ⅰ - 010 Karl, Ⅰ - 011 Fanji, Ⅰ - 011 Karl, Ⅱ - 006 Fanji, Ⅱ - 009 Fanji, Ⅱ - 010 Karl, Ⅲ - 006 Fanji)
US Brain Project (Ⅰ - 007 Karl, Ⅰ - 008 Fanji, Ⅱ - 006 Fanji, Ⅱ - 010 Fanji, Ⅱ - 010 Karl)
Human Connectome Project (Ⅲ - 002 Fanji, Ⅲ - 006 Fanji)
Allen Institute for Brain Science (Ⅲ - 002 Fanji, Ⅲ - 005 Fanji)

Contents

Recommendation ... i

Introduction 1 ... iv

Introduction 2 ... vii

Introduction 3 ... ix

Preface ... xi

Table of Contents by Subjects ... xvii

III- 001 Fanji ... 1

"Subjectivity" is the most difficult problem in both consciousness and artificial intelligence studies; translation and Machine Translation; anthropomorphic and Eliza effects; computer and the brain.

III- 001 Karl ... 7

The social brain; translation and understanding; the computer and the brain are essentially different two systems; the rules of logic apply only to a small part of the various words we say every day, and are basically used only in the axiom system applicable to propositional calculus; rationality and emotion.

III-002 Fanji ... 17

The question of "how" is an important issue in science; individual brain and social brain; Machine Translation; computer and the brain; human connectome program; Alan Institute for Brain Science.

III-002 Karl ... 22

The descriptive and explanatory parts of scientific studies; in order to produce complex behavior, there is not necessarily an agent computing all the time; brain networks and accumulated wisdom will be able to recognize a single brain; reafference principles and closed loop control.

III-003 Fanji ... 30

Intelligence; free will; circular causality and linear causality; to achieve the same function as digital technology, there may be simpler and more energy efficient methods.

III-003 Karl ... 36

AlphaGo; face recognition; automatic translation; the next breakthrough needed by artificial intelligence: lower energy consumption and small sample learning; language learning; priming effect; skill is not equal to intelligence.

III-004 Fanji ... 44

There is no intelligence in the Deep Blue chess-playing system; strong artificial intelligence and weak artificial intelligence; general artificial intelligence and artificial consciousness; Robots with their own mind and will; social consequences of artificial intelligence; the need for large amounts of data and high energy consumption are bottlenecks in the further development of artificial intelligence.

III-004 Karl ... 49

Paradigm shift is taking place in the field of artificial intelligence; from deep blue to deep learning; parallel computing.

III-005 Fanji ... 58

"Intelligence" is not "the ability to solve specific problems", but "the ability to get the ability to solve specific problems"; the classification of artificial intelligence; The Allen Brain Observatory; big science.

III-005 Karl ... 61

Gehlen's Handlungskreis; social institution and behavior; the definition of intelligence.

III-006 Fanji ... 71

Intelligence and learning ability; Human Connectome Project completed the first phase; The HBP entered the "operational phase"; brain projects of various countries.

III-006 Karl ... 78

Big science; Parkinson's law; intelligence and skills; Moravec paradox.

III-007 Fanji ... 86

The definition of intelligence; the mind of others; neuromorphic chips.

III-007 Karl ... 92

Dive into the meta level; scientific writing culture; neuromorphic chips; new industrial revolution.

III-008 Fanji ... 100

Machine Translation, AI is just a tool for the foreseeable future.

III-008 Karl ... 106

> *Machine Translation; interpersonal communication; deep learning.*

III-009 Fanji ... 117

> *Machine Translation; free will.*

III-009 Karl ... 122

> *Free will; linear causality; circular causality; Machine Translation; China's New Generation of Artificial Intelligence Development Planning.*

III-010 Fanji ... 132

> *AlphaGo; New Generation of Artificial Intelligence Development Planning.*

III-010 Karl ... 137

> *Facts speak louder than words; the rise of AI in China; the future of AI.*

Epilogue ... 147

Translation Postscript ... 164

Acknowledgements ... 166

Ⅲ-001 Fanji

24.08.2015

Dear Karl,

My heartiest congratulations! Congratulations for your promotion to grandfather. ☺ Please also give my congratulations to Adi, Jelka, and Jürgen. Also, to Alex of course, if he can understand what my congratulation means ☺. Thanks also for your promising to let me know your observation about how a young brain develops, very interesting indeed!

I totally agree your statement that "*And although we will never know exactly what is going on deep in the other person's mind, my hypothesis is that this subjective pleasure must be pretty similar in your mind and in my mind.* ☺" I think that the reason is that we as human beings have similar brain structures, and both of us graduated in science and technology and worked in interdisciplinary fields almost all our lives, so that I guess that the same stimulus will evoke similar brain activities, although they will never be identical, owing to our different genome, different connectome, different experience, and different social and cultural environments.

Your words remind me of the "subjectivity" problem again, which is the most tricky problem, both consciousness studies and artificial intelligence face. We have discussed a lot about the subjectivity of consciousness, although we cannot get a consensus if the problem about how such subjectivity emerges from a physical brain can be solved with physical laws someday, or just acknowledge it as an irreducible emergent property of some kind of brain activity of human beings. If a person behaves like a normal person, we believe that he or she must be conscious. However, for a machine, even if it behaves like a normal person, I am still not sure if it is conscious. Of course, if a machine is made so sophisticatedly, that no one can distinguish it from a human being, then people would think it is a human being, and, thus, think that it is conscious.

However, at least at the moment or in the foreseeable future, there is no such machine, and not any machine has any mental activities, although people have used so many anthropomorphic terms to describe its behavior. For example, sometimes people call natural language processing "natural language understanding", although the machine doesn't "understand" the language at all. As I told you before, I was annoyed to read some translated version of science books owing to their poor translation. I found all the errors are owing to the fact that the translator did not understand what the author wanted to express. They just did literal translation word by word, just like old machine translation based on grammatical rules and a dictionary or followed wrong examples. If a machine translation system learns samples like such translation, no matter what kind of advanced deep learning is used, how could the translation be correct? I don't think that any deep learning system can "understand" the real meaning, it can only learn from large data samples.

This reminds me of the fact, if I learn to speak English with friends all speaking in Chinglish, then the result must be me speaking Chinglish! Sometimes, people declared that they developed a robot with emotion, although it has only emotion-like expression but without feeling. People also use "thinking", "perceiving" etc. to describe a robot, which leads to confusion. Journalists like to express it this way, I can realize, as such an expression grabs readers' eyeballs. However, even many scientists also like to express it in this way, which may even confuse themselves at last. Sometime, anthropomorphization is so powerful that people would believe an unconscious machine has its mind. As a matter of fact, even in the early days of AI, MIT computer scientist Joseph Weizenbaum developed a chatterbot ELIZA in 1966. ELIZA parodied a Rogerian psychotherapist, largely by rephrasing the patient's words as questions:

"Human: Well, my boyfriend made me come here.
ELIZA: Your boyfriend made you come here?
Human: He says I'm depressed much of the time.
ELIZA: I am sorry to hear you are depressed.
Human: It's true. I'm unhappy.
ELIZA: Do you think coming here will help you not to be unhappy?"

Weizenbaum discovered that, patients, even if fully aware that they are talking to a simple computer program, will nonetheless treat it as if it was a real, thinking being that cared about their problems — a phenomenon now known as the "Eliza Effect." Weizenbaum later wrote, "*I had not realized ... that extremely short exposures to a relatively simple computer program could induce powerful delusional thinking in quite normal people.*" And we can see Eliza Effects here and there today.[1]

Therefore, it seems to me, that the problem of subjectivity is not only so important for consciousness studies, but also for AI research. Anthropomorphization is very powerful for popularizing AI to the public, but we must be clear that is also apt to induce misunderstanding.

Therefore, "singularity is near", robots have their own will to liberate themselves and destroy mankind is so popular. As if the danger were near. However, as we discussed earlier, even for human beings ourselves, we still don't understand how these inner properties emerge, or as I argued, what is the sufficient and necessary condition for them to emerge? Now we know some necessary conditions for consciousness emerging in human beings, such as Dehaene's "signatures", or some necessary conditions in general, such as Edelman and Tononi's integration and differentiation or Tononi's integrated information. However, we still have no idea about sufficient and necessary conditions. Tononi and Koch thought that nonzero integrated information is sufficient and necessary for being conscious. However, this will lead to a panpsychism conclusion. We don't have any machine with such an inner world. It seems that there is still a long way to go for developing such machines, or as John Searle called, "strong AI". In addition, should we develop such a machine is also a problem.

By the way, I just had a glimpse to Kurzweil's "*How to Create a Mind*". In the last chapter "Objections", he criticized Searle's "The Chinese Room" thought experiment. He argued that although the man in the room does not understand Chinese, the whole system — the man plus the rulebook does, otherwise how can the system answer questions in Chinese correctly? Therefore, from his view, Watson understands what he said. His arguments are wrong, it is the compiler of the rulebook in the Chinese room that understands Chinese, not the rulebook itself or "The Chinese Room".

In short, I cannot deny the possibility that someday we may have a conscious machine,

---

[1] https://en.wikipedia.org/wiki/ELIZA_effect

anyway the brain itself is a physical system, and it does have its inner world, there is not any logical reason to exclude the possibility that other physical systems may also develop inner worlds, if they are complicated enough to meet the sufficient and necessary conditions to be conscious. However, as we know little about the sufficient and necessary conditions under which such inner worlds may emerge, to declare that the development of AI will create supper intelligence to dominate our planet is still too early. This is just like the statement that everything will have an end, so does the earth. Of course, this statement is correct, should we worry about the ruin of the earth right now, just like some cult believers? In summary, although we should know that there is such possibility or risk and should pay attention to it, we need not worry about such possibility too much, at least at the moment.

I like your discussion about "discussion" very much, which is absolutely right. People are apt to think along a habitual track. Every established idea seems so natural, it is rare to doubt and think in another way. During discussion, the opponent's word may challenge such habitual thinking and push one to consider it from another view or angle. Of course, the premise is that the person should have an open mind as you emphasized in your last letter. Otherwise, it is just like the old Chinese saying: "Playing a violin to an ox".

The habitual thinking is very stubborn, just as you analyzed in your last letter:

"Sometimes we are misguided, not because we are dedicated followers of a specific school, but because we carelessly use certain terms, concepts, or views just as everybody else does. ... thinking about the brain as an information processing device, is an example in modern neuroscience."

Just like you,

"I feel ashamed that I, until lately, and also in the course of our correspondence used this information processing perspective quite thoughtlessly just as if it would be a matter of course. But it isn't, and I simply didn't think about it, like most people don't, and only in the course of our discussion realized that it is a very misleading, idealized, *perspective which logicians and mathematicians imported to a biological subject of which they only had a vague idea.*"

Of course, I know the book "*The Computer & The Brain*". As a matter of fact, I have had one copy on my book shelf for more than half a century, the paper of which is already yellow and fragile, however, I am ashamed to tell you that I have never read it through. I bought it in June, 1966, just at the beginning of a special age, thus I had no chance to read it then and even in the next ten years. After that period, I was busy with other things and just left it untouched. After reading your letter, I took the book carefully off of my book shelf and had a quick reading. Yes, you are right, even at the very beginning, von Neumann found the big difference between the computer and the brain, if he was not stopped by his death, it is very likely that he would have dug deeply into this problem. In his time, the nick name "electric brain" had already been very popular in the media, and even today in China. However, he already noted that the brain did not work digitally. Inserting or deleting a spike would not change the information carried out by the impulse train very much, however, if the information is represented by the impulse train as a binary string, their values would be radically different!

I also know von der Malsburg, I visited him once at his office in Ruhr University, but this was my only chance to meet him. However, I am also ashamed to acknowledge that I know little about his negative attitude against the popular information processing paradigm about the brain. I only know that he is one of the first to suggest that synchronized oscillation may play a key role in solving the "binding problem". Maybe you could tell me more about him, especially his thinking about the analogue characteristics of the brain instead of digital.

As for the Turing test, although I have had the question I asked in my last e-mail for a long time, recently I read a paper written by Dr. Pei Wang[1], he raised the same question and thought more deeply. Wang criticized that using the passing of the Turing test as the definition of intelligence is too human-central. He argued that passing the Turing test is only a sufficient condition for being intelligent, but not necessary. He

---

[1] https://cis.temple.edu/~pwang/

argued that in the history of AI, the mainstream is to make a computer doing something that only the human brain could do before, but not to make the machine indistinguishable from human beings. He said that even Turing himself understood that if a machine could pass the Turing test then it had to learn how to pretend to be foolish or to lie, otherwise it would not be like a human being owing to something, for example, the speed of its calculation being too fast. Unfortunately, in most papers talking about the Turing test, it was rarely mentioned about this point.

Yes, I have translated Koch's book *"The Quest for Consciousness"* with his former Ph.D. student Xiaodi Hou. We have asked him more than 70 questions when we did not understand the meaning of some sentences, he kindly answered all our questions. In a preface for the Chinese translation, he wrote:

"Translating any text is a very difficult and demanding undertaken since the thoughts underlying the ideas put down on the printed paper need to be understood before they can be rendered into a different language. And if the translator has done his job perfectly, he will and should remain completely invisible, so that nothing intrudes between the reader and the author."

It reminds me of my question mentioned above, how machine translation can "understand" *"the thoughts underlying the ideas put down on the printed paper"*. He and Susan Greenfield had a debate if consciousness is a global property or localized a few years ago. It is very interesting. And I also found something very interesting about the works in the Allen Institute for Brain Science led by him, and the Human Connectome Project (HCP) which has just passed its first phase this year. However, as my letter is already too long, I would have to discuss these problems later.

Best wishes.

Fanji

III-001 Karl

07.10.2015

Dear Fanji,

Thank you for your very interesting letter and your grand-father-promotion congratulations. Well, I have to admit that I wasn't exactly keen on becoming a grandfather. But what happened to Adi and me was exactly the same thing that happens to all grandparents. As soon as Alex had arrived we simply fell in love with the boy and really enjoy it. But this experience nevertheless changed my perspective because it made me think about my new role and about the role of grandparents in general.

Of course, you are familiar with the concept of the "grandmother cell" which stands for the assumption that objects like the face of your grandmother can be identified by a single neuron in the brain. This theory has many flaws and was substantially and rightfully criticized. That's why I don't like grand-mothers (and fathers) being ridiculed by the connotation with such an erroneous assumption and believe that they have deserved to be honored by more qualified theories. Therefore, I have developed a theory of "relevance of grandparents" in the evolution of mankind. And you are the first one to hear about it! Many biological textbooks about evolution are gene centered and everything goes about DNA and mutations. This narrow perspective supports the idea that after you have invested half of your DNA to your offspring you have done your biological duty and that was it. While this may be true for Drosophila and species that don't care of their progeny it is not appropriate for humans.

DNA can't transfer ontologically acquired experience and the collective knowledge of a species. Everything that is relevant to successfully maneuver in this world has to be installed in a young brain from scratch again and again. And while parents have to feed, fight, protect, and make a living grandparents do the bulk of the job of systematizing, aggregating, and filtering the knowledge and do the teaching of the relevant know-how and skills.

A species, so much depending on learning, needs more than one parent generation to do the job properly and that's why it provided an advantage in evolution when groups of humans had members who lived longer than one reproduction-cycle and helped do this job. To have experienced wise old men and women in the family, who also contribute to the social cohesion of the group, provided a huge selective advantage. And that's why grandparents are so important! I wonder why it took me so long to realize such an obvious circumstance.

I'm not sure whether I'm the first one who had the idea. But I'm dead sure that if someone else had it before, it must have been a grandparent. ☺

I would like to discuss the relevance of social matter, which we have already briefly touched some time ago in the context of memes and institutions, in more detail, if you like. My impression is that most neuro biologists and psychologists are too much focused on single brains. They tend to neglect the fact that a human brain can only work properly as part of a network of brains that all have to be primed with an enormous amount of content.

We all live with the feeling that our knowledge and skills are an integral part of ourselves and that we own them in a subjective way. For some parts, this perspective is appropriate but not for others.

The thing we experience as our personality is actually a fabric woven from many threads. Some of them we have spun ourselves, such as the ability to walk, swim, or ride a bicycle. But other threads were spun by others and handed over to us by our parents, friends, teachers, or writers of books. We use them as programmers use prefabricated functions from a library. Language consists of a whole bunch of such threads. Letters and numbers are elementary threads and complicated thread is the realization that each circle is in a certain relation to its circumference which someone long ago called Pi. Sometimes we know the inventors of a useful element and honor them by giving it her name. But often we don't know to whom we owe things that have proven to be extremely useful elements and building blocks. So, we know neither the inventor of the spoon, the comb, the wheel, or the ship. Nor do we know who we owe the know-how to make metal, beer, or paper. A single person could perhaps invent all this, but a human life is not long enough. Just as a good programmer could program all parts of the libraries he uses, but that would never end.

If you look at an adult brain from this point of view, by far, the largest part of its content is not produced by its owner but is inherited or borrowed and imported from others.

Well, it is maybe a little more complicated in reality, but I hope that the example makes clear that it is impossible to understand the network-system just by analyzing a single brain.

But this is a whole new field and, for today, I would rather come back to your considerations about the similarity of subjective feelings between two individuals and the difficulties related to translations, which are both right on the spot in my eyes. We may not agree on the question whether qualia can be reduced to the underlying physics, chemistry, and biology. But, I fully agree with what you say about the problems of ascribing consciousness or mental activities to machines and the dubious, and often careless, use of anthropomorphic terms in this context. To call natural language processing "natural language understanding" and to say that robots "think" and "perceive" is not just sloppy language, it is an unacceptable mistake.

And I also agree with what you say about the need to comprehend first before you can do a good translation. I'm not an expert in the field, as you are, but often have the impression that translators and journalists reporting about scientific findings or technical achievements don't exactly grasp the matter they are dealing with.

But I also believe that all of us are in similar situations very often in our lives. What you say about the relevance of "understanding" is not just a problem for robots or AI systems, it is also a problem among humans. It is hard to tell whether your neighbor sees the same color of blue as you do. But it is also hard to tell whether a scientific author has really grasped the core of an argument when he talks about "quantum leaps", "Schrödinger's cat" or the "Chinese room". These are highly complicated matters, for sure, which are not easy to comprehend. Therefore, it is striking how easily people are using those terms. "Quantum leap" became almost a household term and is often used by journalists when they want to express that a very big revolutionary step forward was made. Not all seem to realize that a quantum leap is about the smallest step possible in the physical world. So actually, it is just the opposite of what they want to tell us.

But our brain doesn't bother because it is able to somehow "smell" the intention or meaning. Sometimes it can also toggle sense to nonsense and has no problem to live and deal with it. Every digital computer will crash when you try to divide a variable by another variable with the value of zero. And programmers have to be very careful to avoid contradictions and logical hiccups because von Neumann machines don't forgive mistakes. Not so for our brain, which is very forgiving when it comes to mistakes. Actually, it hops from mistake to mistake with ease and just guesses around. It is the vagueness and fuzziness of our perception and the relatively solid interpretation which our mind derives from it which is so treacherous. But at the same time this is what allows us to act in a very robust and persistent way. Unlike von Neumann machines, our mind has no problem to stand firmly on loose and shaky ground and live with contradictory and confusing input, or, at least, has the illusion to do so. Von Neumann was already fully aware of this, maybe more than many of his contemporary successors.

I was to make the point that neither our perceptions nor our intentions are based on logical reasoning or consistency and very often we don't foresee the consequences of our actions. The rational monitor in our mind, actually it is a story-telling process, wants to make us believe that we follow a rational storyline and supplies us with supportive arguments that help us to maintain the illusion of acting consistently.

While scientists are trained to spot for logical flaws and the detection of contradictions, such things play a minor role in everyday life. When you listen to, otherwise, quite intelligent people you can detect that they may contradict themselves logically within seconds. But this seems to be no problem at all. Not for them and not for the listeners. As long as the flow of the story is emotionally exiting and in line with an expected meaning, logical flaws are simply filtered out like noise. And there is a good reason. The rules of logic are only applicable for a very small subset of the variety of statements we produce every day. Basically, they are only useful in axiomatic systems where the propositional calculus is applicable and where proofs of true and false can be made, like in math. And even there are limitations as we know since Kurt Gödel. The majority of our statements and arguments, however, fall in a class where terms like "right" or "wrong" or "contradictive" are of little help because they consist of subjective preferences. Here even the most elementary rules of formal logic like the one of transitivity aren't applicable.

You find people that prefer chocolate ice over vanilla ice and vanilla over strawberry ice. But when you give them the choice between strawberry and chocolate they may opt for strawberry. A debate about whether this is foolish is about as fruitless as the debate about whether blue is a more beautiful color than red.

And it gets even worse when it comes to the universe of normative statements which deal with the statements related to obligations and permissions. Generations of philosophers and logicians have tried to build something called "deontic logic" which might help us to make logical operations on normative statements and therefore would be a means to tell others what is allowed and what is not, derived from higher principals on a logical basis. It was all in vain and the search for it is about as promising as trying to invent a machine with perpetual motion or to find a proof for the existence of God. Nevertheless, it is tried again and again. But formal logic is only applicable for a very tiny subset of all the statements and arguments we have to deal with. Our brain knows this and therefore cares little about such rare events. Normative orientation (that is, the rule-set of what we should do and not do) our mind does not find by logical derivations but through inner emotions, impulses and desires which we control by observing what others do and what actions are rewarded or sanctioned.

Acting like this may seem weird, inconsistent, and contradictory and often when we rethink about what we did we find that we should have done otherwise as we did in the heat of the action. But our conscious mind is, by far, too slow to make rational decisions in most interactive situations. There simply isn't enough time to consider and understand everything in full detail when you are in the flow of action. But our brain is very good in acting "without thinking" in situations, where we have no complete information or full comprehension of causality or of the consequences of our actions. It is just the normal operational mode of a brain. Rational and scientific reasoning are very rare and artificial kinds of thinking, which you have to learn the hard way, like playing the piano.

I struggled with this rationality principle for many years when I was working as a postdoc at Karlsruhe University. At the time, it was something like a cornerstone in many theories of human action and decision making, but to me it looked like a total failure. So I made this problem the theme of my habilitation defense in 1980. The title was "*Zur Frage der Angemessenheit des Rationalitätskalküls in den Handlungs-und Entscheidungstheorien*" ("On the question of the appropriateness of the rationality calculus in the theories of action and decision making")[1]. And with the typical exuberance and zeal of the young scientist I attacked and completely destroyed the idea of rationality as the core-principle of human acting and decision making.

It had escaped my notice however that two smart guys in Israel, Amos Tversky and Daniel Kahneman, had already realized that something is wrong with this paradigm and worked on alternative and more realistic theories which earned Daniel Kahneman a Nobel Prize in economy in 2002. Meanwhile Kahneman's differentiation between two levels of human thinking (System 1: Fast, instinctive and emotional; System 2: Slower, more deliberative, and more logical) became the new paradigm in economic decision theories. And after he published his best-selling book "*Thinking, Fast and Slow*" (2011) his often quite funny examples for the many situations in which we act all but rational became part of the popular-psychological knowledge to which other sciences often refer.

We may need to differentiate more than two systems to comprehend how our mind works, but Kahneman, in any event, helped us to better understand why we have no problem easily working with concepts which we haven't fully understood. And he also helps to explain why we are often distracted or fascinated by aspects of a thing that are actually irrelevant. Schrödinger's famous cat is a perfect example. Sometimes such hard to comprehend and mysterious concepts seem to be especially attractive, and this poor cat seems to be the prototype. Schrödinger's cat became the darling of philosophers as one of their favorite metaphors and it is a must in modern science literature. I wonder how many of those who cite it, often as a proof for the validity of the Copenhagen interpretation of quantum mechanics, are aware of the fact that Schrödinger invented this thought experiment only for the purpose of demonstrating the absurdity of the

---

[1] Schlagenhauf, Karl (1984). Zur Frage der Angemessenheit des Rationalitätskalküls in den Handlungs-und Entscheidungstheorien, in: Lenk, Hans (Hrsg.) (1984) Handlungstheorien-interdisziplinär Bd. III, Zweiter Halbband, Wilhelm Fink Verlag, München, 680 – 695.

Copenhagen interpretation where a cat can be both dead and alive until you have a look at it.

Same thing with Searle's famous Chinese Room, which probably is one of the most cited metaphors in science right after Schrödinger's cat. And as you can see from Kurzweil's interpretation, which, as you correctly say, is absolutely wrong, if not to say foolish, it is meanwhile cited by everybody to demonstrate whatever they want. I believe that the enormous success of both metaphors has nothing to do with their inner quality and original meaning but a lot with their emotional impact. It is this amalgamation of something mysterious with something very familiar and well-known that makes these terms so impressive and sticky to our memory. While no-one really understands what this Heisenberg uncertainty and Bohr's complementarity mean or what quantum superposition and entanglement really are, everybody knows what a cat, a glass tube, and a hammer are. And while not everybody may know what exactly makes hydrocyanic acid so dangerous everybody understands that it is a deadly poison. The drama around the destiny of the poor cat is so emotionally loaded that our brain is just fascinated by it.

Our brains were not designed to do math and abstract physics. They evolved as object-recognition, meaning-detecting, story-recording and story-telling organs. And can you think of a more thrilling story than the one about the sophisticated murder of this cat? So, our brain absorbs it with ease and, actually, it's the perfect and literal example of what the Americans call a "no-brainer". We don't have to learn it; we just hear it and it is burned to our memory with the same ease as e.g. the message that a famous actor had an affair with another actor's wife.

Searle's Chinese room doesn't have such dramatic dimensions but its popularity benefits from the same mechanism. For people in the Western hemisphere, the Chinese language is a synonym for the most distant and hardest to understand language of all. The pictography of the Chinese script looks enigmatic and with the rich and complicated culture linked to it, it seems so exotic and mysterious that no-one can really understand it. That's at least the myth rooted deeply in the mind of most people in the West. So, when Searle chose Chinese as the language that had to be translated by the agent in this automated translation room he went to the max in terms of difficulty of the job and emotionally priming and touching the readers. I bet if he had chosen the example of translating a Java program in Python and called his thought experiment "the translation room" no one would talk about it today. And if Schrödinger would have chosen a coin to be flipped after a radioactive decay occurred instead of the cat to be killed no one besides the experts would have ever talked about it. But Schrödinger and Searle were smart (or lucky) enough to enter the minds of millions of people not via the very narrow rational gate but by the wide open emotional gate and were rewarded with the highest ranks in the citation index and will probably remain there for eternity. ☺

Of course, such metaphors represent two separate problems for a translator because they highly depend on the cultural background of the various languages and on the knowledge level of the readers in focus. The level of previous knowledge that can be expected from a reader defines the level of additional information that has to be supplied. The other difficulty is to find the appropriate metaphors for culturally related elements that may not exist in the world of the target-language. The more specialized the matter is the more elaborate language and standardized terms will be at hand. When a French author writes an article on a specific mathematical problem in string-theory it will probably be less difficult to translate into Chinese than the work of a French philosopher on "free will". At least as long as the Chinese translator understands string-theory. The point is, that the long and dreadful work of learning the principal elements of a discipline has to be invested on both sides. There is a nice citation which illustrates this problem in one of those Einstein movies. When a society person is introduced to Einstein and asks him: "Professor Einstein can you explain to me the principle of relativity in simple words" Einstein says: "No!" Needless to say, that Einstein's answer wasn't snobbish. It's just plain impossible indeed. And it is impossible not just in such complicated scientific matter. It is also impossible when it comes to transmitting the fascination of a baseball or soccer game to someone who has never seen one.

If the core objects about the matter are not installed in the mind of the receiver, even the best translator is at a loss. You simply can't build up complete libraries in a mind just on the fly, while reading an article, in the same way as you can't teach how to play the piano in 20 minutes. You can explain the keys on the piano and how to stroke them in 20 minutes and then you have to practice for 20 years, every day. The same applies to all kinds of sport, math, physics, and just everything. If the library isn't already primed in the mind, to a certain extent, there is no real comprehension from reading a book. You may be able to pass the next exam just by pure remembering of the wording (if the professor isn't Richard Feynman), but that's not comprehension.

This is the core of the problem you are rightfully complaining about. But there is no easy solution to it because the old rule that there is no free lunch also applies for intellectual food. The sciences themselves can provide help when they standardize their matter and, indeed, the history of the natural sciences is a history of norming terms, methods, and tools and of standardization. It is much easier to find orientation in such standardized territory than in an open chaotic landscape. Typically, you can tell the stage of development of a science by its degree of standardization. Some people even say that before a discipline doesn't have such a system of coordinates; you can't call it a science. Physics, they say, only turned into a science after Newton provided the necessary framework. Under this perspective you can hardly call neuro-biology a science yet. Therefore, the more modest approach of the US BRAIN initiative seems to be a more reasonable step towards making neuro-biology a science while the original HBP-approach, to me, looks more like a mild version of human hubris.

Thank you for telling me about Dr. Pei Wang and his thoughts about the inappropriateness of the Turing-Test. He is a very sharp thinker and I liked what I found on his homepage. I didn't know him but when I read about his "*General Theory of Intelligence*" it somehow rang the bell. The arguments he uses reminded me of what Ben Goertzel has published some time ago. A little research revealed that Wang and Goertzel are colleagues at Temple University and have co-published a book "*Theoretical Foundations of Artificial General Intelligence*". I came across this kind of idea in 2005 after Goertzel had criticized Jeff Hawkin's book *On Intelligence*. What Wang and he say isn't mainstream but very interesting and inspiring indeed.

I don't know which of the articles on Pei Wang's you have read and whether "*Three Fundamental Misconceptions of Artificial Intelligence*" was among it.

In case you haven't, you may find it interesting. There is a download-link to it in the section "The Possibility of AI".[1]

I'm not sure whether this NARS (Non-Axiomatic Reasoning System) which Pei Wang

---

[1] http://www.iiim.is/2010/05/questions-about-artificial-intelligence/

has developed is the ultimate solution to AI but, at least, the critique of mainstream AI looks substantial and helpful to me. And of course, I like Wang's general position that computers and brains are essentially different systems. So this was a very helpful hint — thank you! I'm looking forward to discussing the arguments with you.

Well, you are the modern Scheherazade now when you tell me that you found out interesting things about Christof Koch's work at the Allen Institute for Brain science and in the Human Connectome Project (HCP) but will only tell me what it is in your next letter. ☺

Of course, I'm very interested in learning more about it and can't wait until you tell me what the news is. In any event, it's good that you know Koch since you have translated *"The Quest for Consciousness"*. So, in case it is something really thrilling we may ask him for more details. And you know what? It so happened that I know Paul Allen because I once did some joint technology development with one of his companies. It had to do with alternative, non-verbal programming methods and databases. However, as my letter is already much too long, although I haven't even discussed all the interesting matter you have raised, I will rather tell you more about this in my next letter if you like. ☺

Best wishes.

Karl

III-002 Fanji

22.11.2015

Dear Karl,

You are a real scholar, I have to admire. Even being promoted to grandfather makes you think about scientific problems ☺. It is really an honor for me to be the first to hear your "relevance of grandparents" theory. As for the "grandmother cell", I don't doubt the experimental fact, the problem is that to recognize one's grandmother depends, not only on one "grandmother cell", instead, there must be a circuit or even a system responsible for recognizing. However, we know almost nothing about such a circuit or system. It is not only the tip of the iceberg which broke the Titanic, it was mainly the body of the iceberg under sea which sunk the ship! Or as an old Chinese joke: One guy ate a cake, and was still hungry, so he took the second, and on and on, until he finished the seventh cake. Now he was full and said: "If I knew that earlier, I would have only eaten the last cake, so that I could have saved most of the money!" Connectomic studies may solve this problem someday.

To say it is the grandmother cell which makes the subject recognize her/his grandmother reminds me of a story told by American neuroscientist, Michael S. A. Graziano in his new book *"Consciousness and the Social Brain"*. *When a father and his son were watching a magic show, the magician's trick was so tricky, that the father asked his son: "Jimmy, how do you think they do that?" The son replied: "It's obvious, Dad." The father was so surprised and said: "Really? You figured it out? What's the trick?" The son said: "The magician makes it happen that way."* The grandmother cell is the magician! But how does it do it? Although I am afraid that I don't agree with Dr. Graziano's main ideas expressed in his book, I appreciate this story very much. As a matter of fact, the story pointed out the flaw of many research works, such as many results of brain imaging. Of course, I don't mean that such results are meaningless, but that it is not enough. Knowing who makes the trick is important, but more important is telling how he makes the trick! As for the grandmother cell, I am sorry to say, it is even not the magician, but only a member of a team of magicians.

Your theory about the role of grandparents playing in evolution is very interesting. I totally agree that besides natural selection, passing on knowledge and experience to others, especially to the new generation, is very important for human beings, in which grandparents may play an important role. This makes human beings strengthen their ability so quickly and dominate over other species at last. Grandparents contribute a lot in this process ☺. And social interaction also plays an important role in it. It reminds me of a quotation from Emerson M. Pugh: "*If the human brain were so simple that we could understand it, we would be so simple that we couldn't.*" His paradox puzzled me for a long time, and then I realized that, on the one hand, his remark seems reasonable on the fact that the human brain is so complicated that one brain is difficult to know itself. However, on the other hand, it is not only one brain trying to learn itself, but as you said, it is a social network of evolving brains trying to learn a single brain. A social network of brains is much more complex than a single brain! Therefore, although I don't expect that someday we could declare that "Aha! Now all the mysteries about the human brain are discovered!" we can still approximate that goal gradually, even if we could never reach it.

As I told you in my last letter, when I check my draft of translation, if I find something doubtful, I consult the original. Most of the time, I find there is a mistake owing to not understanding the author when I translated. Now when I read translations of popular science books, I often find something hard to understand, consult the original books, most of the time, I find the cause is the same — the translator did not understand what the author was talking about. He might just make a literal translation word for word, or after another's similar translation, just like machine translations. Even worse, occasionally, I could not guess why the translator translated like that! Thus, it seems to me, for daily life translation, as there are large quantities of good translation examples, and its grammatical structure is not so complicated, generally speaking, there will be no serious problems for machine translation. However, for contents which are rarely talked about, with complicated grammatical structures and need real understanding, I doubt if machine translation can beat qualified human translators. As I am a layman of AI and you are the expert, I would like to hear what you think about machine translation.

I like your remarks about the differences between the human brain and the electric "brain", some of which von Neumann, the father of the electric brain, mentioned even more than 60 years ago! Your remarks also remind me of the following words by Nobelist Gerald Edelman and Giulio Tononi in their book "*A Universe of Consciousness*", which I translated more than a decade ago, they even wrote a special section titled "The Brain Is Not A Computer" in this book.[1]

"... the brain has special features of organization and functioning that do not seem consistent with the idea that it follows a set of precise instructions or performs computations. We know that the brain is interconnected in a fashion no man-made device yet equals. First, the billions and billions of connections that make up a brain's connections are not exact ... At the finest scale, no two brains are identical, not even those of identical twins. ... These observations present a fundamental challenge to models of the brain that are based on instruction or computation ...

Another organizing principle that emerges from the picture we are building is that in each brain, the consequences of both a developmental history and an experiential history are uniquely marked ... The individual variability that ensues is not just noise or error, but can affect the way we remember things and events ... No present-day machine incorporates such individual diversity as a central feature of its design, ...

If we compare the signals a brain receives with those of computers, we uncover a number of other features that are special to brains. First, the world certainly is not presented to the brain like a piece of computer tape containing an unambiguous series of signals, Nonetheless, the brain enables an animal to sense the environment, categorize patterns out of a multiplicity of variable signals, and initiate movement ...

We have also shown that the brain contains a special set of nuclei with diffuse projections — the value systems — which signal to the entire nervous system the occurrence of a salient event and influence changes in the strength of synapses. Systems with these crucial properties are typically not found in man-made devices, ...

Finally, ... the most striking special feature of the brains of higher vertebrates is the

---

[1] Edelman G M, Tononi G. A Universe of Consciousness: How Matter Becomes Imagination[M]. New York: Basic Books, 2000.

occurrence of a process we have called reentry ... It is the ongoing, recursive interchange of parallel signals between reciprocally connected areas of the brain, and interchange that continually coordinates the activities of these areas' maps to each other in space and time ..."

Of course, I don't think that they have listed all the differences between the brain and the computer. Late Jen Matsumoto listed two other differences: The brain compiled its own program and searched for data by itself. In addition, the brain "trains" itself by active interaction with its environment, maybe there are more differences you can list. Anyway, Edelman and Tononi proposed something deserving to be thought over, especially their idea that the brain is a selective system, but not a computing system. As you know well, Jeff Hawkins also emphasized that the brain is not a computing system, but a memory and prediction making system. Anyway, the computer metaphor of the brain seems quite doubtful.

I am glad that you like Dr. Pei Wang's thought. I don't know him personally either. I just came across several of his papers in a Chinese web journal *"Mr. Science"* and am fond of them very much. As these papers are written in Chinese, so it is easier for me to read and understand as a layman. I have downloaded all his publications including these papers in Chinese which could be downloaded from his homepage (https://cis.temple.edu/~pwang/)[1], and will read them later. I agree with all of your comments about him and am glad to discuss his ideas with you someday.

Wow! You know Paul Allen himself! He is also a critic to Kurzweil! Allen's critiques in his paper *"The Singularity Isn't Near"* has annoyed Kurzweil so much that he defended himself against Allen with four fifths of space in the chapter 11 titled "Objection" in his book *"How to Create a Mind"*.

OK! Now it is the time to come back to discuss about the Human Connectome Project (HCP) and the Allen Institute for Brain Science.

---

[1] In this website, you can find many interesting and instructive materials, Especially the "Suggested Education for Future AGI Researchers" part:
https://cis.temple.edu/~pwang/AGI-Curriculum.html
This is the best list of this topic we could find anywhere. It is nicely sorted according to the initial knowledge of the readers, step by step all the way up to the highest level of up to date science. We strongly recommend readers interested in AI to read it.

As you said in your previous letter, at first, I also misunderstood the goal of the HCP. Only after reading the related materials I understood that the goal of the HCP is to map the connections between brain areas instead of neurons. The task is much easier and practical, even so, they limited their task to map a healthy young adult's brain in the first five years and leave the developing and aging healthy brains, and the brains with a variety of disorders to the future. An important task of the project is to parcel out each hemisphere into more areas on a more solid foundation than Brodmann's division of the cortical cortex into 52 areas based only on cellular architecture. This will give them a solid foundation for further works. They also used four different MRI methods to map anatomical and functional connections between these different areas. [1]

The Allen Institute for Brain Science was founded by Paul Allen and his wife in 2003, and now Christof Koch is the president. In November 2011, they launched an Allen Mouse Brain Connectivity Atlas to map a 3 dimensional high resolution connectome in the mouse brain, as a first step, only in some areas in the visual system. They also launched another project — the Allen Cell Types Database to categorize neuron types in the mouse brain.

Compared with Markram's HBP and the original description of the BRAIN initiative by President Obama, the goals of the above projects are very limited and practical. They focus on important basic data collected with relatively matured approaches. Although their working capacities are huge, new technologies should be developed to make work more efficient and better, experts with different backgrounds should be organized into one team, just as Sebastian Seung said that such tasks would be successful, at last, if enough financial support was given, their goals were held firmly and the team was well organized. These projects may give good examples of big science, team science, and open science. However, I think this may not be a universal way to organize all scientific studies, especially those which need creativity and insight.

As "the daylight has come", I have to stop here.

Best!

Fanji

---

[1] http://humanconnectome.org/ccf

Ⅲ-002 Karl

09.01.2016

Dear Fanji,

First of all, Happy New Year to you and your family!

Thank you for your subtle and witty comments on the possible existence of a grandmother cell. I very much liked your old Chinese joke about the guy who ate seven cakes and believed that it was just the last one to fill him up and therefore thought that he could have saved most of the money. And I even more liked your concise and crisp reduction of this insight to the metaphor *"It is not only the tip of the iceberg which broke the Titanic"*. In this case I'm (almost) sure that it is not an old Chinese saying. ☺

If it was you who has invented it, I ask for you permission to use it in my future writing because it is perfectly applicable in so many cases when the causality chain of human reasoning brakes. Of course, in case you are willing to grant me permission, I will only cite it with reference to your priority!

With this metaphor and with your reference to Graziono's *"the magician makes it happen that way"* you are touching a crucial point both in research methods and in the inner mechanics of our mind's reasoning and its inbuilt longing for causal explanation.

I believe that this longing for causal explanation is our mind's favorite criteria for finding orientation and that it also defines how we do science. The descriptive part of science and research where objects and phenomena are named, categorized, systematized, and measured is something totally different from the explanatory part. In the first part, we deal with the "what" and "where" questions and in the second part with the "how" and "why" questions.

The weight distribution between these two fields is very different among the various scientific disciplines. And typically, it is so that the less we understand how a system works the more emphasis we put on the descriptive part which is somehow driven by the hope that the more we learn about a phenomenon and the more precise we can describe it the closer we get to understanding the puzzle.

Sometimes this helps, indeed, but quite often this kind of scholastic diligence just keeps us busy, gives us the illusion of understanding, and may even prevent us from cracking the enigma behind the phenomenon.

I share your opinion that this magician story points "*out the flaw of many research works, such as many results of brain imaging*". And as you say, such descriptive research isn't meaningless but not enough. Let's hope that the BRAIN initiative is not just out for producing ever more data for the first category. Of course, it would be nice if we would have a taxonomic system for brain locations and activity events similar to the one Carl Linnaeus has introduced for comparative botany long ago.

But even if biologists would have managed to describe all plants in the world and perfectly characterize and sort them into Linnaeus' system it wouldn't have revealed the enigma behind the miraculous variety of all those plants, the function of DNA, or the role of evolution. And the same is true for the dance of the "*team of magicians*", as you have rightfully called them, doing the trick in our brain which supplies us with consciousness.

Actually, we don't even know whether the trick is really so complicated or whether we just overlook quite simple principles, while the solution may lay right ahead of us.

Valentino Braitenberg has shown us with his wonderful artificial creatures, the, meanwhile famous, "Braitenberg Vehicles", that a seemingly complex and hard to understand behavior can be caused by very simple or even primitive internal mechanisms. An outside observer may find it very difficult or even impossible to understand the internal coding of these machines although its internal "intelligence" is very primitive and based on just a couple of telephone relays, simple sensors, and motors. An investigator of such a vehicle trying to understand its internal logic may be, nevertheless, deviated to create enormously complex theories about a mighty internal agent controlling such a machine, although there is none, and the solution of the puzzle is rather trivial. [1]

---

[1] http://en.wikipedia.org/wiki/Braitenberg_vehicles

Our mind seems to love stories where cause and effect rule and where agents and their intentions are responsible for what's happening. Braitenberg shows us, however, that you don't need calculating agents to show complex behavior. His vehicles perform surprisingly human like behavior without the use of supercomputers and heavy numerical calculation. Braitenberg vehicles are nothing else but technical thought experiments. They help us to understand that there may be quite intelligent machines or animals that can successfully survive in a complex world without heavy calculation or a central control agency. Insects, especially ants, are good examples for this possibility in the real world. They have very small brains and nevertheless are an extremely successful species building even complex forms of social systems.

Braitenberg, a very independent thinker and border crosser between neurology and AI has heavily influenced AI engineers, especially Rodney Brooks with his insect-like robots. To me, he was one of the wisest men in the business although he got little credit for it. Probably because he did most of his thinking with the help of paper and pencil and never asked for billions. ☺

I'm grateful that you didn't declare my "grandparents are important" theory as pure nonsense. You added an interesting argument to it of which I didn't think before when you cited Emerson M. Pugh's paradox: *"If the human brain were so simple that we could understand it, we would be so simple that we couldn't"*!!!

I also came across this puzzle but never struggled with it as intensely as you did. I was quite thoughtless about it and regarded it as a pseudo problem. My solution was that I asked the rhetorical question: "can a faint human being, merely able to lift 100 kg, solve the puzzle of how to lift 10,000 tons?" The answer is yes, because we can make machines that outperform humans in almost everything we and our bodies can do. This is also true for most tasks our brain can perform. Computers are better in memorizing, faster and more reliable in doing calculations, and technical sensor-systems outperform the abilities of humans in the physical and chemical range by far. Actually, there is only a very small portion of our mind's capabilities left which machines can't do yet. So why shouldn't it be possible to do the rest as well? But your answer to the paradox is much more subtle and better of course.

You're right; the solution is that the power of a single brain and mind can be exponentially improved by aggregating the capabilities of many brains and minds. And when you do this, not only in parallel at a given point in time, but also sequentially over eons you aggregate the intelligence and knowledge of many generations (where grandparents play a major role). And this gives a perfect answer to Pugh's paradox. Many brains and their accumulated wisdom will be able to understand a single brain.

Fanji, what you did here is just great!!

Now when I thought about it, I found that the major flaw of the paradox is that Pugh overlooks the fact that a naked, empty brain can do almost nothing while a brain loaded with the right knowledge can do almost anything. The point is that minds can share their content and continuously aggregate, focus, and hone their knowledge. This is exactly what happened in technology. And it is the reason why our engineering-competence has exploded over a few hundred years after the ramp-up of self-organization of matter to this starting-level took some billion years.

In any event, it was you who has solved Pugh's paradox. You may think about publishing it and I would be happy if you could mention the grandparent relevance theory that inspired you to find the solution.

So, we may both become famous and maybe rich in the end. ☺

Your friend Walter has taken considerable effort to deal with this cause and effect mechanism which seems to be so deeply rooted in our brain, and also with the role of causality and intentionality in the building of consciousness. He worked on it for decades and the results of this huge effort are laid down in his publication "*Consciousness, Intentionality, and Causality*". I discovered it just recently after you had directed my interest to his work. Meanwhile, after I was more acquainted with his perspective I found that what he laid out here to be the very core of his theory and that what he has to say is quite substantial and relevant.

In the summary of the article he makes these statements:

"*According to behavioral theories deriving from pragmatism, Gestalt psychology, existentialism, and ecopsychology, knowledge about the world is gained by intentional action followed by learning. In terms of the neurodynamics described here, if the intending of an act comes to awareness through reafference, it is perceived as a cause. If the consequences of an act come to awareness through proprioception and exteroception, they are perceived as an effect. A sequence of such states of awareness comprises consciousness, which can grow in complexity to include self-awareness. Intentional acts do not require awareness, whereas voluntary acts require self-awareness. Awareness of the action/perception cycle provides the cognitive metaphor of linear causality as an agency. Humans apply this metaphor to objects and events in the world to predict and control them, and to assign social responsibility. Thus linear causality is the bedrock of social contracts and technology."* [1]

What Freeman expresses here defines a huge playground giving room for more than one PhD-thesis. The article is very condensed and not easy to read because you have to be familiar with a few concepts not only in neuro-biology, but also in physics, logic, chaos-theory, cybernetics, philosophy and many other fields. So, these 36 pages gave me a hard time in doing related reading and learning and I'm still not sure whether I can get my arms around everything Freeman is out for. To me, it seems like a distillate of his life-long research and his legacy to future generations of researchers. And it also includes an attempt to contribute to a general theory of self-organization in the body, the brain, and the mind.

For Freeman, as you can see in all of his work, the interaction of the body with the material world is the starting point. Here he puts great emphasis on the principle of reafference. It is based on the idea that with every intended action an expectation is compared with the resulting sensory input. In the event of a difference, the brain-body-system tries to reduce the difference between the expected value and the realized sensory input after the action in subsequent action cycles until a minimum has been reached. The principle of reafference, (you don't find the term in Kandle's bible, where it is called "efference copy"), has been developed to explain how the motoric system of muscles can control and stabilize its movements (e.g. grip and move an object or improve gaze stability when moving) and also how the brain can filter self-induced signals from inflowing sensory-signals. The most popular mentioning of the principle is in the context

---

[1] Freeman W J. Consciousness, Intentionality and Causality[J]. Journal of Consciousness Studies, 1999, 6(11 – 12): 143 – 172. Freeman gives credit to Erich von Holst and Horst Mittelstaedt for introducing the principle of reafference already back in 1950.

of explaining why we cannot tickle ourselves while other people can, simply because the internal copy of the outflowing (efferent) motoric signal is the same as the incoming sensory signal. What we feel is exactly what we have expected, and we cannot surprise ourselves. You not only can't tickle yourself you also can't tell yourself a joke with great success. ☺[1]

You know all this, of course, and the idea that organisms keep internal models of the outer world isn't new. But what I found remarkable is how Freeman used this very fundamental principle as a starting point of a complete concept of human behavior and intentional action, and not just for explaining the movement of muscles, all the way up to the level of consciousness.

You know this field much better than I do and, therefore, I would like to hear your opinion about this concept and whether I'm wrong with my opinion that Freeman's ideas would make a very promising theoretical framework for future research both in the HBP and also for the BRAIN initiative. It would be interesting to learn what Christof Koch thinks about it.

What makes it so attractive to me is the familiarity it has with similar thinking in the world of cybernetics, not in the sense of digital von Neumann computers, but with the principles Norbert Wiener has developed in his theories about closed loop control and the subsequent developments of automata theories. The problem with these developments is that almost all of them were realized digitally while neurons and the brain are analogue systems. In this context it may be important to remember that Wiener developed his ideas around cybernetics and closed loop control for anti-aircraft gun analogue computers.

In any event, Wiener's idea that a system has a desired state which it wants to reach by acting in the physical world and that after an action, the sensory input is compared with the internal expectation and the difference to what is reached, is minimized in sequential loops has been extremely successful in the technical world. It is the backbone-idea that controls all the machinery we see in robotics and AI.

The point is whether this adaption has to be done by numerical calculation (which is

---

[1] https://en.wikipedia.org/wiki/Efference_copy

very energy consuming using today's digital computers) or whether there are better means.

Daniel Wolpert from Cambridge, who is an expert in computational biology and sensorimotor control, made the point that the principle of reafference can also be described with the function of a Kalman-Filter. [1]

This is an interesting hint because the Kalman-filter is one of the most widely used algorithms in modern digital signal processing. Originally, it was invented to improve the communication between low-power satellite transmitters and the earth and meanwhile it is the engineer's preferred workhorse when it comes to telling the difference between signal and noise in all kinds of control systems.

However, filtering doesn't have to be done by digital calculation. Analog radio and TV systems were full of filters that separated all kinds of signals by means of combined passive elements like condensers, electromagnetic coils, and resistors. This filtering happened very fast and cost (almost) nothing, very much like mechanical sieves separate a finer kind of gravel from the larger ones. You could do the same separation using a camera, a computer, and a robot plus an enormous amount of energy. We are so much used to the concept of digital calculation and algorithms that we often tend to forget that there are much easier, cheaper, and sometimes much faster solutions available since long. Mirrors are a good example. Classic glass-silver-mirrors reflect the incoming light with the speed of light, and no additional energy is needed. Today, people use smartphones as mirrors. Here the camera with a CCD-sensor-chip is used to digitize the picture, send it to the CPU, apply some algorithms on it, and then send it to the display with some delay and after the use of a considerable amount of energy. A glass-prism that separates a beam of sunlight into the various colors of the rainbow is another example. It works with the speed of light and costs nothing. You can do the same with arrays of sensor-chips, a CPU, fancy algorithms, a color-TV and lots of energy.

I'm not sure whether the idea of the brain as an analogue percolation system to filter meaning from noise is the ultimate approach. But when I compare the energy consumption of the brain (20 Watt) with the chips IBM fellow Modha calls

---

[1] Daniel M. Wolpert et. al. An Internal Model for Sensorimotor Integration. Science (269) 1995, S. 1880 – 1882.

neuromorphic, which may need gigawatts to do the same job someday, it makes me think that something with this digital calculation approach is fundamentally wrong.

A friend of mine, Bernd Ulmann, is a leading expert for analog computers (and has gathered one of the largest collections of analogue computers in the world [1]), at least didn't object to my idea. ☺

It's good to know that we, Freeman, and Jeff Hawkins are not the only ones who doubt that the brain is a computer. Thank you for the citation of many good arguments from "*A universe of Consciousness*" by Edelman and Tononi which are all valid in my eyes. And what you report about what Jen Matsumoto says comes pretty close to Freeman's position.

So, Edelman's and Tononi's book you have also translated! Wow, once more you impress me with such a revelation. Is there any relevant book in this field which you haven't translated or an author you don't know? ☺

I would love to come back to your question about the possibility of machine translation which is a good example for one of the remaining weak spots of AI. Why we are still doing so poorly in this field has to do with the fact that language is a thing of content and meaning. It is condensed culture and nothing which can be detected in the hardware of a naked brain. The way it is installed in our brain belongs to the most subjective part of the mind you can think of. Language actually is a perfect example for this serial aggregation of knowledge I was talking about and, therefore, it is the very core of the puzzle and probably one of the last ones we will be able to solve.

But, this is a totally new field and while this letter, again, is so long, although I didn't even talk about Paul Allen, I would rather start this debate in a future letter maybe after you have told me whether my enthusiasm for Freeman's article is well founded or maybe exaggerated. Of course, I hope for the first and send my best to the admired master of the field in the land of middle.

Good night!

Karl

---

[1] http://www.analogmuseum.org/english/

III-003 Fanji

24.02.2016

Dear Karl,

Thank you very much for your kind season's greetings. In return, Happy Lunar New year to you and your family!

It will be a great honor for me if you cite my iceberg metaphor in your writing. Of course, it is not an old Chinese saying, but a new one by myself ☺.

You are absolutely right to distinguish the descriptive part of scientific research from the explanatory part: "what" and "where" vs. "how" and "why". For a very long time, biological studies were mainly limited to the former part, and the situation remains the same, even today, in some fields of biological studies. At the very beginning, to know "what" and "where" is important, but to know "what" and "where" does not mean to know "how" and "why" at all. The reason is obvious; however, many people seem to have forgotten such a trivial truth and are satisfied with the former. Your story about Braitenberg Vehicles is a good lesson, which I did not know before. If we don't know how it works, we might think it "knows" what is good for it, we might think that it has a value system to guide its behavior. Thus, to judge the inner world of an agent just based on its behavior is very questionable. Unfortunately, many reports on AI in the media do just this.

Your story about Braitenberg Vehicles also raises the problem: "What is Intelligence?" This story shows that you cannot judge intelligence just by its behavior. Your story reminds me of what Pei Wang said in a Chinese web magazine *"Mr. Science"*, in which he emphasized that you must distinguish skill from intelligence. In his opinion, intelligence is the ability to adapt when there is not enough knowledge and resources. Intelligence is the ability to improve skills with experience. Braitenberg's vehicles may be very skillful to follow light, but it cannot improve such ability with its experience. Therefore, according to Wang's criterion, Braitenberg's vehicles are skillful, but not intelligent.

Your story about Braitenberg's vehicles reminds me of a story in J. Henri Fabre's classical book "*Souvenirs Entomologiques*", in which he described the behavior of wasps. He found that wasps could build a nest for their offspring with mud sophisticatedly. The nest is just like a jar to reserve its eggs and captured insects as the food for its larva. It killed an insect and took it to its nest and laid eggs on the top of this insect, then flew away to catch another insect and put it on the top of the first one, on and on, until there are ten or more insects in the nest, which would be enough to feed the larva until it grows to be a wasp, then it would close the opening for safety. In this way, the larva would eat the oldest insect first, and the freshest one at last. You may admire how thoughtful the wasp is for its offspring. However, Fabre played a trick to see if the wasp was really intelligent. Every time when the wasp took an insect into its nest and flew away, he just took the insect and the egg on it (if there was an egg) away, however, the wasp never noticed that its nest was empty, and just went on and on, until it closed the opening at last even when the nest was empty! Therefore, although the wasp seemed intelligent and had high skills to build its nest, its complicated behavior was just its instinct, a fixed procedure without the ability to adapt the change depending on the situation. It had no intelligence.

I especially appreciate Pei Wang's idea to distinguish skill from intelligence. No matter how complicated a skill is, if it cannot be improved, and is just fixed at the same level, then it cannot be called intelligent. Jeff Hawkins emphasized that "*Prediction, not behavior, is the proof of intelligence.*" "*Intelligence is measured by the capacity to remember and predict patterns in the world ...*" [1] I used to think that intelligence is the ability that an agent can still solve a problem which it has never met. My idea may coincide with the above definitions. Of course, if a problem the agent has never met, i.e., its knowledge and resource about solving the problem is not enough, yet, the agent could still predict the result of its action, it must be intelligent! Fabre's poor wasp was not intelligent, as it could not predict the result of its action when there was a tricky scientist who took its insects and eggs away, a situation which it had never met!

Although Fabre's wasp gives an excellent example how a living creature can show

---

[1] Hawkins J & Blakeslee S. On Intelligence[M]. New York: Levine Greenberg Literary Agency, Inc, 2004.

complicated behavior without intelligence, we still don't know what happens in its brain. Braitenberg's vehicles give an even better example, as we can know the simple mechanism inside it thoroughly!

I am very glad that you agree with my comments on Pugh's paradox. As a matter of fact, I've heard of a similar paradox long ago. Someone said that we could not understand our own brain, just as we can't pull ourselves out of the swamp by our own hair. Of course, this new version is even more feeble, as only another person could lift you away from the earth by your hair in principle. Your comments on the social network of minds struck me to give an explanation to these paradoxes. Therefore, if my explanation could be published someday, the credit of the discovery of this explanation should be shared by the two of us.

Last Jan 30th is Walter's 89 birthday, in Shanghai; we usually cerebrate 89 birthdays as a big event, so I sent him an e-mail with an electronic birthday card. He is still working hard and published several books and papers with his former students and colleagues, although his health is not so good after his wife passed away last year.

Yes, his paper "Consciousness, Intentionality, and Causality" is very condensed and not easy to read. My feeling is just almost the same as yours, *"I'm still not sure whether I can get my arms around everything Freeman is out for."*.

The argument which gives me the deepest impression in this paper is his emphasis of circular causality in intentionality.

"Intentionality cannot be explained by linear causality because actions under that concept must be attributed to environmental and genetic determinants, leaving no opening for self-determination." *"Awareness and neural activity are not acausal parallel processes. Not does either make or move the other as an agency in temporal sequence. Circular causality is a form of explanation that can be applied at several hierarchical levels without recourse to agency."* *"The most important, with wide ramifications, is the assumption of universal determinacy, by which the causes of human behavior are limited to environmental and genetic factors, and the causal power of self-determination is excluded from scientific consideration. ... It is absurd in the name of causal doctrine to deny our capacity as humans to make choices and decisions regarding our own futures, when we exercise the causal power that we experience as*

free will." [1]

In this way, he gave a reasonable explanation about free will, which puzzled Susan Blakemore and almost all the scientists she interviewed so much: "*To be frank I had rather expected, before I began, that nearly everyone would intellectually reject the idea of free will while finding it hard to live their daily life without any such belief ... As Samuel Johnson put it so memorably 'All theory is against the freedom of the will; all experience is for it.'* " [2]

It is Walter who pointed out where this contradiction comes from. Just as you once said, it is a great pity that Blakemore did not interview Walter before she wrote the book. By the way, when I discussed the causality problem with Dr. Hans Liljenström, he told me that our common friend Herman Haken was the first to propose the problem of circular causality, which was ignored by most scientists, who are still insisting on the principle of linear causality and the world view of classical determinism.

The other important view I felt, but could not express as clearly as Walter did, is his remark in this paper: "*Two things distinguish humans from all other beings. One is the form and function of the human body, including the brain, which has been given to us by three billion years of biological evolution. The other is the heritage given to us by two million years of cultural evolution.*"[3] It is the latter which makes us develop technology and change our lives so fast. This is one of the main topics we have discussed in our previous letters.

I totally agree with your remark that "*Freeman's ideas would make a very promising theoretical framework for future research both in the HBP and also for the BRAIN initiative.*" However, I am not sure if these two giant projects have considered the importance of circular causality. In my eyes, their approaches may mainly still go along the way of linear causality in the microscopic levels and upward, however, they seem to be paying little attention to "*the global state downwardly organizes the activities of the individual neurons.*" This is only my impression, I am not sure if it is really the case,

---

[1] Freeman W J. Consciousness, Intentionality and Causality[J]. Journal of Consciousness Studies, 1999, 6(11-12): 143-172.
[2] Blackmore S. Conversations on Consciousness[M]. Oxford: Oxford University Press, 2005.
[3] Freeman W J. Consciousness, Intentionality and Causality[J]. Journal of Consciousness Studies, 1999, 6(11-12): 143-172.

you may tell me if my impression is reasonable.

Your arguments about if technology should always develop only by numerical calculation is very instructive. The energy consumed in the switching between high and low voltage levels makes such machines very energy-consuming. Your many marvelous examples indicated that there could be a simpler and more energy-saving way to realize the same function than the digital technique could do. When thinking falls into some basin of an attractor, it is often difficult to jump out ☹.

Your examples remind me of Dr. Ramachandran's mirror box to cure the pain of phantom limb. Some amputee may feel his lost fingers digging into his palm so tightly that it makes him feel unbearable pain. Dr. Ramachandran knew vision often wins in multi-modality competition. He thought if he could find a way to let the patient see his lost fist loosening, the pain might be gone. Of course, now we have a variety of virtual reality technologies, but it was very expensive in 1990s, maybe all the grant money he had was not enough to buy such an instrument. Thus, he just bought a big mirror, put it in the middle of a big box to separate the box into two halves, and he opened two holes in the front side of the box on both sides of the mirror, so that the patient could put his arms into the holes. Then the patient could look at the image of the healthy hand in the mirror as if this were his lost hand. When he let the patient loosen his healthy fist, the patient noticed the image also loosening as if his phantom fist opened, and the pain was gone!

Of course, you may know the story, however, I was so excited when I first read this story in his classical book *"Phantoms in the Brain"*, which I also translated into Chinese. Ramachandran is a genius to use low-tech to solve difficult problems! By the way, as there are 12 sentences which were too difficult for me to understand, I wrote to Walter and asked his help. He immediately gave me a detailed answer, although his sight was not so good, and his health was also not so well. He is such a generous and kind gentleman!

I am looking forward to hearing your further remarks on language, Paul Allen, and other stories. However, it is already dawn and I have to wait for your next letter.

Good morning!

Fanji

P.S. You must notice that it was announced on January 27, there will be a competition between the Go-playing system "AlphaGo" developed by DeepMind with the former world champion Mr. Lee Sedol next month. People are arguing if the AI machine can beat the top Go-player all over the world, although it had beaten the European champion, Mr. Fan Hui, in October last year. I am very curious about the result, what do you think?

III-003　Karl

20.03.2016

Dear Fanji,

Thank you for your kind letter, your season's greetings, and the allowance to cite your great iceberg metaphor!

I've been thinking about your many arguments and the inspiring input you've given, but under the impression of the latest news, I'd like to start with the topic you mentioned in your PS, the contest between Googles AlphaGo AI system and Go champion Lee Sedol. Meanwhile the showdown happened and AlphaGo won four out of five given games. Once again, a machine has defeated a human being, after IBM's Deep Blue computer won against chess world-champion Garry Kasparov in 1997 and later, in 2011, Watson with the "Jeopardy!" — champions.

Well, not many people in the computer and AI business were really surprised and most expected this result, including me. Actually, I wondered how the champion was able to win a game at all and found the whole operation more to be a huge marketing show with little proof that a new quality in artificial intelligence was achieved. Or as one of my young friends with a PhD in informatics, who wrote his thesis on AI, put it: "more of a home game advantage for a computer wasn't possible". He was right because axiomatic games with a limited rule set are paradise for computers running algorithms on millions of parallel kernels and a nightmare for a fuzzy single human brain. And this advantage gets even bigger the larger the event spaces of the possible constellations in a game becomes. It would also be bigger on a chess board with 100×100 fields instead of 8×8 as in regular chess. And while Go has a much deeper event space than chess the advantage for AlphaGo was much bigger than for Deep Blue 20 years ago. With databases at hand where all games ever played are stored, a whole known universe of events is defined and fully accessible for the machine. In such schematic applications controlled by simple rules, but running on vast and deep playgrounds, raw computer horse-power really helps, but this has nothing to do with intelligence in the sense you and Pei Wang are talking about.

A really intelligent, surprising (and also humorous) move of AlphaGo would have been if it would have proposed his programmers not to waste hundreds of millions of dollars and burn gigawatt-years for such a pointless demonstration and rather offer the Go-champion 10 million in case he loses. This would also have given the programmers a chance to introduce AlphaGo to the next level of intelligence: rules of ethic. The PR effect would have been the same and the money could have been spent on relevant problems. ☺

So, although I believe in the long-term success of AI and even strong AI, as you know, I wasn't very happy about the staging of this fairground show. The resonance however shows that it is a huge marketing and PR success, because all media people hopped on it. The sensation was fired up when the unavoidable chorus of admonishers once again called that the end of humanity is near. This time because machines and robots will take control. Deriving such a consequence from a rather silly game-victory is rubbish, of course, but it helps to make everybody believe that Google and DeepMind have developed an AI system so powerful that it is a threat to mankind. Smart goldmakers know that it increases their chances of collecting money to accomplish the promised wonder when experts warn of the devastating consequences of such technology.

If they say we must not make artificial gold under any circumstances, because otherwise the economy will collapse, it's good for the gold making business. This type of alarm increases the popularity of the goldmaker and draws the discussion's attention only to the bad consequences. The claim that gold (or AI) can be made, which hardly anyone can judge, becomes irrelevant. For the public and potential investors, the emotional message of such alarm calls, practically confirms the rational claim. Humans are escape animals and run away in panic without thinking when one of the flock calls alarm. This reflex works again and again as reliably as in your wasp-example. And, therefore, every-time someone rings the "the end is near" alarm bell the gaze is focused on the apocalypse and no-one will ask any more whether it is possible to make gold (or whatever wonders promised) at all. So, a clever goldmaker is always accompanied by alarmists and admonishers. And so, are AI-gurus, especially when they know that they can count on the help of fellow Silicon Valley business people, who are all happy to help and willing to warn of the horrifying consequences of AI.

Of course, Google is making some progress in applying neural networks. But to win a Go game against a Go-Champion after IBM's chess demonstration (and 20 more years of Moore's law in place) is no better testimony for the superiority of the machine over the best mathematician than a computer calculating Pi to a million decimals in less than a second. Axiomatic logical games are a very small subset of the real world and here recurrent neural networks and dedicated hardware are on their home-turf. It's one domain where Kurzweil's method of ever more and quicker hardware really works. While it was Google's PR-guys who took the victory in this case, there has been much more relevant progress in the application of neural networks that happened almost unnoticed recently.

For many years one of the biggest challenges in AI was face recognition. Even recognizing different sorts of animals was a really difficult task for a machine. For many years, I used the example that computers and algorithms have big difficulties to tell the difference between a cat and a dog as proof that there is little progress in AI. This has changed recently after Facebook began analyzing the countless images people store there. Now after they have trained their networks with hundreds of millions of photos they are pretty good at not only telling the difference between a cat and a dog but also in identifying human individuals. They don't use recurrent neural networks (RNN), as Google does, by the way, but convolutional neural networks (CNN), invented by Yann LeCun, who also heads Facebook's AI research group (FAIR). CNNs are a smarter variant of the early perceptron, the first kind of artificial neural network, specially designed for pattern recognition.

The progress Facebook made is much more relevant, in my eyes, and really amazing because the possibility of identifying people in a video is a real breakthrough while a computer winning a board game is not. Interestingly, FAIR is also experimenting with CNNs to do automated translation, something Google also does with RNNs.

There is much room for improving machine translation indeed, and there is a huge field of promising applications ahead of us. What Google and Microsoft have provided here, so far, with their translators for written language is sometimes quite helpful when it comes to simple text and trivial matters like understanding what an internet-site is about. But used on complex texts or even literature, the results are often rather poor, misleading and useless or even outright ridiculous.

However, this will be the playing field on which the AI contest will be decided because those who can provide a machine that can read and listen to written or verbal language and really understand will change the game.

Some people say that China and its AI players like Baidu and Tencent have an advantage in this race, simply because they have more users and therefore larger amounts of data at their disposal to train their neural networks. This may be true at least as long as it really takes hundreds of millions of pictures or documents to teach a neural network the difference between a cat and a dog. This teaching is done by calculating statistics out of the variations between elements of such pictures, and the more examples you have the better the statistic gets. It really works, sometimes even amazingly well, while in other cases it fails. But in any event, it requires an incredible amount of hardware and energy.

And when I think of a 2 year old child that can solve the same task after maybe 10 cats and dogs it makes me think, however, that something is wrong here. Maybe it's the same kind of wrongness as in the case when supercomputers, even in theory, have to burn gigawatts to match the power of the original brain running on 20 watts. To me it looks like something is even fundamentally wrong with these algorithmic networks, so generously called "neural" as if they would be mimicking our brains. Everybody who has a closer look can see that they don't and as nice as it is to have such tools because they are much better than what we had earlier, we have to be aware that, in reality, they are only a first small step.

The problem is that this step may lead in the wrong direction, as in the case of the steam-engine which I mentioned some time ago. The big problem is that it is very costly and time consuming to correct such mistakes. And when you have started to build an industry on a bad principle it may be almost impossible to correct such a mistake for a very long time.

To me Google's pompously staged Go-victory is primarily a marketing campaign with the intention of attracting the AI programmers in the world to their AI-platform Tensorflow. Here they compete with Facebook, Apple, Baidu, Microsoft, IBM,

Amazon, and many others and this is a very relevant race indeed. It is comparable to the battle of the PC operating systems in the 1980s, and the ongoing struggle for the most popular programming language. When IT-behemoth IBM who had dominated the computer-business for half a century like a monopolist lost the operating war against Microsoft, it never recovered from it. In the old days, the IT-center of a company was called the IBM-department and the people running the machines were educated and trained on IBM's operating systems and IBM-programming languages and felt and behaved like a priesthood belonging to a mighty church. Their Latin was COBOL and their catechism was the manuals describing the famous IBM – 370 – mainframe architecture.

Microsoft, which looked like IBM's successor as the dominator, for a while, soon lost control over operating systems and programming languages after they overslept the advent of the internet with the related relevance of search-engines, social-networks, and the mobile-phone revolution. Now Google and Apple are defining the operating systems and the related languages.

In all those innovations and revolutions, languages, platforms, and the related tools were always a key-factor. The first computer language a human brain learns seems to be as defining and formative as the learning of the mother tongue. Switching from a programing language to another with a different logic is as difficult as switching from your mother tongue to a foreign language.

Most people always run their next language through an emulator on top of the former language. When object oriented programing became popular I have seen many good programmers struggle with the new concept and never become really familiar with it because they always tried to emulate the new logic on the old basis they were accustomed to.

There seem to be sectors in our mind's architecture that behave like a write-once-read-only memory chip, at least, when it comes to learning languages. The same mechanism seems to be responsible for the phenomenon that most of us never manage to get rid of the characteristic dialect and sound of our mother tongue. You can hear, in seconds, whether it's a German or a French speaker giving a talk in English. Germans typically never learn to pronounce English "Vs" in the correct way and use "Ws" instead.

Typical French can't pronounce a German "H" correctly and I guess you can hear from what region in China a speaker comes, even when he tries to speak pure mandarin.

It is possible to correct this and hone verbal abilities, but it is an enormous effort and therefore most people prefer not to make this effort and rather stay with the principles with which their minds and language were originally prepared. The fact that it is so extremely difficult to change basic principles of the first language we have learned is one of the reasons why I'm very skeptical against Chomsky's idea of a generative, universal grammar rooted somewhere deep in the lowest levels of our brain.

In any event, Google's managers, as everybody in the IT business, know about this priming effect. And while it is so important to be the leader and the first to prime the minds of a young generation of engineers now when a new era of AI starts, Google makes enormous investments to take the lead and become the priming institution number one. And from what I can see in the surrounding of my young AI-friends, it really works because many are willing to hop on the Tensorflow bandwagon. From an engineer's standpoint, it's understandable, of course, that people want to be with the party that wins. They hope to learn what will be the new standard and don't want to waste their time learning frameworks, tools, and languages that may be obsolete soon. But times are gone when you could be a happy COBOL-programmer in the IBM department for 40 years. And when I'm right and the modern neural network architectures are all based on shaky ground, the real innovation is still ahead of us and there will be no established standard of the actual kind because there are much better things to come, maybe very soon.

And in the position of Chinese AI makers I wouldn't count on the advantage of having close to a billion users and related huge amounts of data. I would rather look for solutions that, as the 2 year old can do the job with fewer amounts of data-sets and especially with less energy. I wouldn't be surprised, now after we have seen that pattern recognition and understanding and translating of language is feasible for machines, to see solutions, soon, that will be better not only on a linear scale but better by orders of magnitude. And this leap into the future doesn't necessarily have to come from one of the big research labs of the industrial heavy weights with all their money and resources. In this domain, it can be a small group with the right idea about a fundamentally different method or algorithm. This happened often in the past and it is especially likely to happen again when the industrial mainstream is betting on principles that are so obviously inappropriate. In any event, there is an exciting track ahead of us and I am curious to see what will surprise us next.

I have to admit that my view results more of a bad feeling and I also can't propose a better solution. But I'm happy that you, Walter Freeman, and John von Neumann seem to share my point of view, that something is wrong with the computer-mind similarity.

I really liked how you chose the Braitenberg Vehicles as a starting point to reflect on the elements of intelligence. A glance at the literature could give the impression that there are as many definitions of the term intelligence as authors who write about it. It is in vain to ask for the "correct" definition of intelligence because, in open systems, all terms about phenomena are variables and not axioms as in closed systems like mathematics or logic. A definition may only be less or more helpful or fruitful in relation with the empirical testing of a specific theory of intelligent behavior. This process of operationalizing a variable is a difficult, and often, cumbersome operation that must always be adapted to the respective behavior. This is an almost trivial insight but not everybody in the neural sciences seems to be aware of it.

I also liked that you quoted Pei Wang in this context and what he adds to the discussion. The aspect of adaptability, indeed, adds a relevant dimension which helps to differentiate substantially different kinds of intelligent behavior. And Henri Fabre's famous wasp experiments which you have cited give a striking example. And I especially liked Wang's idea to distinguish skill from intelligence as you did. There is a problem, however, when you want to measure pure, "skill-free" intelligence in insects or in humans. When you test a system that can learn, regardless of whether it is a biological system, or an artificial system based on silicon chips, time and experience come into play. Intelligence, over time, improves the skills of a system that can learn. In the case of the wasp, evolution had to do the job and behind every wasp stand millions of wasps who helped to finally filter out what the modern wasp can do. We humans are much better off because with the help of language (and grandparents of course ☺ we can accumulate, transfer, apply (or sort out) the formerly acquired know-how of our ancestors and our fellows.

Time is a major factor not only in this process but also in my writing because again I ran out of time and space although I haven't responded to all of your questions and not even told you about Paul Allen. ☹

I apologize for being such an undisciplined pen pal and hope that you can forgive me.

Best and good morning/evening.

Karl

III-004 Fanji

28.04.2016

Dear Karl,

I am very sorry to tell you sad news. I just heard from Prof. Leslie Kay, a former Ph.D. student and long-term cooperator of Prof. Walter Freeman that he passed away, peacefully, at home in Berkeley on the 24th of this month. It is a great pity that I cannot go there to take part in his funeral. I had to ask her to transfer my deepest condolences to his family. The neuroscience community lost a great scientist and thinker; I lost a mentor-like friend.

He was always ready to help others. As I told you before, I have asked him for help many times, and he always gave me advice in time, even for very trivial matters. For example, once I asked him what the difference was between the words "neural" and "neuronal", which makes me confused when I read literature. He told me, these two words are almost the same, and only when you want to emphasize something closely related to a neuron, then you only use "neuronal" instead of "neural". Although this is a very simple question, it is difficult for me to ask someone in my country. Neuroscientists, who are not English native speakers, may feel the subtle difference, but may not be so sure to give me a definite answer; Professors of English language and literature do not understand neuroscience at all, I am afraid. As I know from Leslie and other colleagues, Walter was working almost up to his last day! It is said that a new book co-edited by him will be published soon. What a great man has gone! He will live in his friends' hearts forever.

As you mentioned in your marvelous letter, IBM's Deep Blue computer beating the chess world-champion, Garry Kasparov, in 1997 might have been the first thunder-clap for the third rise of AI. Your words remind me of the remarks by Deep Blue chess playing system designer Dr. F. H. Hsu in his book "*Behind Deep Blue: Building the Computer That Defeated the World Chess Champion*". He declared that the game was played between two kinds of persons who played different roles: one was the player, and the other were the makers of the tool. Kasparov played against Deep Blue twice, the results were different. In the 1996 game, the person as the player won; while in the 1997 game, the person as the tool makers won. As for if the Deep Blue is intelligent, his answer is that Deep Blue has no intelligence at all, it is only a sophisticated tool, which can only show some behavior as if it were intelligent in some limited field, which you pointed out as "*axiomatic games with a limited rule set*". It is Kasparov who is the player with real intelligence, although he was the loser in the game. As an example, Kasparov complained several times that there was some cheating during the competition; Hsu said that Deep Blue could never accuse others out of thin air in a similar way. I think that Hsu and you have explained the matter clear enough (I would not like to cite your comments here as you know them much better than me ☺), why should some person still declare that the AlphaGo AI system had beaten mankind! Although to beat a Go champion is much more difficult than to beat a chess champion, there is no essential difference in principle, I think. So, I am also not so surprised by AlphaGo's victory. The only thing that surprised me is that the day came so much earlier than many people, including me, expected!

By the way, in your letter you said that you "*believe in the long-term success of AI and even strong AI*", what do you mean "strong AI" here? As we know that there are two different understandings of the term "strong AI", i. e. general AI or AI with consciousness. Generally speaking, I agree with your expectation for both. Now there are some people, such as Dr. Pei Wang are working on general AI (I am sorry that I haven't read his original research papers on this topic), although I do not know if there is any successful application with general AI now. It seems that all the successful applications of AI belong to weak AI, which can only solve a very special problem, such as playing the Go-game, self-driving car etc. Although some people are now talking about artificial consciousness and there is even an international journal titled "*Artificial Consciousness*", I am very skeptical if now is the time to study such a topic when people cannot even give consciousness an operational definition. It seems to me that to reverse engineer something you don't understand at all is absurd. In addition, I don't understand why people should invent some intelligent robot with its own mind and will? To develop AI to help human beings to do something that they cannot do is one thing, to develop an AI robot with its own mind and will is another thing. Should it be banned as chemical or biological weapons?

In China, the response to AlphaGo's victory is similar to what you described in your letter. Many people believed that "*the end of mankind is near*" and "*machines and robots will take over control*" soon, which you explained in your great letter that clearly this will not be the case. However, a general impression has been made that "*Google and DeepMind have developed an AI system so powerful that it is a threat to mankind*", which made Google occupy a very advantageous position against its competitors.

The second reaction is that people are worried that after AlphaGo, there will be Alpha-physician, Alpha-lawyer, Alpha-bookkeeper, Alpha-taxi-driver etc. and many people would lose their jobs, although similar worries have repeated again and again in different forms since the first industry revolution. Of course, many old jobs would be lost, but new jobs would appear. In Shanghai, many small shops have closed owing to Alibaba or Jingdong; however, workers delivering goods which are ordered by customers from the website are needed in large quantity, especially during the holidays. Of course, I am not sure if the positions for new jobs are more than old jobs. However, in history, the situation seems just like this, and it gave people more leisure time to do creative works or enjoy their lives. If this will be the case this time again, I am not sure, and the government and community should pay enough attention to this potential social problem in advance.

The third reaction is from some biologists, they said machines can never have creativity, insight, mind and consciousness. Although this is the case at the present moment and in the foreseeable future, I am not sure if their arguments will be true forever. Anyway, human brains are also kinds of, although special, physical systems, they do have such properties, why should other physical systems not be able to have similar properties in principle? As for if we should develop such a system is another problem.

The fourth reaction is that there would be a new species, a mixture of human beings merging with intelligent machines, which will be a kind of superhuman. They argued that there were already people implanting some chips to strengthen their abilities. I don't understand why people should risk implanting some foreign chips in their body to strengthen their abilities if they are healthy; we have already used a variety of machines to strengthen our abilities, which makes us as if we were superhuman in the eyes of our ancestors. Anyway, to me, the goal of science is to make the majority of people happy, not to make them superman.

Just like you, I would not like to look down on the progress of AI in recent years. Only a decade ago, comparing computers and human brains, authors used to say that although computers can do calculation and logical operations much faster and accurately than human brains since the computer was invented, a brain of a 3 year old child could beat any computer in object recognition and action with ease, such as to recognize its mother's face from a crowd of people or to cross a busy street. Now people cannot say that. As you know that I am not familiar with technology and am a layman of AI, I am not sure if my impression is correct. I think that it was no accident that the breakthrough happens today, it could not happen a decade ago owing to the fact that there were no computers as powerful as today, and there was no internet sharing huge data. Without these conditions, deep learning would still be only a theoretical achievement. Some people said that deep learning is the victory of mimicking brains, however, as we discussed many times before, at most it is an achievement inspired by brains. The organization of CNN is quite different from brains! Deep learning gives the artificial neural network the ability to learn by training with hundreds of millions of data to solve practical problems, not like its ancestors such as perceptron which could only solve much simpler problems in a toy world. Neural networks with deep learning do have some ability to improve their skill, they are intelligent, according to Pei Wang's definition, but such improving has some limit, and it would reach to some saturated level sooner or later. Brains do not have such limitation; they can not only improve some special skill, but also develop new skills. So deep learning is still a kind of weak AI technology. It does not belong to general AI, let alone AI with an inner world!

You are absolutely right, at the present moment, AI machines have to be trained with mass data and consume a huge quantity of energy, while human brains can learn with only a small set of data and consume only 20 watts. It is still too early to say that AI has defeated human brains, even in weak AI fields. I am not sure if, nowadays, AI has led us in a wrong direction, they can solve many practical problems, do their jobs even better than the top human experts, and it seems that there is still broad space for it to develop. However, as you said that if people think that AI should develop along the way it is on now forever, it may be totally wrong. And *"the big problem is that it is very costly and time consuming to correct such mistakes"*.

Your words about the priming effect remind me of Walter's explanation why people are not willing to change their field even they know there might be a new field with more perspective. He wrote the following words in his preface for the Chinese version of his book *"Neurodynamics: An Exploration in Mesoscopic Brain Dynamics"* which was translated by my colleagues and me:

"... Why is the field (neurodynamics — citer) so sparsely occupied if it has such great promise? There are three reasons to consider. ... A second and more cogent reason for the sparseness of practitioners of brain theory is that established investigators have large intellectual and financial investments in the disciplines for which they trained. They also undoubtedly have strong and deep personal commitments as well, which are necessary for success in any field of science, and they find great satisfaction in continuing their pursuit of results along already productive lines of research using familiar technologies. There is little incentive to change fields unless there is clear and compelling advantage. These considerations give reason to suppose that the best hope for rapid growth of activity in the new field of mesoscopic brain dynamics into a new tide lies with recruiting scientists who are not already firmly committed to an existing point of view. Young people who are still in their formative years and are not yet burdened with a great deal of knowledge about an existing discipline may have a substantial advantage over mature scientists in this respect."

Although you have never read his above words, as I am afraid that you don't read books in Chinese, your comments and Walter's are so similar!

Look forward to hearing from you more about these open problems.

Fanji

III-004 Karl

27.05.2016

Dear Fanji,

I was very happy to the find your e-mail in the inbox this morning but the bad news that your friend Walter has passed away made me very sad. It is always terrible to lose a friend and in this case, it is even worse because you also loose an intellectual counterpart and many open questions can never be answered. Although I never met Walter myself I feel very familiar and sympathetic with his way of thinking. I had heard his name before. But only after you introduced me to his work I realized how important his work was and how relevant a thinker, not only in the field of neurodynamics but also in system-biology and many other fields, he was. You often read interesting books written by intelligent authors who have good ideas. But it rarely happens that someone really touches you because he has something to say which enriches your knowledge and widens your scope in such a way as Walter could do it. I have already mentioned this when I talked about this long but immensely rich and compact article "Consciousness, Intentionality, and Causality". It is written like a legacy as if he wanted to document, summarize, and structure the outcome of his 40 years of continuous wrestling with the same puzzle for the coming generation of researchers.

It's not just his mastery in the field and his competences which impresses; it is also his deep modesty. He never pretends to know and comprehend more than he really does, and he can resist the temptation to overstress a good idea which works on one level and to declare it as a fundamental principle of the whole system. He's just the opposite of a loud-speaker or goldmaker pretending to understand everything. He clearly states that there are many aspects of the problem of how the brain works where we don't even have a clue what's going on. The fact that he emphasizes and focuses on the "mesoscopic" level, the one between the microscopic of the molecules and the macroscopic of the brain and the whole body, is so typical for his work. He was aware that there are many more levels to discover and comprehend above the one of the neural network that allows the salamander to find orientation with the help of its olfactory system, which he knew so well, until we will comprehend how the whole brain works and consciousness somehow arises.

I can believe that he worked to his last day because I saw this video of Walter when he gave a talk where he presented a summary of the theory on which he worked for the most part of his life to an audience at the University of California at Berkeley just a few months ago. He acted as lively and intellectually strong as you can be, and it was a delight to see how well he had honed his thoughts and his theory. He used the optimization principle of reafference, which he found to be so relevant on improving his own theory in continuous cycles of approximation. It's unbelievable how much passion, discipline, and persistence this man had.

A giant has gone, and this is a sad day indeed!

But the next generation of researchers is ready to stand on his shoulders and may reach and discover the land he has seen.

And although he knew much better than most that neural networks, used in modern AI and machine learning, have little in common with its biological eponyms, the young generation of AI engineers may benefit more from his work than he had possibly expected.

There is a paradigm shift going on in this domain and this time it's very unlikely that the hype and euphoria about AI will lead to the next AI winter as we have seen it in the past.

You have put the spotlight right on this phenomenon when you cited from Dr. Hsu's book "*Behind Deep Blue*".

I didn't know his statement that in 1996 it was the player as a person who won and in 1997 the person as the tool makers who won.

This was absolutely correct 20 years ago, because Kasparov had to compete with the combined forces of many experts. They had put all their aggregated knowledge and expertise into the software and databases. Especially important, was the database containing the "opening book" which allowed the machine to select the best lines of known and evaluated moves during the first part of the game without doing much calculation.

The next step was to look ahead as many moves as possible, evaluate the positions, and choose the best move derived from rational calculation based on approved principles. So, the engineers tried to prime the machine with as much of their own knowledge as they could before the game started. The machine was built like a better informed and quicker human player. It tried to compensate for its lack of intuition by its superior memory of former games and fast rational calculation of more moves and avoidance of errors; especially the trivial ones which sometimes happen even to the best human players. Chess literature is full of examples of deadly mistakes even grand masters have made when they overlooked easy to see dangers or opportunities in the heat of the action and under time constraint. Mistakes even an early chess program would have never made.

Chess, at the time, was something like the model organism for AI researchers, similar to drosophila and C. elegans in biology labs. And the paradigm in place was that one had to make the computer fully understand the event space of a problem and teach it to the level of an expert. It was the time of the ontologies, expert and knowledge-based systems and the focus of the programmers was on semantics and languages. The major problem of AI seemed to be how to tell the machine, in the best way, all that humans already knew about the field.

And twenty years ago, the machine literally only did what it was told to do by its programmers. In times of GOFAI (Good Old-Fashioned Artificial Intelligence) programmers tried to collect everything that could be known about a domain, like chess in this case. The aim was to build a large knowledge base, understand the rules, distill the knowhow from experts and take advantage from the fact that a machine can have access to a large memory.

Especially important, was the "opening database" where all known opening moves were stored including the information about the winning-likelihood of a move. This is similar to the method human chess players use to prepare for a match. A good player knows thousands of constellations (there are more than 1,300 named opening constellations alone!) and can also look ahead for 5 and sometimes many more moves, depending on the remaining number of pieces. A machine can do this much better, of course, and, in principle, can fully calculate the whole game, if time wouldn't be a limiting factor. This was a big problem 20 years ago, but nowadays computers can defeat any human even with software based on GOFAI.

It happens to be that I was confronted with this problem already ten years before Deep Blue played against Kasparov. At the time, he was an advertisement partner of the Atari computer company. My company, ADI Software, had developed a database management program, Adimens, which had been quite popular at the time. It was sold by Apple, DEC, Hewlett Packard, Nixdorf, Siemens and other manufacturers. After two of my students had made an adaptation for the new Atari ST computer, it was one of the first database programs with a graphical user interface and became kind of a standard on Atari machines. And so, it came that the German director of Atari made contact with the Kasparov team and, especially, his friend Frederic Friedel. He had the idea to build a new kind of chess program and a chess database containing stored games and was looking for the best tools to store positions and moves. PCs were relatively slow in those days and had very limited working storage in the range of half a megabyte of RAM-space (even todays smartphones have 8 thousand times larger capacities). So, the idea which we discussed with Friedel was about a very specific notation of a chess move so compact that as many as possible could be stored in the database and loaded quickly to the computer's memory to do fast calculations. At the time I couldn't think of a better way to overcome the limited performance of the computers, and so the notation-method Friedel proposed seemed quite convincing to me.[1] I didn't realize at the time that there are much more promising ways for AI which had to wait for thirty more years. I was too blind to see them but other people, like Geoffrey Hinton could. In any event, although I found Friedel to be a very sympathetic and super-smart person I couldn't see any real business in making chess programs and therefore didn't engage in this field. He realized his idea with the help of a very skillful programmer and managed to establish "ChessBase" as one of the most popular chess-programs and databases in the world. He

---

[1] Prof. Gert Hauske who was kind enough to read our manuscript raised the question of whether this notation was derived from the ideas Claude Shannon had presented in his fundamental work on chess programs as early as 1949. As far as Karl remember the notation was a proprietary format, but Karl is sure the developers of ChessBase knew Shannon's famous work. Shannon C. Programming A Computer to Play Chess [J]. Philosophical Magazine, Bd. 41, 1950, Nr. 314.

remained obsessed with chess and when the contest between Kasparov and Deep Blue took place in 1997 Friedel was on Kasparov's team.

But times have changed a lot since and to make clear what kind of paradigm shift we have seen let me explain some aspects of the challenge first.

In the days when Friedel invented his specific chess-notation (which has survived in modified form until today) the biggest problem of a computer was to load large amounts of data from the stabile but slow long-term memory banks to the quick but volatile working memory of the CPU. This is still a performance bottleneck today but back then it was a huge problem. Shifting content from a physical storage area to another makes applications slow and also prone to errors during the process of reading and writing and it can cause logical problems in case changes happen during running processes. The handling of such processes in databases is the fine art of informatics to this day. It is rooted in the design of the dominating von Neumann machine. Here two kinds of separate physical storage devices have to be continuously updated by scooping data plus program code from the slow long term memory to the quick (and energy consuming) short term memory and vice versa.

Speed isn't the only problem that comes with this method. The other is that program code and data are stored in the same space. This is like frying an egg and cooking the soup in separate sectors of the same pan. If the soup boils over it may spoil the egg and therefore the whole meal. Overwriting the memory area reserved for a data-variable may corrupt the program code stored next to it and cause the program to do stupid things or to crash. Actually, this is the most common way hackers use to hijack a foreign computer.

There are other computer-architectures like the Harvard-architecture, which is even older than the von Neumann architecture, where code and data are kept in physically separated parts of the memory. It's used in embedded systems and microcontrollers not only for safety reasons but also for speed reasons because here you can read and write data and code in parallel.

In modern multi-CPU-architectures we see many new mixed forms of von Neumann- and Harvard-architectures optimized to handle the flow of data and program code for a given problem. The management of memory and the bus-structures, through which the data travel, are always in the focus of the engineers who try to find the right architecture for a problem. And there is a myriad of possibilities to handle this with maybe as many possible variants as in biological systems. The major difference is that engineers don't have to wait billions of years but can intentionally shape and continuously adapt their systems; and when doing so, they use the artifacts of the last generation to shape and improve the next generation.

I mention these technicalities here because they have a lot to do with the problems neurobiologists and AI-engineers have in common when they look at the brain and its intellectual performance. The brain too has long and short term memory and neurobiologists always guess how content is shifted from one part to the other. The much different use of energy for the two types of storage is a major issue for sure, but actually we know very little about how the brain does the job.

When engineers and biologists worked, in parallel, over the last 70 years there has always been a tendency to explain the brain and its architecture in terms of the latest computer generation. And so, it was in GOFAI until it came to its culmination when Deep Blue won against Kasparov. It was a huge PR-success for IBM but not much happened after this impressive show.

It was extremely costly to build an expert system merely for winning a chess game and when you went to the next game you had to start from scratch again to build a knowledge base for let's say Go, not to talk about more relevant things like a system to detect different kinds of cancers in tissue samples, do handwriting recognition or identify individuals from a video.

With the principle of a single master unit that oversees and controls the whole process, decides what to do next and tells the subunits what to do, you need enormous computers that soon run out of speed when the event space develops exponentially. Go, the next drosophila of AI-Gamers, after chess is by many orders of magnitude more complex just because it is not based on an 8×8 board but on one with 19×19 lines.

Going parallel with a number of general purpose CPUs was an idea quite popular for a long time, especially in Japan, but never really made it. The overhead in parallel computers to organize the CPUs was too high and it needed a new kind of programming no one was willing to invest in when the advantage was just linear, at best, and could be overcome by the next generation of classic computers.

The true break-through for parallel-computing came, however, but not with general purpose CPUs but with GPUs, the workhorse in gaming computers to deal with pictures and videos. GPUs are about the least intelligent kind of computers. They can do very few things, like shifting a pixel, but this they can do in parallel and extremely quick.

And GPUs are also well suited for doing the type of calculations needed for a very old kind of AI: Neural networks.

However modern AI didn't become so powerful because we have much faster machines and smarter algorithms now but because people have overcome the perspective that the machine has to be primed with all available knowledge before the start.

The neural networks AlphaGo used when it won the Go game against Lee Sedol are a totally different kind of AI-Animal from its predecessor Deep Blue. They aren't even relatives and maybe as different as fish and birds.

The difference is that AlphaGo's deeplearning network wasn't told all the rules, strategies, or tricks of the game by human experts nor did it learn it from databases. It just learned it by doing. Demis Hassabis one of the founders of DeepMind, which was acquired by Google in 2014, describes this method quite impressively using the example of Breakout, a computer game similar to "Pong", the old videogame, we all have played in the 1970s.

In an interview with *Wired*, he shows how the neural network did when it learned the game, with the only objective given to maximize the score of the game:

"*After 30 minutes and 100 games, it's pretty terrible, but it's learning that it should move the bat towards the ball,*" ..."*Here it is after an hour, quantitatively better but still not brilliant. But two hours in, it's more or less mastered the game, even when the ball's very fast. After four hours, it came up with an optimal strategy — to dig a tunnel round the side of the wall, and send the ball round the back in a superhuman accurate way. The designers of the system didn't know that strategy.*"[1]

---

[1] https://www.wired.co.uk/article/deepmind (June 22, 2015)

For people not familiar with modern neural networks it is hard to believe that this is possible, and it is even harder to understand how the program could do this.

More amazing than the fact that it is possible indeed, is the circumstance that the programmers also don't exactly know how the program did the trick.

A neural network is, at the beginning, about as dumb as a rock and (unlike DeepBlue) knows nothing at all of the world and of this particular game. Its entry neurons are started with arbitrarily sets of weights and then it starts moving around, very much like Freeman's Salamander, exploring the world. There is little success in the beginning but after every cycle the program adjusts the (positive and negative) weights of the neurons and thereby learns which variations make the output score better or worse. This is an awful lot of calculations (all partial derivatives) which literally explode in numbers when there are many neurons and many layers. But with enough time or computer-power the system will learn how to play the game perfectly well, just by exploring.

Of course, this works especially well in closed axiomatic systems. But it can also be applied in open systems where neural networks often, very quickly, learn to perform a task such as identifying specific dog breeds out of 2,000 different kinds of dogs. Many data are needed to train the system and lots of GPUs (or dedicated hardware like ASIC-chips).

In this article Hassabis and his friends describe, quite nicely, what makes the difference between old and new AI. Of course, most of these applications are limited to specific problems. And while the know-how hidden in the "hidden-layers" of the network isn't accessible indeed and therefore the programmers really don't know "how the magician does the trick" it is impossible at the current state to transfer such know-how from one application to the next.

But it isn't impossible in principle and the young Mozarts of the AI business like Andrej Karpathy and Ian Goodfellow may find ways to extract such knowledge to build more general kinds of AI and also better methods to train the networks.[1] Sometimes it is just a little idea that causes immense progress like the one Goodfellow had when we

---

[1] Actually transferring knowledge between artificial intelligence systems is one of the most active and promising fields in AI research. See: Tyukin I Y, Gorban A N, Sofeykov K I and Romanenko I (2017). Knowledge Transfer Between Artificial Intelligence Systems. https://arxiv.org/abs/1709.01547.

made two networks compete against each other and thereby made them hone their expertise much quicker and more efficiently. The method is called GANs (Generative Adversarial Networks). Supposedly, Goodfellow had the idea after he visited a bar with friends and programmed the solution overnight. I'm not sure whether this is true, but I like the story and it is true, for sure, that the young man became a superstar of AI overnight.

There is still a lot of work to do, of course, but there is real progress in this field and not just PR-ballyhoo. And the best news is that the brightest young sharks of a new generation feel attracted to this field. And therefore, you are right that I meant General AI when I said that I believe in the long-term success of AI. I'm not sure about the consciousness question yet, which may turn out to be a pseudo problem in the end, which you may not like to hear. ☺

But we fully agree about the relevance of what Pei Wang has to say. Meanwhile, I have read more of him and found him to be really relevant. He too, is an independent and deep thinker who seems to love to stay with the same puzzle for a long time. Again, I'm so grateful to you for drawing my attention to an interesting researcher and his ideas!! I don't know whether you have read his article "*Three Fundamental Misconceptions of Artificial Intelligence*"[1]. If not, I highly recommend reading it, because this is about the best comment on this subject I have ever read. Cristal clear and really brilliant — you will love it and it is worth a whole round of discussions.

But dawn is showing up again and I have to close.

Good morning and best to Shanghai.

Karl

---

[1] Wang P. Three Fundamental Misconceptions of Artificial Intelligence[J]. Journal of Experimental & Theoretical Artificial Intelligence, 2007, 19(3): 249–268.

Ⅲ-005　Fanji

14.07.2016

Dear Karl,

I am deeply moved by your genuine sorrow for Walter's passing away, although you did not know him personally before. As an old Chinese saying says "hearts which have a common beat are linked" and "the wise appreciate one another"! Walter is the very man you described in my eyes. He would be pleased to find another colleague who really understood him if he could have read your e-mail.

Your letter also answered some questions about AI that puzzled me for quite a long time. These questions are: Will the third AI winter come? Is there any essential difference between the approaches used in Deep Blue and AlphaGo? If there is, then what is it? How to evaluate the recent progress in AI? I am sorry to admit that I know little about computer engineering and AI. Making the matter worse, I never played chess and Go. This also badly influences my understanding of the situation. Compared to me, you, my dear friend, already developed Adimens and were even involved in something about developing a chess playing system ten years before Deep Blue played against Kasparov! You are an expert in this field, so it is lucky for me to find the very person to answer the very questions I want to know!

It is very interesting that you said that even the programmers of deep learning neural networks don't know "how the magician does the trick" in the system which they themselves designed. Why? I just read a few essays by Dr. Pei Wang, he may have given an answer to this question. He explained that it is the whole of the hundreds of thousands of parameters of the network that decides which move AlphaGo should make, and these parameters are determined by the whole history of its training process. Therefore, it is very difficult, if not impossible, to elucidate the reason of the move, even if you can review the whole training process again. Maybe this is also the case for the brain action in a similar way, I suppose. Do you think this answer is reasonable?

In addition, another argument Dr. Wang gave in his essays, which I am very interested in, is that *"intelligence" is the ability of learning, but not the ability of solving some specific problem, which is the skill.* According to him, *"intelligence" is not the ability to solve concrete problems, but the ability to get the ability to solve concrete problems.* He classified "intelligent" systems into four categories: a system having a constant skill level and cannot learn from its experience, which does not have intelligence, from his point of view, even if it may have a very high skill level, e.g. the Deep Blue chess playing system; a system having limited learning ability, which can improve its skill at its learning phase, and reaches a constant level after the training phase, AlphaGo is an example of such a system; a system having constant learning ability, which can learn new skills and improve its skills forever, human beings belong to this category; a system having learning ability that can be raised exponentially, which is what Kurzweil and others imagine, but until now we cannot even give one example of such a system.

From his view, most of the nowadays "intelligent" machines, such as Deep Blue, should be excluded outside of AI, although they constitute the main body of traditional AI. In the main stream of AI, artificial intelligence is defined as the ability of a machine to do some work only the brain can do. Wang argued that according to this definition, after a machine does this work, then the work can be done not only by the brain anymore, and it should not be considered as AI again, a contradiction in the definition itself. Though people may argue that the definition could be modified as follows: The AI system is a system which could do something that before only the brain can do, however, according to this new definition, even a system that can only do addition or subtraction should be considered as "intelligent", which most of people will not agree. Machine learning, such as deep learning, belongs to his second category, and artificial general intelligence belongs to his third category. He said that if there was some system belonging to his fourth category, then such system should be called an "artificial God" system instead of an artificially intelligent system. Although his argument to exclude his first category from artificial intelligence may violate the common belief of the public, I think his idea delineates the scope of artificial intelligence in a clearer way. Although people may argue why should a machine only being able to do addition be excluded from AI? You can accept it and admit that such a system is in the lowest level of AI. However, this may be concerned with the definition of intelligence, if so, then Fabre's wasps should be considered intelligent, although there is not any flexibility in their behavior. What is your opinion?

Well, now let me shift to another topic. You may have noticed reports yesterday, which published the achievements announced by the Allen Brain Observatory of The Allen Institute for Brain Science launched four years ago. And a report titled "*Brain-data gold mine released*" was published in today's *Nature*[1]. The Allen Brain Observatory opened their data of neuronal activities of 18,000 neurons in four areas within the visual cortex of 25 mice in response to a variety of visual stimuli, including a clip of the Hollywood movie "*Touch of Evil*" for 360 experiments.

This is the first time in history of opening the data of neuronal population activities, at the cellular level, real-time to show how neurons coordinate their activities while completing a cognitive task. The Allen Institute for Brain Science is a member institution joining the US BRAIN Initiative, this must be good news for the project. The Allen Institute for Brain Science operates in a new way for scientific research similar to industry. They organize experts from different fields to attack a common target in a team and open their data and tools of analysis to the public. Some people call such research "big science", "team science" and "open science". It seems to me, such an approach is suitable and efficient for data collecting, if the amount of work is extraordinarily huge, but the procedure is rather fixed. The Allen Brain Gene Expression Atlas, The Allen Mouse Brain Connectivity Atlas, and the Allen Cell Types Database released by The Allen Institute for Brain Science before, are such examples. The Human Connectome Project may be another example, although I don't know its exact progress. However, all these works were strongly supported by giant government grants or a billionaire's donation. In addition, I don't think such an approach can be extended to most fields of scientific research, for which the main objective is not just collecting data, but strongly depends on insight and creativity. The reason for this argument we discussed many times in our previous letters.

Best wishes as always.
Fanji

---

[1] Shen H. Brain-Data Gold Mine Released[M]. Nature, 2016, 535: 209-210.

Ⅲ-005 Karl

29.08.2016

Dear Fanji,

Thank you very much for your kind letter. Brief letters often come with heavy content and so does your mail. It contains 3 relevant aspects and questions — and all are difficult. ☺

One is the intelligence question defined by Pei Wang, the other is the paradigm-shift in modern AI and its possibly enormous consequences, and the third is the question of organization of BR (Big Research) which you raised after the publications of the Allen Brain Institute.

I'm not sure whether I can give satisfying answers to all of them, but hopefully I'll manage to tell you the "how I met Paul Allen" story after much delay this time. ☺

But let me first tell you how much I liked what you said about "*hearts which have a common beat are linked*". Actually, I feel this kind of resonance when I read the core parts of Walter's work. But there is always some danger involved in worshiping a scientist when you feel this kind of resonance and to buy everything he says. We tend to believe anyway that when a person has achieved an ingenious performance in one area everything he or she does or says is also ingenious.

Einstein, who became the first popstar of modern science, is probably the most popular example. And although he was, indeed, a true genius of physics, he wasn't a genius in many other fields. Whatever poor Einstein said to a journalist made the headlines the next day (and especially) when he was acting far beyond his range of competence and sometimes produced sheer nonsense. He liked to play this game and because he had a good sense of humor sometimes did it on purpose just for his own pleasure. Other celebrities who don't possess the ability to make fun of themselves often can't resist using their popularity and earnestly comment on everything in public even if they don't have a clue about what's going on.

Demis Hassabis made this point quite politely when he commented on the popular warnings about the dangers of AI expressed by many scientists and celebrities like Elon Musk, Steve Wozniak, Stephen Hawking, or even Hollywood actors (see the Wired article which I have cited in my last mail):

"These are people who are not actually building something, so they're talking from philosophical and science-fiction worries, with almost no knowledge about what these capabilities can do."

This is very true indeed and a great example to support my longtime hypothesis about a core feature deeply rooted in the architecture of the mind of all humans:

"The less we understand a topic, the more definite is our opinion on it".

Of course, Walter Freeman was far from acting beyond his area of competence. So just to demonstrate that I try to be critical even with my friends and heroes, I have to tell you that some parts of his writings are quite alien to me. I can't share his sympathy for Heidegger and Merleau-Ponty. This is less relevant because we all have different methods to stimulate our creativity and provoke our minds to generate new ideas and insights, but the point is that he missed a better solution that was readily available, since long, which he probably would have liked even better.

To me, phenomenology, in the tradition of Edmund Husserl, always looked like one of the least attractive parts of philosophy and especially in the Heidegger variant more foggy and obscure than enlightening. I remember that when we had seminars at the university with the leading proponents of critical rationalism, to them "Heideggerism" was the prototype of huff-puff and humbug. In public they expressed their disgust more politely but from the traditional after-work meetings in a nearby wine-restaurant I remember terms like "nonsense on stilts" or "pseudo-intellectual garbage" from quite a few well-known philosophers. As a young scientist who tried to find his orientation in the field I was impressed by such strong statements and got kind of pre-magnetized by it. Later I found that not everything in Heidegger's work is plain rubbish, but also that similar ideas had been formulated by anthropologist Arnold Gehlen in a much clearer and more useful way.

I think Walter became aware of Heidegger and Merleau-Ponty because of Hubert Dreyfus, who made them popular in the US, and amazingly enough, managed to get quite a few American researchers and even leading computer scientists, like Terry Winograd (the PhD supervisor of Google cofounder Larry Page) interested in rather fluffy European philosophy.

If someone would have explained Gehlen's much clearer view to them they would have probably been more impressed I guess. Just have a look at the passage which starts with the term "Handlungskreis" in the German Wikipedia-article on Gehlen. [1]

I'm sorry that this is only available in German but I know that you can understand it.

Here I made a rough translation of this passage which hopefully makes clear that Gehlen talks about circular causality and a process very similar to what Walter Freeman later called "reafference".

"He describes the action circle with the following example: When you try to open a jammed door with a key, you have to move the key back and forth. You will notice whether it works better in one direction or in the other. So you experience success or failure in these attempts, you get feedback. If you respond to this feedback and change your actions, you will experience the intended success, the lock opens.

Gehlen describes this process as circular. The cycle process activates psychic intermediaries, the perceptions, continues via the physical parts, then via the own movements and then into the factual level and back again. As a result, Gehlen does not see the action as a dualism: The process that is taking place cannot be divided into physical and mental things. All parts are inseparable from each other and work together constantly in the same process. He describes his concept of action in the following words: "The action itself is — I would say — a complex circular movement, ..., and depending on the feedback, the behavior changes.""

---

[1] https://de.wikipedia.org/wiki/Arnold_Gehlen

This is from 1957!!

I doubt that much of Gehlen's work, if any, was ever translated to English and so not many scientists abroad took notice. [1] You could make a much better translation of the text I'm sure and I think it would be worthwhile to make a translation of Gehlen's core ideas into Chinese, because it is so immensely inspiring. And young researchers, both in neuro sciences and in machine learning, who are concerned about the "binding problem" at the next higher level, which is from psychology to sociology and anthropology might benefit from his perspective.

What Gehlen is out for is, in a nutshell, an extension to what Schrödinger started to deal with in "*What is Life*"; the puzzle of the principle of self-organization of matter and energy. Schrödinger, the physicist, asked how the cell can perform its major job, which is in his technical words "*exporting entropy*". In more popular words this means how can a cell establish and maintain order in chaos. So, he looks at the microscopic level at the lower end of the brain. Gehlen asks the same question at the upper level of the brain and beyond. He asks which social tools and inventions allowed humans to get so well organized that they gained an advantage over other animals, which are much stronger and physically better equipped.

So Gehlen is interested in the action-relevant content stored in the brain. The part that can be transferred to others: skill and knowledge. While many are yearning for mind-uploading we tend to overlook that our ancestors invented such a technique some ten thousand years ago. We call it language and use it in verbal and written form. Both are quite slow and terribly imprecise processes but in the long run so successful that it was even worth this incredibly long phase of childhood and learning which differentiates humans from all other mammals. This wasn't new insight for anthropologists of course. But Gehlen's detailed analysis of the self-organization on the social level was new and also that he, at the same time, realized how learning, in general, works as the text cited above demonstrates.

Gehlen's objects of interest aren't cells or neural networks but "institutions". That

---

[1] In the meantime I found that at least two of Gehlen's major books have been published in English.
　　Arnold Gehlen. Man in the Age of Technology[M]. New York: Columbia University Press, 1980.
　　Arnold Gehlen. Man: His Place and Nature in the World[M]. New York: Columbia University Press, 1988.

means tools humans invented like the hammer or knife but on a social level. It's like capsuled procedures much more powerful and relevant than what was introduced as "Memes" by Dawkins many years later. Formal marriage belongs to it but also legal systems, parliaments, political parties, policy, military, religions, schools, tax-systems, banks, credit-card-payment, Stop-signs, traffic-lights, and internet-security and even organized crime and corruption. In short, everything that regulates our lives as capsuled subroutines in the real world. They have to be taken into consideration when we want to successfully maneuver in a given environment into which we are born. We may adapt and change them while we pass them on from generation to generation. This allows for a much quicker process of adaptation than the one via modifying DNA, and a process that includes real intentional design, something ordinary evolution can't provide. I don't want to bore you with a deep introduction to Gehlen's anthropology and theory of institutions, but I believe that it could help to close the gap which I call the "*second binding problem*".

We can't comprehend human actions without these rules and norms defined, stored, and transmitted outside of the individual brain. Even if we were able to map a brain down to the very neuron and atom, we would know nothing about these fields of force that guide and restrict our actions. Physiology and psychology are not enough to understand human actions without the knowledge about this type of content which has to be reinstalled in every brain from generation to generation. Neuro scientists, of course, know this, at least, in principle but often tend to forget it after they have banned it to the boundary conditions of their equations.

But, as much as you can't understand the brain and the mind without its embedment in the body you can't understand them without considering the social embedment and the rule-related content that was loaded to it. We experience this knowledge which is basically a set of rules of do's and don'ts as subjective. But in truth, it was acquired from our ancestors or our fellows. This is nothing that would have surprised my grandmother, but it is amazing how many brain researchers act as if it would be possible to deduce the brain functions from bare brains without taking this learned content into account.

This is the reason why I'm very skeptical about the use of an intelligence definition which neglects learned skills and tries to extract the pure ability to learn.

Each new piece of knowledge acquired is always treated and stored in our mind in the light of what we have learned in the steps before. And it is even physically related to it via the connectome.

There is no empty disk in our mind as on a computer and no possibility to check the performance of a raw, empty brain in terms similar to Gigahertz or the RAM-space of a machine. And even with our computers this isn't possible. To check the performance of a machine you have to boot it first from any kind of BIOS. And only after the machine has gone through a short phase of fast-motion "*childhood*" in its boot-process it "*knows*" that it is a Mac, a Windows-PC, a Linux-Server, or whatever machine. And even if it is the same hard-ware, these systems will show very different behavior, even in elementary use of its hardware depending on how much effort the programmers took in writing good drivers. And this goes on with every layer of programs you load to it and the more data you feed e.g. to its databases. There are databases which show the puzzling characteristic that the more data stored in them, the faster they become. If the data is from a limited universe of objects this is quite understandable. An "*experienced*" address database, which already contains most zip codes of a country, needs much less effort when updating index lists (inserting new nodes in key trees) than an empty database, which knows nothing. The more "*experienced*" the system is, the more likely it is that a zip code occurring in a data set is already known and the better the learning advantage over a less experienced system. Jeff Hawkins gives a very good explanation in "*On Intelligence*" that the human brain may benefit from similar effects. The model he proposes says that as long as sensory input is not new it is just quickly piped through the layers of the columns in the neocortex and only if there is something new or relevant a much more costly attention and/or storage process is started.

I understand why Pei Wang is interested in separating learned skills from a "*pure*" learning-intelligence. It would be helpful to check the performance of his NARS (Non-Axiomatic Reasoning System) for general AI. But, as much as I appreciate his perceptive analysis in "*Three Fundamental Misconceptions of Artificial Intelligence*", I do not like this idea of a pure experience-free intelligence.

Trying to explain a phenomenon by the "*correct*" definition, as in the good old scholastic days, is in vain anyway. At best, it can be more or less useful in an empirical sense when it comes to testing a theory. It is interesting, however, to see how many diverse definitions across the various disciplines exist that deal with the performance of our brain.

One of the best and most complete considerations around the problem of "intelligence" I know of also includes a list of about 70 different definitions of intelligence. Shane Legg, one of the founders of DeepMind, and a student of Marcus Hutter and Jürgen Schmidhuber, took the effort to collect them in his doctoral thesis "*Machine Super Intelligence*". His thesis is a very interesting read and impressed me more than the one of his cofounder Demis Hassabis. You find a link to it at Legg's Wikipedia site. [1]

What surprised me among all those definitions listed by Legg is that many researchers don't take account of an aspect which is the most crucial to me: time and speed.

For a good reason all IQ-tests, whatever they may measure, are performed under time constraints. There may be intellectual tasks where time doesn't matter, but in most competitive situations speed counts because if you are too slow someone else will be quicker and you lose.

And to my surprise, my favorite definitions were not on Legg's list of 70 definitions of intelligence.

The first one I read many years ago but don't remember who the author was and I also don't recall the exact wording. It was something like "*in biology intelligence is when an animal realizes that time has come to give up a formerly approved strategy*". So, the poor wasp would be really excluded. ☺

The other one is that in intelligence it is "*incompetence-compensation-competence*" that counts. This is my private variant of a term German philosopher Odo Marquardt has coined. Originally, he meant it as a description of the regrettable state in which today's philosophy is situated. Once competent for everything, today it is mainly concerned with compensating her own incompetence. This is more than a funny term, because in social interaction (and in science) we often have to hover over unknown ground and, quite often, very bright people used to calculating everything accurately to 10 decimals are stuck in situations when they don't have the necessary information to do their usual calculations. In the same situation less intelligent people may perform much better

---

[1] https://en.wikipedia.org/wiki/Shane_Legg

because they are used to acting without much calculation and reasoning.

A third one I owe to my dear friend Aharon Beth-Halachmi from Tel-Aviv. He's one of the wisest and most knowledgeable men I ever met and he highlights the delicate difference between intelligence and cleverness when he says:

"A clever person will never put himself in a situation where only an intelligent person can find his way out." ☺

The final definition of intelligence I have produced after I read too many fierce debates among scientists from hostile camps. I don't claim a copyright to it because I'm sure someone else must have had the idea before:

"No-one knows exactly what intelligence is, but everyone thinks he or she has enough of it while others don't." ☺

A crucial element of intelligence which is often not covered by intelligence definitions, IQ-Tests, and also not in AI is creativity.

I was very pleased to see that in this last publication of Walter's together with Robert Kozma, they also mention Arthur Koestler's *"Art of Creation"* and Huizinga's *"Homo Ludens"*.

The first is dealing with the question of how creativity works and the second with the important role of playing in developing intelligence. The latter I can observe right now, quite intensively, when I see how the connectome in the brain of my grandson Alexander develops from day to day at an immense rate after he started walking, talking, and playing with everything in reach.

I had heard of the genius of Koestler and knew that he was one of these Austrian-Hungarian polymaths like his friend John von Neumann. But I never read one of his books in the original and now took the opportunity to buy "Art of Creation". I'm not done with it yet because it's more than 800 pages. And I'm not sure in how far his theory of bisociation (which is two associations combined) to explain humor, the sensation unique to humans, and the "eureka" effect is still relevant. But what I can say is that it is, for sure, a real gem and a masterpiece of western literature which I can highly recommend for a long holiday and if only for the sake of intellectual pleasure.

In the meantime I have read parts of Freeman and Kozma's "*Cognitive Phase Transitions in the Cerebral Cortex — Enhancing the Neuron Doctrine by Modelling Neural Fields*".

It contains an updated summary of Walter's research and of the work he and Kozma have jointly done. I don't know whether you have read it yet. Much of it will look familiar to you but it also contains "commentaries" of many relevant people of the neuro-community paying tribute to the master and discussing his contributions to the field, your friend Hans Liljenström among them. The book is very much in the tradition of the famous old "Festschrift" of German universities, where former students and colleagues honored a professor when he retired or celebrated a great birthday.

I will tell you more about it when I've fully read it. For today, I will leave it with the recommendation to read the commentary of Paul Werbos (Chapter 19) because there you will find an answer to your question about the sense or nonsense of Big Science projects like BRAIN or what the Allen Brain Observatory does. Werbos seems to be quite knowledgeable and experienced in this big funding business and has a very explicit opinion on it. I mostly agree with what he says on the tendency of science as "leaf counting" but don't want to reveal too much here. First because I think you will like to read what he writes and second because this mail again is far too long already. Therefore, I will also postpone my comments on the newest developments in Machine Learning and AI.

But one thing I have to complete as promised, before you start hating me, the story how it came that I know Paul Allen. At the time, when I was in this database software-business, Paul Allen had already retired from Microsoft and started a new company called Asymetrix. They had a very nice software-tool called "Toolbook" which allowed designing interactive multimedia user interfaces including video and sound in a much easier way than by regular programing. This is quite common today but was a sensation in the early 1990s when no-one knew what the internet and a browser were. Our CTO, August Wegmann, a real computer science genius, managed to combine his brainchild, our database product Adimens, with Toolbook. At the time, it was one of the most productive development tools of that kind in the world, particularly suitable for self-service terminals and ATMs. And so, it came that I got in to contact with Asymetrix and also with Paul Allen. Unfortunately, Microsoft had started a new programming language "Visual Basic" at the time which was more in the style of traditional programing languages and considered Toolbook to be an unpleasant competitor. I don't know how hard Paul tried to work on Bill Gates in the case but, in any event, Microsoft neglected the development and ignored the opportunity in the same way as they ignored the internet and the relevance of browsers shortly later. Ironically, it was IBM who took interest in the solution. At the time Microsoft turned into a serious threat for IBM although both were officially still partners for OS/2, (IBM's version of the PC operating system). It ended up with IBM buying our ATM multimedia solution for banks, however transporting it to OS/2. After the advent of the internet, Microsoft came back to us and we developed a complete Banking solution called Banking NT, by ADI, which Bill Gates and I jointly presented to a group of banks in Frankfurt in 1997. But this is a different story. ☺[1]

Best for today and good morning to Shanghai.

Karl

---

[1] Schäfer, Waldemar. Gates Präsentiert Deutschen Bankensoftware-Partner [R]. Handelsblatt, Nr. 200, 17/18.10.1997, p.22.

Ⅲ-006 Fanji

15.10.2016

Dear Karl,

Thank you very much for your DeepLetter ☺, which I read three times in order to grasp the essence of your deep thinking, some of which I never thought of in your way, and some I felt in some similar way but could not express as clearly as you did. In one word, your letter is very instructive and inspiring.

At the very beginning of your letter, you pointed out that

"But there is always some danger involved in worshiping a scientist when you feel this kind of resonance and to buy everything he says. We tend to believe anyway that when a person has achieved an ingenious performance in one area that everything he does or says is also ingenious."

Although few people would deny the importance of critical thinking and rational thinking, people are apt to do just what you described, including me sometimes. And sometimes, people are also apt to do just the opposite, they would reject every word proposed by a person whom they disliked. Your story about your earlier attitude to Heidegger's work may be an example, however, you are not the person belonging to the people I've just described, you found there was something reasonable in his work, and even more, you found a better expression in Gehlen's work. I am sorry for my ignorance of these philosophers, I even haven't heard of their names before ☹.

It is very interesting that Gehlen pointed out so early the importance of skill and knowledge transferring between persons. And it is also interesting that you compared verbal language and written language with "mind-uploading". Yes, when I read some essay talking about mind-uploading, I cannot help thinking of reading vivid novels, especially autobiographies of some late person. I felt as if the author were still alive and speaking to me about his experience. Is this a form of mind-uploading? I don't think that anyone would agree. The living experience is in my mind, not in the book. Of course, the autobiography is also a copy of the author's thought when he was writing. If this is true, why would copying the connectome of a person into a computer be mind-uploading?

You are absolutely right that mind can exist only in a context of body and social environment or "institutions" as you said and in the interaction between them. Therefore, I am skeptical about the "brain in a box" theory. If an isolated brain is only maintained by some nutrient substance, I am wondering if such a brain can still have any mind in the long run. Think about the stories about those prisoners who were kept in a dark single cell for a long time, most of the poor guys almost lost their minds!

Your words about the process of transferring skill and knowledge, which is a much quicker process of adaptation than the one via DNA, reminds me a remark by Dr. Ramachandran. He emphasized that, thanks to the former process, people learn to kill polar bears and take its skin to warm themselves quickly, instead of waiting to evolve some thick fur, long before that the species may already have been extinct owing to the cold, so that human being can adapt to harsh environments quickly.

Up to now, I almost agree with you on all the remarks you gave in your last letter. However, I would like to argue with your following remarks:

"I'm very skeptical about the use of an intelligence definition which neglects the learned skills and tries to extract the pure ability to learn ... I do not like this idea of a pure experience-free intelligence."

I am sorry that I might have misrepresented Pei Wang's argument in my previous e-mail, in which I wrote something about the difference between skill and learning ability, and emphasized the importance of learning ability for intelligence. As a matter of fact, as I wrote in my e-mail before my previous e-mail that Pei Wang defines intelligence as *"the ability of adaptation under insufficient knowledge and resources"*, in which resources mean time, speed, memory capacity and etc. just as you emphasized in your last e-mail. Thus, I thought that in this point, you and Wang have no difference. The problem is that in other places, Wang emphasized the importance of learning ability of intelligence instead of skill. Maybe this is not his formal definition of intelligence, as the meaning of the two expressions does not seem identical. Maybe he just wants to express that there is some hierarchical structure in intelligence. Those without learning ability are the lowest, then the one with learning ability only in its training phase, and then with learning ability all the time in a linear way, and finally, with learning ability all the time in an exponential way for the highest, although we cannot exemplify any system with the last type of intelligence up to now. As far as the skill is concerned, although it will be easier for the agent to adapt to a novel situation if it has more learned skills, it is hard to define what kind of skills should be essential for AI, just as Wang argued. Similarly, if we define the intelligence as the skill only the human brain has, then it would mean that no other animal other than human beings has intelligence; and if we define intelligence as the skill any brain has, then the wasp would also be intelligent according to such a definition ☹.

I am not sure if you will agree with my argument.

In your letter, you also say something about big science, and last month there were several significant events related to this topic, so I would like to share these pieces of news with you and know your opinion.

The first event is that it was announced that the Human Connectome Project (HCP) had fulfilled its first phase in June of this year. As we mentioned earlier, the HCP only studied the macroscopic connection between different cortical areas and subcortical structures both anatomically and functionally for normal young human adults. According to their report, they have reached the following achievements:

(1) Improved four different magnetic resonance imaging (MRI) technologies (Structural MRI, Task-activated functional MRI, Diffusion MRI and Resting-state fMRI) to observe anatomical and functional connections between different cortical areas and subcortical structures noninvasively. The spatial resolution, temporal resolution, and signal noise ratio have been improved.
(2) Using the above four technologies, one hour's data were collected for 1,100 subjects, including twins from 300 families.
(3) Divided each hemisphere into 180 different areas, much more accurately than the traditional Brodmannian areas. This may have basic meaning for macroscopic

connectomic studies.

(4) Their data and analytic tools have been opened to the neuroscientific community. [1]

The second event is that a Specific Grant Agreement was signed between the EU and the HBP coordinator on the 12th of last month. The EU promised to support the HBP with 89 million Euros for the period between 1/4/2016 and 31/3/2018. That means that the HBP has been allowed to enter its second phase — the "operational phase" at least for the next two years. According to the official HBP website [2], the HBP achieved the following accomplishments in its first phase — the "ramp-up" phase:

(1) A digital reconstruction of microcircuitry in a young mouse neocortex has been made;
(2) An HBP Web Collaboratory has been established, registered users can access its software and data bank;
(3) Neuromorphic computing systems have been developed in the member laboratories in Heidelberg University and the University of Manchester.

Compared with the landmarks of the end of this phase in the original proposal — Establishing ICT platforms, using them to process some important experimental data and beginning to serve the neuroscience community, the achievements seem to fulfil the primary goal. However, if it is examined carefully, it seems that it has just passed the threshold. Although the HBP has called for neuroscientists all over the world to use their platform, including its brain simulation tools, visualization software, and remote-accessed supercomputer, it is not clear what the response of the community will be. If a platform is finally successful depends on if it can make some breakthroughs in science or application. The landmark achievement the HBP has made in the first phase is a paper[3] published in the Journal "Cell" last year, which reports a simulation of a tissue of 1/3 mm^3 in a young mouse cortex with 30,000 neurons and 40 million synapses. 82 scientists cooperated for 20 years for this paper. They used 207 different

---

[1] http://humanconnectome.org/ccf
[2] http://www.humanbrainproject.eu/
[3] Markram H, et al. Reconstruction and Simulation of Neocortical Microcircuitry [J]. Cell, 2015, 163: 456-492.

kinds of neuron models, 13 different kinds of ionic channels were considered. Owing to the lack of experimental data, synaptic connections were hypothesized; the activity pattern within this microcircuitry was simulated, which is similar to experimental data. This work was praised by Christof Koch:

"The digital reconstruction of a slice of rat somatosensory cortex from the Blue Brain Project provides the most complete simulation of a piece of excitable brain matter to date." [1]

Markram declared that he realized his promise at last. However, he obviously forgot or was unwilling to mention his promises in 2009 and 2012 respectively that he could simulate the whole mouse brain in 3 years instead of just a micro-circuitry! In addition, as some scientists pointed out, their micro-circuitry model has no meaningful function! Using a much simpler model can almost repeat the phenomena they simulated.

Not all scientists praised the paper. Your country-fellow neuroscientist Moritz Helmstaedter criticized that *"There are no real findings. Putting together lots and lots of data does not create new science."*

Britain neuroscientist Peter Latham remarked:

"This was an amazingly huge amount of work, but it teaches us nothing new about how the brain works,"

He agreed that their work may be helpful for testing some hypothesis, however

"Do you want to spend a billion euros on this? That's the question." [2]

Of course, it may be still too early to give any conclusion about the perspective of the HBP, especially after the dramatic reform. It seems that now that the HBP may become some permanent international infrastructure with IC technical platforms focusing on

---

[1] Koch C, Buice M A. A Biological Imitation Game[J]. Cell, 2015, 163: 277-280.
[2] http://www.sciencemag.org/news/2015/10/rat-brain-or-smidgeon-it-modeled-computer

studies of brains, cognitive neuroscience and brain-inspired computing, which is far away from Markram's goal — creating a whole artificial mouse brain, or even a human brain in the foreseeable future.

The third event is that many powers all over the world have launched or are preparing to launch their own big project on brain research. International cooperation and coordination has also been considered. In April this year, academies from 14 countries release a joint declaration to call for international cooperation on brain research [1].

Last month, Rockefeller University, Columbia University, the US National Science Foundation (NSF), and the Kavli Foundation convened a meeting to discuss how to coordinate the brain projects all over the world. Representatives from the EU HBP, US BRAIN Initiative, Allen Institute for Brain Science, China brain project, Japan's Brain Mapping by Integrated Neurotechnologies for Disease Studies (Brain/MINDS) program and Israel's Brain Technologies (IBT) attended the meeting. They introduced their aims, achievements, measuring 10—15 years impact, challenges and Efforts to address reproducibility, data availability and resource sharing. This meeting gave scientists a chance to review the big brain projects and explore the possibility of international coordination and cooperation. However, except for the EU HBP, US BRAIN Initiative and Allen Institute for Brain Science, other projects are just launching or preparing to launch, it is still too early to give these big projects a conclusive evaluation. If just the ongoing projects are considered, it seems that all these projects are focusing on data collection works such as making a variety of atlas and new technology development. These works may provide an important basis for breakthroughs in brain research, but not a breakthrough itself. In addition, the amount of work for making various atlases is huge, if not daunting. Spending 6 years, the HCP has only made a database for a macroscopic connectome for healthy young adults, Much more work is expected for other stages in the lifespan, and a variety of patients. And maybe this is the easiest task for similar data collection!

It seems to me that the problem how the brain research should develop is still open. To organize big projects may be only one of the ways, and what role they should play in the progress of brain research should still be thought about. Even for these big projects,

---

[1] http://sites.nationalacademies.org/cs/groups/internationalsite/documents/webpage/international_172183.pdf

how to organize them is still a problem. The HBP has given people a serious lesson, if the reformed project will work well, let us just watch!

An essential problem is how to evaluate the state of art of brain research, has brain research attended an industrial age or is it still mainly in some handicraft way, just as American Physicist Michael Roukes said that "*We're still in the neuroscience craft era, everyone has their secret sauce.*"[1]?

As for data sharing, The HCP and Allen Institute for Brain Science may give good examples, however their data are atlas like and both are strongly supported by government or billionaire donations. Could "craft" neuroscientists provide their data too? At the end of last century, US scientist S. H. Koslow proposed a US Human Brain Project to call for data sharing all over the world, and the term "neuroinformatics" became popular to the scientific community. It was quite popular then and an international organization was built to promote and coordinate corresponding activities all over the world. However, it declined quickly and is not mentioned any more when people talk about neuroinformatics today. I don't know what happened to this project and why it declined so quickly. Maybe data sharing is easier for big science, but not for craft science. Nowadays, when people are talking about neuroinformatics again, should the experience and lesson from Koslow's HBP be summarized to avoid similar failure? I don't know, do you have any idea?

There are so many problems to be discussed about the big science, however, the letter is already too long, and I have to stop here.

Best regards.

Fanji

---

[1] Reardon S. Global Brain Project Sparks Concern[J]. Nature, 2016, 537: 597.

III-006 Karl

18. 11. 2016

Dear Fanji,

Thank you for your kind letter and especially for your thorough report on the latest developments in the big science projects in the field of brain research. Some of them I had seen but not others. I was more focused on the developments in AI over the last weeks, where a lot is going on at the moment, and so your summary was very welcomed and helpful. And it led to the discovery of a link between the two fields which are very important to me, and therefore I am particularly grateful for this impulse. It has to do with new concepts of "neuromorphic chips" that are now appearing everywhere. Actually, they are not as "neuro" as the name suggests but some of them contain rather smart ideas. In this case, I was surprised to see some analogue parts in the hardware of the BrainScaleS computer which was especially built to run the brain emulation software of the HBP.

What you mentioned in your report on the HBP in point (3) "Neuromorphic computing systems" invited me to have a closer look at the architecture of the machine and I was pleased to find that people at nearby Heidelberg University and at the Technical University in Dresden were involved in making the VLSI – CMOS wafer which builds the heart of the machine. I had taken a look at it some time ago, but not closely enough. Mostly because I wasn't impressed by the HBP for the reasons we have discussed and approached it with a negative attitude. But that was inappropriate, not to say arrogant. ☹

Sometimes we tend to doubt everything that comes from a particular source, which can make us blind to interesting things. Very much as you have described it the other day when you said that worshiping a person for everything he says is a fault but denying everything he says is another one. I try to avoid this mistake as much as possible, but it happens anyway. Thanks to you, I took a closer look at it and got inspired to a few ideas which may turn out to become relevant in a project for a new hybrid concept of an analogue-digital machine which I'm discussing with two of my friends right now. And you made me aware of a relevant kind of know-how people have here in my direct neighborhood. Thank you!! Isn't it funny that the reference to it had to come from Shanghai! ☺

When it comes to the results which you have summarized from the three big science events I agree with your sober analysis and summary and can't see any real progress, especially in the HBP, when you compare it with the promises made. We have to be fair with the people who did diligent work for sure, but Peter Latham's question

"Do you want to spend a billion euros on it?"

is the one to ask.

But, as you have described in the third part of your survey brain research became a darling of many governments and billionaires and there is a continuous flow of money into this field.

Sometimes waves are forming, like the famous "La Ola" in a football stadium, and everybody has to be engaged in atomic physics, space programs, cancer research, nanotechnology or whatever the hype of the day is. Now it's "Neuro"-time and it looks like no government of an industrialized country with any self-respect can afford not to follow the band-wagon. ☺

It feels like a new bonanza-time and, as always, in such times too much money can flow into a presumed goldfield and attract too many people.

So, the question we always have to ask ourselves in the context of scientific progress is: would it help to call $1,000$ doctors to the bedside of a seriously ill person — and how much better would $10,000$ be?

When you ask for lessons learned from these movements or for an alternative, I have no answer. At least not one leading to a practical solution. Such movements happen like natural events, e. g. avalanches or tsunamis. You can describe and understand them (afterwards) but not predict or control them. Like waves, they start somehow and also come to an end but you don't know why and when. The only thing you can predict is that the money always comes with a swell in bureaucracy and a decline in productivity as it is described in "Parkinson's Law", the only other law in the natural sciences that is as reliable as the second law of thermodynamics. In short form it is:

"Work expands so as to fill the time available for its completion."

When the young generation hears the term "Parkinson" the only connotation is the well-known terrible neural disease still not curable. Fifty years ago this was different and everybody linked it to the name of Cyril Northcote Parkinson, the great researcher of organizations and bureaucracy. He had summarized his observations how bureaucracies swell and how bureaucrats make sure that their budgets and the number of people working for them grow steadily, in a series of humorous essays published under the name "Parkinson's Law".[1] It was very popular at the time. Not only in the West but also in the former Soviet Union where people had fun seeing a system described, they knew so well. A huge bureaucracy that kept trying to control everything in the economy, with the only consequence that the more they tried, the less it worked.[2]

To me there's more wisdom in the work of Parkinson than in some scientifically beefed up trivialities worth a Nobel Prize in economy we have seen over the last decades. But Parkinson's Law is almost forgotten for whatever reason. I have seen modern researchers with a PhD in economy and management who hadn't heard of Parkinson's Law. It's hard to understand why this jewel of human insight and knowledge has been so forgotten.

Maybe because it is so clear and easy to understand?

Maybe because it's an inconvenient mirror for researchers who don't like to realize that their research has become a huge bureaucratic event?

Maybe bureaucracy will win, in case bureaucrats are the majority and if everybody is a drug-addict who will protest against drugs?

As I said, you can't do much about this. But if there is one aspect in Parkinson's Law that should be obeyed by those who really want to explore new frontiers, it is the first of the two forces Parkinson has identified as growth factors of bureaucracy:

---

[1] Parkinson C N. Parkinson's Law: the Pursuit of Progress[M]. London: John Murray, 1961.
[2] https://en.wikipedia.org/wiki/Parkinson%27s_law

(1) *"An official wants to multiply subordinates, not rivals"* and

(2) *"Officials make work for each other."*

The exclusion of rivals or competitors is the most deadly danger, since it also excludes controversy and discussion. This is like blocking the error-correction (or backpropagation) function in a deep-learning network.

That's why I don't think that the right lesson learned from Markram's autocratic failure is to assign each child its own safe place in the sandbox protected from competition. To do so would be an even bigger blunder because the consequence would be almost a guarantee for little or no increase of knowledge.

I can see a tendency to do this in the way a smart and mighty big science organizer like Christof Koch is ruling the new empire he has silently acquired. Much smarter than the aggressive and overly ambitious Markram, he gives every child a fair share of the cake and invites everybody to the party to celebrate in harmony. For those who want more than a predictable career, publications in prestigious magazines, and invitations to congresses in nice places, it is recommendable to stay away from those big science castles made of poisoned chocolate.

It won't be many, of course, who will be able to resist.

It was a couple of people 120 years ago and it will be a couple of people today.

Well, we have discussed this many times and because I'm probably hopelessly romantic in this question I still have no better solution than hoping for the young students doing crazy things and not following the plans of the committees and their professors. ☺

And, of course, I have to admit that somebody has to do the data collection, the leaf counting, and the less thrilling part of the business. A university full of Einstein's and Gödel's would also be hard to live with. And it is also nothing wrong when today more than 50 % of a year's age-group graduate from high-school and can apply for college (which most of them do) in Germany while it was between 1 and 2 % in 1900.

However, there are alternative solutions to the problem that all parents wish social advancement for their children and see it best guaranteed by an academic career. Ephraim Kishon, the popular author and satirist proposed a solution to the problem already many years ago which was equally ingenious, simple, and inexpensive.

He proposed to allow the names "Doctor" and "Professor" as first names in the birth register! What a brilliant idea and if only they had listened to him. How many lesser talented students in my country would have been spared an unhappy scientific life and how many very talented young people would have enriched our society as happy cooks, carpenters, mechanics, and gardeners? ☺

Well, let's see how the HBP develops under its new rules of harmony.

Thank you for calling my last e-mail a "DeepLetter", although I'm not sure with how big a dose of irony the compliment came. ☺

Adi says that one of my (many) faults is that I always have to teach people even on matters they already know or even know better than myself. ☺ You are such a polite person who would never protest, so I apologize for any attempts to explain things to you that you understand better than I do.

To my apologies I should add that this method of explaining things is the best way to improve my own comprehension of the matter and so does the discussion with another individual, especially when it gets controversial.

Therefore, I very much liked your objection to the skepticism I expressed against a definition of intelligence that tries to separate learned skills from the pure ability to learn. First of all, I have to admit that you are probably right when you protect Pei Wang. From what you have cited I can see that he seems to be aware of the problem I wanted to address. What I said wasn't related to an original writing of Wang (which you probably got from a Chinese article of him) but to what you said in your letter from 24.02.2016:

"*I especially appreciate Pei Wang's idea to distinguish skill from intelligence. No matter how complicated a skill is, if it cannot be improved, and is just fixed at the same level, then it cannot be called intelligent.*"

I can understand why Pei Wang and many others like Jeff Hawkins, whom you have also quoted in this context, put their focus on this higher element of intelligence above mere automated behavior or formerly learned skill. It is in vain, as I said before, to catch any natural phenomenon by a "correct" definition as the scholars in the scholastic times tried to do. A definition is only less or more useful when it comes to testing a theory by empirical means. Therefore, I don't want to argue about the correctness of an intelligence definition which Pei Wang needs for his theory (which I think is a relevant theory in its own right). My point is, that in every definition you always import parts of theoretical assumptions.

In the case of an intelligence that goes beyond mere skill, it is the aim to declare some kinds of intelligence belong to a higher or superior kind. In history, this has often been done to declare humans a unique species with such superior intellectual abilities that any other species is excluded from it. There is a parallel in the history of religion when a similar problem arose with the question whether other living beings have souls besides humans.

Much of the consciousness discussion develops in a similar way. To many people it seems almost unbearable that animals also have a soul or that robots might be conscious. I remember a discussion with a friend, who is a scientist and also a strong religious believer, about this matter. We had a debate, in which he insisted that humans possess souls but lower species don't. When I didn't buy his arguments, he stopped, looked at me and said "you are arguing in a way as if we would be nothing else but other kinds of cats and dogs". To him this was so unthinkable that reducing my position to such an absurdity was equal to a proof that I was wrong.

Of course, I don't compare your arguments and the one of Pei Wang with this religious believer, and as I said, an intelligence definition with the focus on the non-skill part of learning makes sense in the framework of his theory.

To make sure that I didn't do Pei Wang wrong, I read your letter from July 14th again very carefully and I stumbled across a term in your citation of Wang that didn't fall into my eye when I first read it. You wrote:

"*According to him, 'intelligence' is not the ability to solve concrete problems, but the ability to get the ability to solve concrete problems.*"

The important part in this quote is "*ability to get the ability*"!

I like this "ability to get the ability" idea very much! It is more than a definition, because it implies a theoretical idea. And it comes close to my own view that we are continuously improving our cognitive abilities by experience. It is a dynamic process that improves our learning and comprehending abilities over time. It is important to understand that an experienced brain is smarter than a naked brain, not only because it has collected more factual knowledge, but because it has improved its ability to grasp things quicker; very much as a trained artificial neural network is smarter than a lesser experienced neural network of the same kind. There is another puzzle related to this intelligence problem and the differences between natural and artificial intelligence which I want to mention here. The traditional focus of AI researchers and engineers has mostly been on the higher functions of intelligence and this led to a practical problem, already in the early days of AI, which is known as the "Moravec Paradox".

When engineers like Hans Moravec, Rodney Brooks, and Marvin Minsky started to build intelligent systems and robots in the 1980s they had expected that the higher intellectual functions of humans would be more difficult to emulate than the more basic functions. But to their own surprise, they soon learned that this wasn't the case. The hard part was easy and the easy part was difficult to solve-hence the term paradox.

Moravec writes:

"*it is comparatively easy to make computers exhibit adult level performance on intelligence tests or playing checkers, and difficult or impossible to give them the skills of a one-year-old when it comes to perception and mobility.*"

As you can see on the nice Wikipedia page [1], Moravec found good arguments to explain this odd situation. He makes the point that:

"*Encoded in the large, highly evolved sensory and motor portions of the human brain is a billion years of experience about the nature of the world and how to survive in it. The deliberate process we call reasoning is, I believe, the thinnest veneer of human thought, effective only because it is supported by this much older and much more powerful, though usually unconscious, sensorimotor knowledge. We are all prodigious*

---

[1] https://en.wikipedia.org/wiki/Moravec%27s_paradox

olympians in perceptual and motor areas, so good that we make the difficult look easy. Abstract thought, though, is a new trick, perhaps less than 100 thousand years old. We have not yet mastered it. It is not all that intrinsically difficult; it just seems so when we do it."[1]

I've been struggling with a similar thought since long and was happy to find this citation of Moravec. I think it can be linked with Gehlen's idea of the relevance of human inventions like tools and ideas (crystalized in the form of institutions that function like a guidance-or support-system).

While Gehlen, unlike Heidegger, was ignored in the English-speaking world, his ideas could spread under the headline "Extended Mind" which covered some of Gehlen's arguments but not all and, in general, represent a step backwards compared to the anthropological insights Gehlen had already gained more than 50 years ago.

I'm not through with my homework of bringing all this together and may give it a try in one of my next letters in case you are interested.

For today I stop with this subject and rather close with a reference to an intellectual pleasure that you might also have fun with.

Right when I was dealing with Wang's arguments about the relevance of logical reasoning in the tradition of Gödel, von Neumann, and Turing for practical AI systems I, just by accident, watched the movie "*Ex Machina*". It is an independent science fiction thriller from 2015 which addresses exactly the theme of conscious androids and of how to find out whether a subject is a human being or a machine. The plot goes that a programmer is hired to do the ultimate Turing-test with an android. I don't want to reveal too much here because the end is surprising. The theme is not new and has been treated quite often since the days of Star Trek, but I'm sure you will like the way how your specific interest in the problem of subjectivity is dealt with. So, in case you have a chance to see it I recommend it to you and if you can't, Wikipedia will reveal the trick. ☺

Best for today and good night to Shanghai.

Karl

---

[1] Moravec, Hans. Mind Children[M]. Cambridge, Massachusetts: Harvard University Press, 1988, p.15-16.

Ⅲ-007 Fanji

23. 12. 2016

Dear Karl,

I am very pleased and excited to read your great letter.

I have the same feeling as yours. Although we know that "*worshiping a person for everything he says is a fault but denying everything he says is another one.*" And "*try to avoid this mistake as much as possible, but it happens nevertheless.*" Just as the old Chinese saying says: "It's hard to put into practice what is known", so do I. Yes, we must try our best to be fair to ideas from others whom we do not agree with for many points previously. We should not have a prejudice against anyone, although it is sometimes hard.

Thank you very much for telling me about Parkinson's law. Although I do not belong to the "young generation", I am sorry to tell you that the only connotation related to "Parkinson" for me is the disease with the same name ☹. I have never heard of Parkinson's law, although we can find the phenomena the law described here and there.

I apologize for making a fool of myself in trying to be humorous in calling your previous letter a "DeepLetter". You know that I appreciate "Deep Blue", "Deep Learning" and "DeepMind" quite a lot, I just wish my letter would have been vivid and expressed my appreciation of your letter, no irony at all! As a matter of fact, the many things you talked about in your previous letter make me think more deeply, especially for those philosophers I have never heard of. I am sorry for being so ignorant, compared to you ☹.

Thank you very much for pointing out the flaw in my statement "*I especially appreciate Pei Wang's idea to distinguish skill from intelligence. No matter how complicated a skill is, if it cannot be improved, and just is fixed at the same level, then it cannot be called intelligent.*" Although I shall still insist on the first sentence, the second may be too absolute and it may not be Pei Wang's original idea. You are right that intelligence has many "elements" or aspects, one should not equate intelligence with one of its elements, say the ability of learning. I just had a look at the Wikipedia (https://en.wikipedia.org/wiki/Intelligence), in which it is written:

"Intelligence has been defined in many different ways including as one's capacity for logic, understanding, self-awareness, learning, emotional knowledge, planning, creativity, and problem solving. It can be more generally described as the ability or inclination to perceive or deduce information, and to retain it as knowledge to be applied towards adaptive behaviors within an environment or context."

It also lists several other definitions and none of them could be an accurate and generally accepted definition as no one can. The above statement lists a lot of elements, including capacity of learning and problem solving. Maybe "*some automated behavior or formerly learned skill*" should still be considered as intelligent behavior. As a matter of fact, in the mainstream AI, people are just making machines to do some such skill. In fact, in the Pei Wang's classification of intelligence, the lowest one is also the one without learning ability. However, there is still the question I asked in my last e-mail, what kind of skills could be considered as intelligent. Obviously not all skills could be called intelligent; otherwise, the wasp is also intelligent. If only the skills that only human brains can do previously are intelligent, no animal's behavior would be intelligent. So, the problem is still there. Of course, from a point of view of application, it may not be important at all if a machine can be called intelligent or not, only if the machine can solve the problem we need it to solve. Just as you said in your last letter:

"It is in vain, as I said before, to catch any natural phenomenon by a 'correct' definition as the scholars in the scholastic times tried to do. A definition is only less or more useful when it comes to testing a theory by empirical means."

Maybe we have to be satisfied with a rough idea about intelligence at the present moment, just as we do for the concept of consciousness. Just as Francis Crick said about consciousness, we can still study consciousness even without a generally accepted definition. As for the problem of definition, let philosophers do it. I agree with your argument that the ability of learning is a higher element of intelligence, no matter how important it might be, it could not be the entirety of what intelligence means. Even so, Wang's idea to divide intelligence into four different categories according the ability of learning is still quite interesting for me. There seems to be something in it, but I cannot explain clearly enough.

Thank you very much for recommending the movie "*Ex Machina*" which I do not know, I shall try to watch it. As you know me so well, I am always interested in the problem of subjectivity, which I think is the key hard nut to crack both for neuroscientists and AI engineers. I am convinced that neuroscientists could elucidate the neural mechanism of behaviors sooner or later, and a machine could pass the Turing test someday. Even so, we may still not know how subjectivity emerges, and have to treat it as an unreducible emergent property, maybe. A machine could have emotional expressions and intelligent behaviors, but we still don't know if it has feelings, intentions, desires, consciousness and the like, we still don't know if it has an inner world. Maybe it has, only we don't know, maybe it is just a zombie. While the "other mind" is still an open problem for animals, perhaps also for our fellow human being in a strict sense, how could we expect to solve this problem for machines in the foreseeable future?

In your letter, you also mentioned big science, neuromorphic chips, and other interesting problems. I would like also say something about these topics.

In a special issue of the journal "*Neuron*" published on the 2nd of last month, big brain projects, including the EU HBP, US BRAIN Initiative, China Brain Project, Japan Brain/MINDS, and other national projects of Australia, Canada, Korea were introduced by the organizers of these projects. Related topics, including a paper titled "*Big science, team science, and open science for neuroscience*" by Christof Koch and Allan Jones were also published. The contents of these papers seem to support our ideas discussed in our previous e-mails. I don't think that we should modify our views radically owing to these new progresses.

"Moravec's paradox" you introduced last time is very interesting, I don't know this name, although I know the idea from a paper by John Hopfield in the 1990s (Sorry, I forgot the title of the paper). I remember that until a few years ago, people always used pattern recognition and action control to exemplify the deficit of computers. People used to say that even a three year old child could do such tasks easily, while they are difficult for computers. However, owing to the rapid progress of machine learning, the situation seems to be changing. Nevertheless, to carry out these tasks, computers need to consume huge amounts of energy and it is also time consuming for them to do so. This may be the background that neuromorphic chips have been developing in many labs. Besides the BrainScaleS system developed by your country-fellow in Heidelberg University and other institutions, which you mentioned in your last e-mail, another SpiNNaker system under the HBP has also been developed at the University of Manchester.

In addition, there are also several labs in the United States which are devoted to developing similar chips. Maybe the most famous is the IBM TrueNorth chip with more than a million "*neurons*" and 256 million "*synapses*" developed by Dharmendra S. Modha, the man whom Henry Markram wanted to "*string up by the toes*". This is only a part of a " Systems of Neuromorphic Adaptive Plastic Scalable Electronics (SyNAPSE)" project supported by DARPA.

I suppose that you have noticed that recently Modha and his colleagues published several papers on the TrueNorth chip. The energy which is needed for the operation of the TrueNorth chip is only 70 mW, four orders of magnitude less than a von Neumann computer doing the same task, and the speed is 46 billion synaptic operations per second per watt. The key point here is that the processing units and memory are not separated as in a von Neumann computer, thus the energy needed for continuous transmission between them is saved. In addition, the output of its element is spike, thus the huge energy consuming during the frequent switches between high and low voltage levels in von Neumann computers is also saved. A demonstration was shown that such a system can recognize different objects, such as pedestrians, cyclists, buses, cars etc. in real-time[1]. On the contrary, Markram's blue brain cannot be used to carry out any practical intelligent task and needs huge amounts of energy to run. Of course, Modha's neuron is some kind of integrate-and-fire neuron, much simpler than Markram's, which

---

[1] Service R F. The Brain Chip[J]. Science, 2014, 345(6197): 614−616.

made Markram so furious and called the announcement of Modha a "stunt" or "deception". So now Modha emphasized:

"Let's be clear: we have not built the brain, or any brain. We have built a computer that is inspired by the brain. The inputs to and outputs of this computer are spikes. Functionally, it transforms a spatio-temporal stream of input spikes into a spatio-temporal stream of output spikes."[1]

However, in another case, he said that

"It's trying to get as close to the brain as possible, within the limits of today's inorganic silicon technology".

It seems how to take a trade-off between to be "brain-like" and "brain-inspired" is still a problem. There is also other skepticism about the ability of the TrueNorth system. In 2014, Yann LeCun, one of the three main pioneers of deep learning and the head of Facebook's AI research group, wrote that the chip would have difficulty running tasks like image recognition using a deep learning model called convolutional neural networks. In a follow-up 2016 paper, however, IBM said it was able to demonstrate how convolutional networks could run quickly and accurately on its neuromorphic chip. Anyway, I think that the TrueNorth system is worthy of noticing.

Similar chips known as Neurogrid are constructed by Kwabena Boahen from Stanford University, a former student of Carver Andress Mead, the father of neuromorphic engineering. Maybe his neuron is an improved version of the "silicon neuron" developed in Mead's lab, which uses CMOS circuits to mimic a few ionic channels physically. Although his neuron is more complex than Modha's but is still much simpler than Markram's. Boahen's work is also very interesting to me. Mead's lab has a tradition of using neuromorphic chips to solve practical problems, such as developing artificial cochlea and artificial retina, but not a whole artificial brain. Let's see what Boahen will do the next.

---

[1] http://www.research.ibm.com/articles/brain-chip.shtml

Anyway, there is one common point in all the projects mentioned above, they all use spiking neural networks to save energy, which is inspired from the brain mechanism. And all of the systems are implemented directly by hardware, instead of simulation on a traditional computer. As for the neuron elements they use, it could be as simple as a point neuron, or as complex as a silicon neuron with analog simulated ionic channels. Which would be better? I don't know. Although it seems that to simulate a neuron as accurately as Markram does is not practical for application purposes.

In one of his 2016 papers, Modha reported that they had developed a hardware and software ecosystem consisting of a simulator, a programming language, an integrated programming environment, a library of algorithms and applications, firmware, tools for deep learning, a teaching curriculum, and cloud enablement for their TrueNorth chip. As the architecture of the chip is totally new, they have to compile the total software system from scratch, and they even established a virtual SyNAPSE university to teach others how to use their system. It seems to me that such system may be prospective for real-time pattern recognition, action control, and for being used in situations when energy-saving is the key problem. As if such a system could develop into a new generation of computers, it will depend on how many people will invest their time and energy to learning how to use this new ecosystem. Although the prospect of neuromorphic chips seems bright, I am not sure if it would definitely develop into a successful new generation of computers. I cannot help being reminded of the Japanese 5th generation computer project. In theory, the idea is not wrong, and they also developed some ecosystem, however not many people were willing to use the new kind of computers, and the project failed at last. Will neuromorphic chips develop into a new generation of computers, or just be another Japanese 5th generation of computers? I don't know. However, as you know that I am not a technology guy, you may tell me if my impression is reasonable or not and tell me your opinion.

OK, as my letter is long enough, and in my brain, the music of "silent night, holy night" is sounding, I think I should stop here.

Merry Christmas and a Happy New Year!

Fanji

III-007   Karl

05.02.2017

Dear Fanji,

Thank you for your kind season's greetings; and welcome to the year of the rooster!

I wish you and your family all the best for the coming year. You have already described to me what a turbulent time Chinese New Year is and I hope you all got off to a good start and weathered this storm well. But these are turbulent times everywhere and not all of them have such pleasant causes as a New Year celebration.

I will come back to turbulences of a different kind in a minute but let me first tell you how much I appreciated your comments on my questioning whether you were serious with calling my last mail a "DeepLetter" or whether this came with a flavor of irony.

Actually, I had hoped that you would deny all irony, exactly as you did, not just because I prefer to be taken seriously (who doesn't?) but because it gives me the opportunity to tell you about a nice finding in the book "Intuition Pumps and other tools for thinking" by Daniel Dennett. I came across it just by accident when I was looking for some newer contributions from him to the concept of "Memes".

As an example, for one of those tools our mind has developed in the course of evolution the reasons about the principle of "going meta":

"*thinking about thinking, talking about talking, reasoning about reasoning. Meta-language is the language we use to talk about another language, and meta-ethics is a bird's-eyes view examination of ethical theories. As I once said to Doug (Douglas Hofstadter) 'Anything you can do I can do meta'.*" [1]

---

[1] Dennett, Daniel C. Intuition Pumps and Other Tools for Thinking. New York: W. W. Norton & Company, 2013, p.9.

And humor and irony belong to this category of "going meta". It is one of the last layers in the organization of our brain evolution added to the architecture. Irony is also the last thing humans learn in their childhood. The brain is so busy mastering the basic level that there is no energy (or facility) left to deal with an extra (meta) level in our mind. It takes a very long time to deal with more than one level and to master this additional level of humor and irony. Even many adults never learn it and sometimes even people do not reach this meta-level who display excellent performances of intelligence on the first level.

In order to better understand the development of the mind in the brain of my young grandson I went back to the books of child and developmental psychology which describe how this happens in stages. What I have learned is that we shouldn't be surprised when we talk to even the most intelligent 3 or 4 years old and when they don't understand irony. E.g. when by accident they have thrown one of two glasses from a table and you tell them "why don't you break the other glass as well?" don't expect them to laugh as an adult human would do. It is quite likely that they really take seriously what you said (with an ironic intention) and break the other glass as well.

"Intuition Pumps" is quite entertaining although on memes I found less than I had hoped. What Dennett has to say on memes is more helpful and clearer than what Richard Dawkins, and subsequently, Susan Blackmore made from the concept. But none of them can see their relevance for building "institutions" which Gehlen could many years earlier. But this I would like to discuss in a separate letter if you like.

Here I would like to show you another finding which I found in this book of Dennett's which will hopefully address (and please) the professional translator in you.

"First, there is a culture of scientific writing in research journals that favors-indeed insists on-an impersonal, stripped-down presentation of the issues with a minimum of flourish, rhetoric, and an allusion. There is a good reason for the relentless drabness in the pages of most serious scientific journals. As one of my doctoral examiners, the neuroanatomist J. Z. Young, wrote to me in 1965, in objecting to the somewhat fanciful prose in my dissertation at Oxford (philosophy, not neuroanatomy), English was becoming the international language of science, it behooves us native English-speakers to write works that can be read by 'a patient Chinese (sic) with a good dictionary'. The

results of this self-imposed discipline speak for themselves: whether you are Chinese, German, Brazilian — or even a French — scientist you insist on publishing your most important work in English, bare-bones English, translatable with minimal difficulty, relying as little as possible on cultural allusions, nuances, word-play, and even metaphor. The level of mutual understanding achieved by this international system is invaluable, but there is a price to be paid: some of the thinking that has to be done apparently requires informal metaphor-mongering and imagining-tweaking, assaulting the barricades of closed minds with every trick in the book, and if some of this cannot be easily translated, then I will just have to hope for virtuoso translators on the one hand, and the growing fluency in English of the world's scientists on the other." [1]

The (sic) behind the "patient Chinese" is in the original by the way. ☺

Dennett demonstrates the difficulty of a scientific translator's work and he supports the position you have described some time ago. I therefore, hope that you will appreciate what Dennett is saying, even though I'm aware that in other respects you aren't exactly a fan of his position.

Thank you very much for the excellent summary you gave me about the developments in the field of neuromorphic chips. Collecting such information would have been my duty but I have to say that I have been so busy with other things over the last months that I neglected this field more than I should have. And, as always, in these dynamic fields of IT-engineering a lot of things happen and when you don't read the latest news for 2 months you risk losing contact.

So, I'm very glad and thankful for what you did and I'm also very impressed. You always insist that you are not a technology guy, but you are an excellent researcher! What you found is exactly what the very experts discuss right now. One of my friends who is in the chip-making research business reported to me about similar things as you did, but his analysis was not as broad as yours. He met IBM's Modha in December and was also very impressed by both TrueNorth and Modha as an individual. Others in my environment are (similar to what you have quoted from Lecun) much less positive

---

[1] Dennett, Daniel C. Intuition Pumps and Other Tools for Thinking[M]. New York: W. W. Norton & Company, 2013, p.11.

about the chances TrueNorth or other dedicated AI-chips have in the market. Again, you have identified the crucial point when you said that:

"As if such system could develop into a new generation of computers, it will depend on how many people will invest their time and energy to learn using this new ecosystem."

Bingo! This has always been the crucial factor in IT over the last 50 years, much more relevant than the technology itself and its possible benefits.

And, at least, for the moment my young hipos with their AI startups don't care much for such fancy new devices and run as fast as they can following the Tensorflow-path paved by Google and simply use piles of GPU's to get their neural networks calculated. Google may not be the most powerful technology company, but they are an extremely powerful marketing machine.

But nothing is granted for eternity in this business and therefore marketing soon may not be enough.

In any event your survey inspired me to do more research and also ask my young sharks for advice before I come back to you with a more thoroughly funded opinion. Especially I have to have a closer look at this chip designed by Kwabena Boahen, the former student of Carver Mead, which I did not have on my radar-screen.

Thank you for this hint!

I'm a great fan of Carver Mead (especially because of his refreshing opposition against the Copenhagen quantum orthodoxy) and everything that comes from his school has deserved a closer look.

I also liked what you reported on the special issue of the journal *"Neuron"* about the big brain projects, at least to the point that it doesn't put the two of us in the position of fools.

At least not yet. ☺

However, when I think of the huge effort and the enormous amount of money flowing into such sandboxes and compare it with the possible outcome I'm less happy.

But, as mentioned above, we are confronted with more severe things here right now than a possibly wasted billion euro for an ill-fated HBP.

You have heard, for sure, that Europe is in turbulence after the Arab spring, which not long ago, was greeted with great enthusiasm, but has turned into a nightmare of war and destruction. And millions of refugees were and are heading to Europe to escape the threat and misery with Germany as their preferred destination.

A generation of Europeans living in peace and prosperity has been confronted for the first time in their lives with such elementary horrors and threats that they have only known from books, films, or university seminars. Overnight, people realized how fragile the foundations of our pleasant living conditions are. And how easily what they took for granted can be endangered. And the people also learned that being helpful and supportive of other people in need is one thing, but to accommodate and integrate such a large number of people from a completely different culture is another. It also became clear to many that what we had, so far, in the focus of our interest were often rather negligible bureaucratic matters or what some called "luxury problems".

Especially after Great Britain's voters opted for leaving the EU and the US voters now elected a president who has "America First" on his agenda many people are thinking about the future role of Europe both in economy and in world politics.

And, at the same time, this is a strain on the European community and its administration that triggered enormous centrifugal forces in its various member states which, in turn, led to a destabilization of the political conditions that were not considered possible recently.

Many people in today's Germany know nothing else but an ever-increasing economic standard with a generous supply of excellent food, goods, and all kind of services. Where everybody gets medical and dental care of superb quality and has access to a high-quality education system almost for free. And where the productivity of the industry is so high that ordinary people reach a standard of living that is considered paradise in many places elsewhere.

Leisure and tourism industries all over the world thrive on the prosperity of Europe, and especially Germany that supplies its retired civil servants and teachers so well that the cruise ship companies can hardly supply enough ships to carry them around the globe to the most exotic and pleasant places which most other people only know from Hollywood movies.

Germany, at the beginning of the 20th century, appeared as the power house of modern industries built on research, science, and technology, where diligent people worked hard with proverbial Prussian discipline to improve their standard of living.

Today, according to a popular philosopher, Germany's most remarkable achievement is to have democratized luxury. Well, he wasn't popular anymore after he said that and was covered with a "shitstorm" — no one likes to hear the truth when it's uncomfortable. ☺

Actually, the dominating mindset of the young generation seemed to be that the privileged economic situation of the West is guaranteed anyway by its superior industry based on modern science and technology and the time has come to additionally take into account other parameters such as ecological footprint, fair trade, and other elements of a the road-book towards a better world with responsible use of human and material resources.

And now out of a the clear blue sky, Europe was confronted with a turbulent discussion about very basic life questions about how to stop war and terror right at Europe's doorstep and how to deal with a flood of refugees. The agenda ranges from possible military intervention and fighting terrorism in the Middle East to building walls to stop the flood of refugees, to bringing more people to Europe and integrating them into a modern industrial world.

So, there is a great debate going on about how Europe should position itself in a much more difficult modern world.

Of course, the young generation is mostly concerned about which way to go.

This also has a lot to do with the advent of industry 4.0 and the role AI and robots play in this transition to a new kind of industry which may not be as radical as Kurzweil's singularity but rather thrilling in any event.

I'm discussing such matters more and more these days with my young friends both researches and entrepreneurs. And especially the latter ones know quite well where prosperity comes from and are concerned that Europe, and especially Germany, may be harmed by this transition because the backbone of our big industry is depending on automobiles and the surrounding supplier-ecosystem. They are still doing quite well but with a transition to electric cars and competitors like Tesla at the horizon, there are good reasons to be concerned.

The very problem is that Germany, as an American friend has once put it (with some cynical undertone) *"is making the best 19th century products in the world"*.

What he wanted to say is that we should not miss the boat of innovation. And our big industry guys indeed are not exactly famous for being the most innovative of all.

This is much better in the medium sized businesses where we have many global technology leaders in their fields. What Germany misses is new big pillars of whole industries as they were created in the 19th century but today mostly rise in the US and Asia, especially in China.

After the Second World War, Germany generated only one new industrial heavyweight in the technology sector, and that was the software company SAP SE [1].

In Asia we have seen plenty of them, first in Japan, then Korea, and now China which develops with breathtaking speed. My friends in the US who were once very confident that the US will remain the technology super power of the world for a long time now scratch their heads and ask themselves whether it was a good idea to so actively help to make China the factory of the world.

And now AI, automation, and robotics are in the middle of all of this and everybody asks where the next hotspot for the ultimate technologies will be. Many here believe that it will be China, that the US is running out of steam and that Europe is in real danger of becoming the loser in this game. But as always, in economics this is hard to predict. I remember a similar discussion back in the 1980s. At that time, it looked as if Japan would overrun both the USA and Europe. But it didn't happen, mostly because the Japanese economy almost stalled because of internal problems of the country.

But one thing is for sure: Artificial intelligence, machine learning, and robots will turn many industries upside down again and maybe even more than microelectronics did 40 years ago. And while many, once again, are concerned about people losing their jobs who work in dying industries I'm much more optimistic on this question because I have seen this film before.

---

[1] https://en.wikipedia.org/wiki/SAP_SE

In all those innovative phases we have seen more new products, services, and jobs created than have disappeared. Just think of the example of the PC, the Internet, and the Smartphone. The question however is who will be in the lead and who will be able to afford sending their pensioners to the beautiful sceneries worldwide on luxury cruise ships?

About a year ago I would have said that Europe is, in fact, at risk of losing contact and also told some of my ambitious entrepreneurial hipos to think about moving to the US if they really want to play in the Champions League of technology.

But there are some indicators that Germany may have a chance to participate in the AI-race. At least I can see the advent of a new generation of researchers and entrepreneurs who really want to go for it. When I saw this here, for example I believed that it was from Google [1].

But it isn't from Google, it's from a small spin-out from the University of Tübingen, not far from where I was born. I didn't know that Germany had such advanced technology in DeepLearning.

And the people there seem to be related to those who have created a more serious project called Cyber Valley [2].

This is still not enough, of course, to build big industries and it's not yet Google, Facebook, JD.com, or Tencent's WeChat but at least an encouraging start.

So there is hope for me, my pension, and the global cruise ship industry. ☺

I wonder how the developments in Europe and Germany and their prospects are seen from a Chinese perspective.

Best and good night to Shanghai.

Karl

---

[1] https://deepart.io/
[2] http://www.cyber-valley.de/en

III-008 Fanji

20.03.2017

Dear Karl,

I am "jealous" about your ability to deeply dig into almost everything we discussed, this time about irony and humor. You find that "*Irony is also the last thing humans learn in their childhood.*" The example of the child who breaks the other glass as well is funny, but the lesson is deep indeed. I cannot help laughing and thinking.

Although, as you said that I am not a fan of Dennett, I have to admire his remarks on scientific writing and translation. His words are just what I have experienced as "*a patient Chinese with a good dictionary*". When I was a college student of mathematics, it was not too difficult for me to read basic mathematics textbooks in English or even in Russian (in those days, we only learned Russian as the foreign language in universities. Fortunately, my high school still taught English instead of Russian at that time). Of course, here I only mean that the language for pure mathematics textbooks is easy, not the contents themselves. The most difficult part for understanding is in formulae, which do not need translation at all, but needs background knowledge which the reader should know in advance. The grammar is simple, the total vocabulary is small, and there is almost one to one correspondence between terms both in English (or Russian) and Chinese, almost no ambiguity, very few cultural allusions, nuances, word games, and metaphors. I could not think of any other text to be translated so easily. This may be the text most suitable for machine translation, I suppose.

After my graduation, I entered a new field — biological cybernetics, the thing turned completely different. The vocabulary is terribly rich; the grammar of the text is complex. The words used often have many different meanings, depended on the context, not only in one sentence, but also in a paragraph, a chapter, a whole book, or even the author's cultural background. I could not understand the words just with grammatical analysis and a concise English-Chinese dictionary! Just two years ago, when I translated V. S. Ramachandran's book "*Phantoms in the Brain*", I met the following sentence:

"*The classic illustration of a reaction formation, of course, comes from Hamlet, 'Methinks the lady doth protest too much.' Is not the very vehemence of her protest itself a betrayal of guilt?*"

Although the grammatical structure of the sentence is simple and the words used are not difficult, however, I still didn't understand what it meant, I doubted my translation, until I read the total play of *Hamlet*. Then I understood and gave a note to help my readers to understand. Of course, western people are familiar with Shakespeare's works, and the above sentence is easy for them, but not for the Chinese.

After so many years of practicing, now I may translate much better than before, I totally agree with what Christof Koch wrote about translation in his preface for the Chinese version of his book "The quest for consciousness", which I mentioned in my previous letter. Now I would like to cite it again:

"*Translating any text is a very difficult and demanding undertaken since the thoughts underlying the ideas put down on the printed paper need to be understood before they can be rendered into a different language. And if the translator has done his job perfectly, he will and should remain completely invisible, so that nothing intrudes between the reader and the author.*"

Therefore, according to Koch, the essence of a good translation is the full understanding of the translator and can not only be based on grammatical rules or statistical analysis, which may be the very approach current machine translation uses.

In the hot tide of artificial intelligence nowadays, sometimes I hear remarks similar to the following: "Machine translation is so good that translators will lose their jobs soon". Is it really the case? I don't know, but I doubt it, at least for the present moment. Understanding is a thing in the inner world, not just symbol processing. It seems to me that the present machine translation only does symbol processing, but is not involved in understanding. It will make mistakes when the meaning is complex. To test, I used Google translator, which is said to be one of the best of all machine translation software, to translate the quotation of Dennett you cited from English into Chinese, and then back from the Chinese version into English again. The following is the result:

"First of all, there is a culture of scientific writing in research journals that in fact does insist that non-personal, succinct introductions to these issues are made with the least amount of <u>prosperity</u>, rhetoric, and allusions, <u>which is mercilessly as me One of the doctoral students</u>, neuroanatomist JZ Young, wrote to me in 1965 against some fanciful essay in my doctoral dissertation (philosophy rather than neuroanatomy) at Oxford University, <u>an international scientific language that we should use in our mother tongue Write for English-speaking people Works that can be read through 'A Good Dictionary of Chinese (Morbid)'</u>. The result of this self-discipline is self-evident: whether you are a Chinese, a German, a Brazilian, or even a French, you all Insist on publishing your most important works in English. <u>There is basically no problem with English. You can translate as much as you can</u> and use cultural allusions, nuances, word games and even metaphors as little as possible. The level of mutual understanding reached by this international system is invaluable, but at a price: some of the ideas that have to be done obviously require informal metaphors and imaginative adjustments, <u>and everyone is turning a blind eye to closed thoughts. The trick in this book is that if some translations are not easy</u>, then I only hope that on the one hand there should be a high level of translation, and on the other hand the world scientists are required to be fluent in English."

The above translation seems alright except for the words underlined by me, some of which is not correct even as an English sentence or phrase, some of which seems to be a perfect sentence from a grammatical view, but you cannot understand what the sentence expresses or you find the meaning must be wrong. The incorrect translation is about one third of the total quotation. I used Bing and Baidu translation software as well, the results are similar. As I don't know how this software works, I cannot guess how they make these mistakes.

To have a better understanding, I consulted the "machine translation" item in

Wikipedia [1], in which it is said:

"Therein lies the challenge in machine translation: how to program a computer that will 'understand' a text as a person does, and that will 'create' a new text in the target language that 'sounds' as if it has been written by a person.

In its most general application, this is beyond current technology. Though it works much faster, no automated translation program or procedure, with no human participation, can produce output even close to the quality a human translator can produce. What it can do, however, is provide a general, though imperfect, approximation of the original text, getting the "gist" of it (a process called "gisting"). This is sufficient for many purposes, including making best use of the finite and expensive time of a human translator, reserved for those cases in which total accuracy is indispensable.

..."

Claude Piron, a long-time translator for the United Nations and the World Health Organization, wrote that machine translation, at its best, automates the easier part of a translator's job; the harder and more time-consuming part usually involves doing extensive research to resolve ambiguities in the source text, which the grammatical and lexical exigencies of the target language require to be resolved:

"*Why does a translator need a whole workday to translate five pages, and not an hour or two? ... About 90% of an average text corresponds to these simple conditions. But unfortunately, there's the other 10%. It's that part that requires six [more] hours of work. There are ambiguities one has to resolve. For instance, the author of the source text, an Australian physician, cited the example of an epidemic which was declared during World War II in a 'Japanese prisoner of war camp'. Was he talking about an American camp with Japanese prisoners or a Japanese camp with American prisoners? The English has two senses. It's necessary therefore to do research, maybe to the extent of a phone call to Australia.*

The ideal deep approach would require the translation software to do all the research

---

[1] https://en.wikipedia.org/wiki/Machine_translation

necessary for this kind of disambiguation on its own; but this would require a higher degree of AI than has yet been attained. A shallow approach which simply guessed at the sense of the ambiguous English phrase that Piron mentions (based, perhaps, on which kind of prisoner-of-war camp is more often mentioned in a given corpus) would have a reasonable chance of guessing wrong fairly often. A shallow approach that involves 'ask the user about each ambiguity' would, by Piron's estimate, only automate about 25% of a professional translator's job, leaving the harder 75% still to be done by a human.

One of the major pitfalls of MT is its inability to translate non-standard language with the same accuracy as standard language."

Thus, the problem is that the machine does not understand, does not know the meaning of the text which it is translating, and only does literal translation word for word based on grammatical rules or matches the source paragraph as closely as possible in its database. If the source text is non-standard or rare, the translation would be awful. Anyway, in my eyes, the machine only does some symbol processing instead of understanding and expressing the same meaning with the target language. The machine has no internal world! This is the point, and I don't think that the problem could be solved in the near future. Am I right?

What makes me interested in Boahen's work is that he comes from Mead's laboratory, and he uses silicon neurons in his neuromorphic chips. If my realization is correct, his "neuron" is composed of analog circuits instead of digital processors. As I know, Mead's former Ph. D. Student, late Michelle Mahowald, developed a silicon neuron with CMOS circuits mimicking ionic channels. I am not sure if Boahen uses Mahowald's silicon neuron or it's a variation or a new type developed by himself. Using such a "neuron" may reduce energy consumption. As you know that I am not a technology guy, it is difficult for me to read the original research paper on microelectronics; you may tell me if I am right.

Of course, I have noticed the refugee flood to Europe and also the terrible terrorist attacks, which could be heard or read frequently from TV, newspapers, and websites. However, these matters you described are serious social problems, at a higher level than the brain or individual, which are far beyond just science and technology. No pure scientific and technological approach could solve these problems. We are old and retired, I don't think that we could do anything to help. ☹

I agree with your optimistic view about the influence of AI to society and the bright prospect of your country in this competition. Although many celebrities worried about AI and declared that robots with AI might enslave human beings. It is very difficult to confirm or reject such predictions in the long run, but I am quite skeptical, at least, for the foreseeable future. In my eyes, almost all the successes in the AI field belong to weak AI, not one of the weak AI products has its subjective world. Not one has its own intention, its own desire, and its own will. They are only tools serving human beings. Even more, as we have discussed for so long, we don't even understand how we human beings have our inner world, how consciousness emerges, or what is the sufficient and necessary condition for consciousness emerging. For these matters, we even have no idea. How could we expect that we would create a machine with its own inner world in the near future? Thus, AI will only be a tool to help human beings, at least in the foreseeable future. However, the above worries may not be totally meaningless; at least people should think about if we should develop strong AI in the rigorous sense, although at the present moment, no such studies are on-going.

Recently, I read something about free will, as this is still an open problem; no conclusion could be made after such reading. As you know, I know little about philosophy, I am not sure if my comprehension is reasonable. At the very beginning, I wanted to say something on this topic in this letter, however, it is too long already, I may write on this topic in my next letter, if you like.

Best wishes as always.

Fanji

III-008 Karl

01.05.2017

Dear Fanji,

I'm glad to hear that you, although you aren't a fan of Dennett, liked to read what he has to say about the difficult job of a translator. I couldn't resist sending you the part with "*a patient Chinese with a good dictionary*". He could have chosen any language, of course, but Chinese he chose, for sure, because he wanted to tie in with Searle's well-known thought experiment "Chinese Room". But this again, is an example for a translation problem. Many readers of his book may understand this connotation and therefore may smile about it. And if they don't understand it it's not a real problem, because the essence of Dennett's message is understandable even without catching this fine hint for the connoisseurs. So, in case you had to translate this passage into Chinese one day it probably isn't necessary to make a footnote about this. A reader who hasn't heard of the "Chinese Room" may prefer to focus on the argument itself and may be even annoyed if a translator tries to teach him about a very small additional piece of content in this phrase.

The example you gave with your own reading experience with English and Russian math books is very illuminating and so is your example of a real translator's problem in the case of a reference made in Ramachandran's book "*Phantoms in the Brain*" to Shakespeare's play *Hamlet*.

Actually, all three problems mentioned, the very small and negligible one with the "Chinese Room" connotation, the little one in a highly structured and formalized system like math with few cultural allusions, and the big one with a massive cultural impact like Hamlet can all be projected on the same scale.

And they all describe a problem that is not only restricted to translations from one language to another but plays a role in any kind of communication between individuals.

You were right on the spot of this problem when you questioned (in your previous letter) whether we can be sure about what is going on in the mind of others:

"While the 'other mind' is still an open problem for animals, perhaps also for our human being fellows in a strict sense, how could we expect to solve this problem for machines in the foreseeable future?"

We really can't be sure whether others have the same feeling or understanding for a matter we try to communicate to them. Any human communication has this uncertainty built in. It depends on how similar the object libraries in the minds of the two communication partners are. And of course it is not enough that there is an object in your mind and in my mind that is carrying the same label. When mathematicians correspond, they benefit from the fact that there are many well defined objects present in their minds. But the communication is so easy because they take advantage of the work spent earlier when they learned the matter and now do "know" what decimals, primes, square roots, cosines, or integrals are.

If such objects don't exist in a receiver's mind it is close to impossible to build up such libraries over the course of a conversation. In principle, it is possible to build them from simpler terms, but you will run out of time if you try to do so in the course of a conversation. When you talk to someone about trigonometry, you will immediately recognize if he or she knows what the difference between a sine and a cosine is. Very much as you can see after a few moves in a game of chess whether your opponent understands how to play chess. The more closed a system is and the more limited the number of variables and rules are, the easier it is to discover such competence, but only if you overlook the topic yourself. This means that you have already spent time and effort installing such objects in your own library and thereby have linked meaning to syntax. A rich object-library which you have acquired by learning is not the same as skill! Feynman's Brazilian students had all the necessary objects stored in their mind-libraries but didn't have the skills to use them.

This is the reason why I'm so reluctant to buy the argument that intelligence should be measured in a "pure" skill-free manner. Objects stored in our mind-libraries may not only contain content in the sense of factual memories of events. They can also contain tools or methods that help to acquire more knowledge and to make a more efficient and quicker use of it. Programmers use a similar method when moving a function, such as drawing a square root, into a library of a computer language. If they have succeeded in developing a square root algorithm, e.g. using Newton's approximation method, they can place this routine in a library and call it "sqrt". From now on they can easily use the square root extraction function in all other programs on any number. Such thinking-tools can act like an enzyme in a biological process or a turbocharger in a combustion engine. The richer and more experienced such a system of acquired tools is, the better it will perform in learning and solving problems faster. A "quick mind" isn't a quick mind because of its superior physiology, which may help to a certain extent, but mostly because of its library of methods and routines it has built in the phase of learning of how to acquire knowledge and to deal with it. This kind of content is only accessible when it is run through the subjective monitor of the given individual. The subject is the owner of the unique code of this specific brain. You will not detect it when you X-ray a brain to the very atom very much as you won't be able to see whether it is "*Gone with the Wind*" or "*Casablanca*" when you do the same with a video DVD.

Communicating with others typically starts with sending "pings" to the mind of the other to find out about the content in the library of the receiving correspondent. This is not just detecting knowledge, but also gathering information about interpretation, meaning, and emotional connotation. While there is a common pool of words and phrases in everyday language the connotation with meaning and feelings in the individual's brains aren't the same at all but highly subjective. However, humans are pretty good at scanning this part of the subjective landscape in another person's mind.

After a while, we know which objects we can "call" in the receiver's mind and even when we can't be sure that we connote exactly the same meaning to them, we, at least, can cooperate quite successfully. The more elaborate the matter of communication is the easier this gets.

But, only if transmitter and receiver both made the effort to build up the necessary libraries and also align them to full coherence. Coherence doesn't, of course, mean harmony because although we understand someone else's opinion it doesn't necessarily have to be ours.

When it comes to understanding others there seems to be a similar paradox as the one which Moravec already discovered in the early days of AI. It is quite easy for a Chinese and a German to communicate about neuromorphic chips after the participants took the effort to build up the very specific libraries. Communicating about the most basic cultural and social aspects is much harder because the libraries that need to be adjusted here are much less specific.

And the field is so wide and vague that you can never be sure whether the two of us associate the same meaning when an object carries the same name.

It is the same problem as in the case of your *Hamlet* example where you had to transmit an enormous amount of content and meaning. It is close to impossible to install such rich content in the mind of a conversation partner or a reader in the course of a conversation or while reading an article. I have already mentioned an example to this very problem. Einstein's answer "No" to the question "*Professor Einstein can you explain to me your theory of relativity in simple words*". And while the person who had asked this question may have taken the answer as arrogant, it was not, because it is indeed impossible. Sometimes Einstein couldn't resist but giving funny answers to similar questions. Those were subsequently (and often pompously) cited and people who didn't have a clue what relativity, in Einstein's sense, was about, projected their own subjective interpretation into it and enjoyed the illusion of having grasped what the thing is.

But this problem doesn't only exist with complicated scientific matter and it has nothing to do with intelligence or superior knowledge. The vagueness and uncertainty of the interpretation of terms or whole concepts you find also in everyday language and even when it deals with the most basic elements of life. Love, happiness, justice, or healthy food fall in this category where you have a universe of possible interpretations all coming with a cultural flavor as broad and cloudy as a comet's tail. Our brain has learned to deal with this and sometimes is just happy to link a name to a vague meaning even if it's not sure whether it is the same meaning as in the mind of the corresponding partner.

And when lovers have been assuring themselves of their eternal love for centuries, they never knew exactly what that meant for each other. And modern lovers and psychologists don't know either, but it is certainly something different today than it was

150 years ago.

The more elaborate and standardized a field is, the more elaborate are the terms and the language to work on it. But, as I said, you can only take advantage of this when you communicate with another individual that has also taken the effort to establish such a library during a time consuming phase of learning. Sometimes, such knowledge is acquired along with the use of technical devices like computers or smartphones that kind of guide or even force us to learn the basics and obey the rules.

When a Python programmer communicates with another Python programmer they both can communicate easily on a problem about convolutional neural networks regardless of whether they are from Shanghai or from Karlsruhe.

There is another paradox however, which, to me, is even more puzzling than the one of Moravec. When humans are so unique and their minds are made of so extremely different and subjective libraries how can it be that in many situations they act so uniformly and highly predictable?

There seems to be a mechanism, maybe rooted very deep in the old parts of our brains, which causes something like social "phase locking", a phenomenon similar to the one Christian Huygens had discovered when he observed that pendulum clocks positioned nearby soon tend to synchronize.

I'm not sure whether swinging in the same rhythm was really the first step to communication and language, as Walter Freeman seemed to believe, but to see "La Ola" arise in a football stadium all over the world when the spectators create a wave moving around the rows when people raise for a short moment and sit down again is an impressive event.

There is no language needed, no one is commanding or controlling it and nevertheless it works when a small group has started to overcome the first barrier and induces their neighbors to do the same and rise with a little time-delay. And so, the wave starts to travel very much like an action potential travels along the axon of a neuron and also through networks of neurons. When people have seen it in one stadium on TV, it is much easier to start in the next. You don't have to understand what is going on and you don't have to think about what you do. You just do what your neighbor did. It's an automated reflex and while some people can inhibit it this will not prevent the building of the wave. Very much like some kind of music makes people whip their feet or swing their head along with the rhythm.

There has been some buzz around the wonders of swarm intelligence and that it may be superior to rational thinking in some situations. I have already mentioned that I'm not a great fan of this idea because when I think about such collective phenomena and what kinds of catastrophes we owe to this mindless "*do what your neighbor does*" behavior I would rather call it the dumbness of the swarm. It may be very helpful indeed, for fish or birds to jointly build objects or perform movements which make it more difficult for attackers to catch members of the swarm, but for humans not using their mind and relying on paleo cerebral functions of their old brain to me, rarely seems to be a good strategy. Humans are escape animals, which because of their inferior supply of biological means had to run away when a dangerous predator appeared. There isn't much time to verify whether this is really a lion or a crocodile approaching or to debate what to do. You simply run, or you are eaten.

And it was clever to run when your neighbor ran, or one member of the flock signaled alarm. But there are smarter ways to deal with threats.

The problem is however, that you have to install rich libraries in the minds of people to make them able to deal with problems in a rational way. This means education and practicing which needs time and discipline. That costs energy but in the end, it is decisive when it comes to surviving and success.

The more closed and limited a field is the easier it is to establish knowledge and rules that allow people to find orientation and successfully maneuver in it.

It's no accident that young people all over the world are attracted to sports and computer games. Both represent highly structured, well defined and limited micro-universes with extremely reduced entropy.

They are ideal to apply rules and practice rational strategies. The same is true with the sciences, especially math and with technology and engineering which all have a high degree of standardization and solid rules and norms. In sports, games, science, and engineering the framework for action is highly standardized and routinized, which gives the players and actors a high degree of certainty. It also makes communication easier and cultural differences are less relevant.

A Chinese mathematician will easily be able to communicate with a German or American colleague about the importance of a new method of "backpropagation" in artificial neural networks, while they may find it much more difficult to deal with cultural and social issues related to AI.

On the other hand, we have the world of emotions where we have unlimited possibilities of interpretation.

And when a translator has to translate a book it very much depends on what ingredients are on the pizza. If it's about highly structured and closed subsets of the world with extremely reduced entropy, like science and technology with low culturally flavored content, and where you can expect a reader that has taken the effort to load the basics to his library, it's relatively easy.

On the other hand, we have literature soaked with cultural content or, even more extreme, poetry with rhymes, a case that asks for a translator who is able to redo the original artist's job.

Therefore, I believe that Christof Koch's advice that a translator "*should remain completely invisible*" is only applicable for the easier part of translations, which is in science and very specialized fields with well-established and standardized libraries.

Your test of the translation quality of Google Translator by translating Dennett's text first to Chinese and then back to English is very impressive and the results are kind of funny.

I have had similar puzzling experiences. But sometimes I was also impressed by how helpful the result was, as in the case of a PC mainboard from ASUS where the workshop manual was only available in Chinese. In other cases, as in your example, I found the result disappointing, useless, or even outright wrong and misleading.

In American-English for example, the efficiency of automotive engines is defined as the "gas mileage". It expresses how many miles a car can drive with a gallon of gasoline, which means that a high gas mileage is a good thing because it indicates a high degree of thermal efficiency. In Germany, this kind of efficiency is measured by how many Liters a car consumes to drive 100 km. The measure L /100 km is called "Benzinverbrauch" (fuel consumption) which means that a low fuel consumption value is good because it indicates a high degree of thermal efficiency.

Many years ago, I noticed that both Google translator and Bing translator translate the English phrase "*The car has a low gas mileage*" to the German expression "*Das Auto hat einen geringen Benzinverbrauch*" (the car has a low fuel consumption).

So, the meaning of the translated sentence is exactly the opposite of that of the English original.

I checked it again a minute ago, just to see if someone has cured this elementary weakness.

It would have been quite easy to correct, but it's still there after many years. Well, maybe managers of both Google and Microsoft don't care much about the gas mileage of their company cars. ☺

But to be fair with the translating machines I have to mention that there are similar trivial pitfalls that human translators keep stumbling upon.

It's the so called "false friends". A common mistake of that kind in English-German translations is confusing billions and trillions. In German, a thousand millions make a "Milliarde" and a thousand "Milliarden" make a "Billion". But for whatever historical reason, an American billion is a German "Milliarde" while a German "Billion" is a US trillion. So, it's very tempting for Germans when they read the word billion in an English text to translate it into a German "Billion". It's no big deal and everybody should know about this pitfall. But you won't believe how many German journalists make that mistake and report about "Billionen" dollar events in the US, which they unintentionally inflate by a factor of 1,000.

Your demonstration and my own experiences with poor machine translations seem to support the skeptical analysis which you have cited from the Wikipedia article on "machine translation". However, I'm not sure how long the arguments presented there will remain valid.

Many of the arguments in the article relate to good old fashioned AI (GOFAI) based on syntactical analysis, grammar rules, dictionaries, ontologies, and semantics.

These methods indeed, try to "understand" what is in the text and this is often hopeless.

But artificial neural networks (ANNs) that turned out to be so successful in visual deep learning applications, like face recognition or telling the difference between cats and dogs, use a totally different approach. They don't want to "understand" but learn which elements of objects are relevant to make them similar. And the way how these algorithms learn to make ever better guesses over many "hidden layers" is difficult if not impossible to understand. Even their inventors don't understand how the tools they have created do the trick. The old paradigm that a computer can only do what it was told to do isn't true anymore in this new world.

It is the data that these networks absorb which represents the knowledge and not the algorithm!

The biggest problem with today's ANN applications is the enormous amounts of data they need in order to be trained. But if they are fed with the right kind of data they can do surprising things and even outperform humans in tasks that seemed to be impossible to solve for machines until recently. I have mentioned the example that machines are doing better now than humans when it comes to differentiating two thousand different breeds of dogs. Face recognition is another example for the astounding power of such deep learning techniques. But in both cases, we don't understand how the neural networks do the trick.

Here is an article where Aaron M. Bornstein from the Princeton Neuroscience Institute deals with the interesting question of the interrelationship between predictive accuracy and explainability of the various machine learning techniques.[1]

And the answer to this "What versus Why" question is:

"*Modern learning algorithms show a tradeoff between human interpretability, or explainability, and their accuracy. Deep learning is both the most accurate and the least interpretable.*"

---

[1] http://nautil.us/issue/40/learning/is-artificial-intelligence-permanently-inscrutable Dietmar Harhoff kindly made us aware of a recent and very rich overview of the different methods used in AI: Dominique Cardon, Jean-Philippe Cointet et Antoine Mazières. La revanche des neurones. L'invention des machines inductives et la controverse de l'intelligence artificielle [M]. Réseaux, 2018/5 (n° 211), p. 173 – 220. DOI 10.3917/res. 211.0173

So, the trick seems to be to let the machine do what it does best and not to try to get it do the job the way humans would do it. And again, this is the method that was so successful when engineers built vehicles, airplanes, or telescopes whose performance soon surpassed all their biological models by orders of magnitude.

The downside of course is, as Bornstein points out, that it's an unpleasant feeling when even the engineers don't understand what's going on in the systems they have created.

This for sure, is a relevant issue when it comes to risk-critical applications where mistakes can cost human lives or cause other heavy damages.

From the point of view of system security, it might be useful to combine the various methods. And processes of understandable methods can supervise others and perform plausibility checks which help to exclude "stupid errors".

The article also supplies a nice graphic with the many methods applied in the field which makes clear that there is much more available than the popular deep learning.

Right now, we see hype in deep learning, and as always engineers tend to use the most productive tools and try to solve everything with them.

However, my guess is that after a while, we will see more hybrid systems and mixed systems that are well adapted to a given problem.

And we will also see more powerful machine translation systems because your examples show that today's systems are not yet built on state-of-the-art technology. Google was very successful in building a good Go-playing program, but its machine translator is not on the same level as this painting program from deepArt that makes these nice paintings e.g. in van Gogh style from any kind of photos.

A machine translator of that kind based on convolutional neural networks might not only be able to make good translations, but also give them a touch of well-known literature like Qian Zhongshu, Goethe, Balzac, or Hemmingway. You may not like the idea, but I'm convinced that it is not as difficult as it sounds.

Bahdanau, Cho, and Bengio published an interesting conference paper at ICLR 2015 entitled "Neural Machine Translation By Jointly Learning to Align and Translate"[1]

---

[1] https://arxiv.org/pdf/1409.0473.pdf

which is pointing in this direction.

This is a highly technical article and you don't have to read it to the very detail. The most interesting information I drew from it is that what state-of-the-art machine translators don't do yet. So, the message is that the revolution we have already seen in image processing is still ahead of us when it comes to dealing with text. So, we might be prepared to see real improvement in machine translation. Or as one of my young hipos in the field has put it:

"What we have seen so far was just a slight scratching on the surface"

Dedicated hardware like the various neuromorphic chips now appearing everywhere can accelerate this to a great degree, especially, as you rightfully say, when it comes to reducing the enormous amounts of energy today's systems consume. This is especially true for stand-alone applications living on batteries like in IoT (Internet of Things) applications or drones.

Boahen's silicon neuron application is one of the many and I can't tell you from a distance where it is positioned right now. As far as I understood, the novelty is not in a new basic design of Mahowald's silicon neuron but in a more sophisticated architecture on the next level.

But as you have rightly said, the problem with all those new hardware concepts will be how many people will be willing to invest time and energy to switch to a new programming eco-system. As long as great progress can still be made in the old environment, as it seems to be the case right now, the probability is not so high. This is why I believe that we will see a paradigm shift only after a while.

I found it very interesting what you wrote about the turmoil in Europe, and also what you said about the economic perspective and the role AI will play. But now when this e-mail is already so long I will rather postpone my comments to my next letter.

In any event, I'm very much looking forward to learning about your ideas on free will!

Best for today.

Karl

15. 07. 2017

Dear Karl,

Thank you very much for your great letter, in which you gave a deep analysis of the essence of translation and the perspective of machine translation. Your arguments are forceful, and I have to re-consider my previous arguments. You are right, although there are few cultural allusions in a pure math textbook, and there is a massive cultural impact in Ramachadran's book, there is one thing common in both texts. To understand the text depends on the corresponding libraries in the reader's mind, although the libraries needed in these two cases are quite different. Although the grammar structure in pure math textbooks is simple, readers without a proper math background can still not understand, just as I didn't understand Ramachandran's remark before I read Shakespeare's play *Hamlet*. Even the grammar structure of Rama's words is not complex, and the words he used are also ordinary. So, it seems that to translate correctly is not equivalent to understanding correctly. A translator may still translate correctly, even if he or she does not understand.

As a matter of fact, when I first translated Rama's words, I didn't understand what he meant, I searched where the quotation came from, the original play *Hamlet*, and then I consulted the expression in the corresponding section in its Chinese version, translated by a famous translator, so that my translation was correct even though I didn't understand why the queen said so, and why Rama should use her words as an example of reaction formation. I suppose that machines could also do this well. Thus, I have to admit that my examples are not proper, these examples, on the contrary, show that even when translators do not understand, they might still translate correctly. The main point is if the translator has a library rich enough as the original writer, just as you hinted. Maybe the poorest library is just a concise dictionary and a grammar handbook, then a better dictionary with plenty of idiomatic usages and collocations, so that the words in the original text can be translated according to the context in a phrase; and then even considering the context in whole sentence, whole paragraph, whole chapter, whole book, some background beyond the book itself such as all the contents in Wikipedia or even more.

The richer the library, the better the translation. Just as you emphasized in your wonderful letter, a rich library is necessary for translation, no matter if the agent is a human being or a machine. However, even so, I still wonder if a rich library can guarantee that all translations are correct. Understanding is rather subtle, I don't think that machines can understand in the foreseeable future, and a translation made by machines may still need a human being's review who understands, if the translation is expected to be accurate and without mistakes.

Although the Google translated example I mentioned in my last e-mail is poor, as well as the translation given by the Bing translator and Baidu translator, which I haven't shown, this does not mean that machine translation can't make noticeable progress in the foreseeable future. You are right, even human translators such as me make mistakes frequently and I often complain about such translations. It is possible that machine translation can reach the standard of an average human translator, although to make the translation perfect, maybe having it reviewed and revised by a human expert is needed. However, those poor human translators also need such reviews. ☹

Even so, the machine translation can help us a lot, at least for the tourists traveling all over the world. Maybe I should not be too picky with machine translation.

I completely agree with your following remarks: "*So, the trick seems to be to let the machine do what it can do best and not to try to make it do the job in a way humans would do it.*"

Now let me shift my discussion to the topic about free will.

As I wrote in my previous letter, after interviewing 21 leading scientists of consciousness studies, Susan Blackmore remarked:

"*To be frank I had rather expected, before I began, that nearly everyone would intellectually reject the idea of free will while finding it hard to live their daily life without any such belief... As Samuel Johnson put it so memorably 'All theory is against the freedom of the will; all experience is for it.'"*

It seems to be a difficult dilemma to solve. I would like to tell you what I think about and am anxious to know your opinion.

Of course, the problem about free will is difficult; I could not touch every aspect of this problem like a review. In this letter, I only want to discuss the dilemma Blackmore proposed.

Different people have different ideas about what free will is.

First of all, let me argue that free will is not equivalent to free action. No one can do everything he or she wishes. There is always something that people cannot do owing to some external or internal restrictions. Therefore, for me, "Free will is the ability to choose between different possible courses of action unimpeded."[1]

Next, the problem is, if such choice is conscious or unconscious? Does free will only mean conscious choice? People are apt to think that the answer is "Yes". However, the classical Libet's experiment hinted that this may not be the case. Libet asked a subject to decide if he or she should flick his or her wrist at a moment he or she chooses, while watching a moving spot move around a round screen. At the same time, Libet also recorded the subject's EEG. The experiment showed that even half a second before the subject was aware of his or her decision, a special "readiness potential" could already be recorded. Therefore, it is not a conscious order or inner speech in the brain that commands starting the movement. If one insists on that free will must be conscious, then Libet's experiment seems to hint that free will is only an illusion. However, if we consider free will as the ability to choose, then his experiment only shows that such choice does not seem to be necessary to be conscious. In addition, this experiment doesn't hint that consciousness doesn't play any role in decision making, as all similar experiments only consider simple actions, you cannot extend the results to long-term planning, such as planning for your career or marriage when you are young. In summary, for simple action, there is some event happening in the brain, which makes a decision from several options to start the action, and makes the subject aware of this

---

[1] https://en.wikipedia.org/wiki/Free_will

decision a second or so later; as for the long-run plan, you cannot find such readiness potential, you are aware of the plan first, maybe months or years ago, then you execute the plan, although we still don't know the details of such decision making.

In daily life, we are not sensitive to such tiny timing differences, we are apt to think that our awareness of the chosen action is ahead of the action and mistake such awareness as the cause of the action. Libet's experiment just shows the last statement is not correct, however, the cause is still some event happening in the subject's brain, which chose the given action starting at the given moment. As for the function of consciousness, it is still an open problem, it may be the cause for the long-term plan, but we are not sure.

Third, even so, an objection to free will is that according to determinism, every effect is caused by a previous cause, so that you can find a causality chain back to the "Big Bang", and everything is decided at the very beginning, even when something seems to be chosen. This argument seems strong, as it is based on some physical law. However, I have two points against the above arguments. First, we know that deterministic law does not hold at a microscopic level such as the laws in quantum mechanics, so that deterministic law is not universal for all phenomena. Here, I don't mean that free will could be explained by quantum theory as some scientists have suggested.

My argument is that as the regular macroscopic body is composed of microscopic particles, while the Laplacian deterministic laws hold for the former, and the probabilistic laws hold for the latter, that means that in a hierarchical system, different laws may hold for different levels. Considering the fact that the mind is much higher in the hierarchy, why should free will have to observe Laplacian deterministic law, which holds for lower levels, but higher than the microscopic level? In addition, the causality chain mentioned above sounds to be based on linear causality. As a matter of fact, for complex phenomena such as the mind, we must consider circular causality instead of linear causality, just as Walter Freeman emphasized, which I cited in my previous letter on 24/02/2016. With circular causality, the effect can also influence the cause, not only the lower level in a hierarchical system contributes to the higher level, but the higher level can also influence the lower one. So, you cannot trace to the very origin along the imagined causality chain. The structure and interactions in the brain are so complex; it is hard to imagine that you can explain its higher function with linear causality. Just as Freeman said that

"The determinants of human actions include not only genetic and environmental factors but self-organizing dynamics in brains ..." [1]

Such complexity and circular causality leaves some room for free will.

In summary, if free will is considered as the ability to choose, then a conscious will is not necessary, unconscious free will may also exist, but be only limited to a short time scale and simple actions. Deterministic law and free will maybe co-exist, considering free will as an emergent property at the higher level of hierarchical systems with circular causality. However, at the same time, I would like to also emphasize that the degree of freedom of free will is limited. The option that free will could choose is restricted by the external and internal constraints. This is what I think about the problem, what is your opinion?

As the letter is already quite long, I have to stop here.

Best wishes from Shanghai.

Fanji

---

[1] Freeman W J. Consciousness, Intentionality and Causality[J]. Journal of Consciousness Studies, 1999, 6(11-12): 143-172.

III-009 Karl

02.09.2017

Dear Fanji,

Thank you, thank you, thank you!!

You won't believe how much I liked your letter and your brilliant demonstration of how helpful the concept of circular causality is, when it comes to shedding light on this mysterious debate about the "free will".

I never liked how this discussion developed and had many debates about it with our friend Hans Braun. He and I don't like the metaphysical clouds that surround Libet's experiments and consider the far-reaching conclusions that some people draw from them to be much exaggerated and inadmissible. I think all three of us agree that this simple experiment and also the later fMRT-experiments of the Haynes group may mean many things but have nothing to do with the question whether humans possess a free will. Hans wrote a very helpful article about the confusion in this debate where physiologists debate with philosophers about determinism, randomness, and free will. It's titled *"Determinismus und Zufall in der Neurophysiologie — Die Frage des freien Willens"* (*"Determinism and coincidence in neurophysiology — The question of free will"*) [1].

He included all his expertise regarding the physiological basics of signal processing in neurons down to the level of ion channels. Hans has worked on the problem of noise and chaos in neuron communication for a lifetime, maybe as intensely as Walter Freeman has worked on the olfactory system and the phenomenon of "reafference".

You can feel this when both talk about the brain and neurons in the same familiar way as if they were talking about the furniture in their living rooms. Both have seen thousands of brains and have done countless signal deductions from neurons with all kinds of

---

[1] https://www.hss.de/download/publications/AMZ-87_Homo_Neurobiologicus_02.pdf

methods from EEG to patch-clamping. And from both of them speaks this kind of competence that you only acquire when you dive deep into a matter.

So, not many people in the world are as deep in the matter as Hans is and what he has to say is for the connoisseur who really understands. His article is a true gem, but I doubt whether many of the people he addresses, who superficially talk about the matter, are able to follow him. But this debate has long since developed a life of its own that is no longer determined by those who are familiar with the matter, but by those who speculate about it and draw wondrous and far-reaching philosophical conclusions from a few meagre experiments. The discussion seems to be enormously ideologically charged, and my impression is that many people who play in this arena are characterized by the syndrome I have already cited the other day: *"they seem to have the most determined opinion of the things they understand least"*. ☺

But this is not uncommon for scientific debates, even in theories that are much easier to test empirically. Man-made climate change is a popular example.

And while this free-will discussion is hard to come by on the level of the physiological details, your argument is much simpler and stronger at the same time and very convincing to me.

From a higher-level perspective of a long term reasoning mind the pathways of signals and primitive motoric control loops (which we still don't understand) are just not the appropriate model for our acting in the world. Planning a career or a marriage is a different kind of game than lifting a finger or pressing a button after a signal is given, just as you say.

And you are right on the spot when you insist that linear causality is not enough to explain the continuous monitoring process by which we adapt our actions to approach moving targets over long periods of time. And while many processes on the lower level, like the motoric system, may be hardwired, this is not the case on the higher level where soft-libraries come into play which act as reasoning tools or apps. Those tools may be self-developed, modified, or imported from others via education or just acquired by reading.

Our material worlds of objects which may also be built on quite fuzzy atomic microelements are astoundingly stable and predictable on a macro level. And so is our acting on a social level. It's stabilized by rules and institutions (in the sense of Gehlen) which act as stabilizers like rails holding a train in the track.

And the artefacts built by engineers use this principle to the max. Their machines are nothing else but intentionally materialized ideas honed, in the continuous process of circular causality you are describing.

Of course, someone may raise the sophisticated argument that ideas in our minds and also our rational reasoning are based on physiological processes and the molecular biology of our brain. And those may be "caused" by earlier states. But this is also true for a computer running an algorithm. And here, probably no one would have the idea that the result of the algorithm would be caused by the material state of the molecules in the silicon chip. Of course, the result is defined by the inbuilt logic. And what many philosophers, who aren't familiar with the latest development in software technology, especially machine learning, don't comprehend is the fact that such inbuilt logic can modify itself in circular processes of learning.

The old paradigm that a computer can only do what it is programmed for is obsolete.

Neural networks, after they have seen enough data, can perform tasks in a way the programmer who created the system and invented the algorithms neither understands nor has foreseen.

Very much as children invent things their parents don't understand and would have never dreamed of.

The point is, that we are able to capsule ideas and thoughts and make them run on logical machines both internal ones (our brains) and external ones (computers). Our internal ones may not be as reliable as the external ones when it comes to remembering and calculating decisions, but they are subjectively intentional in any event.

And while much of our internal rational reasoning and decision making is disturbed by moods and various physiological micro-states even the slightest influence of such subjective reasoning breaks the simple chain of materially defined linear causality, and at the same time, makes the whole system very stable and predictable on the macro level.

It seems plain ridiculous to me to take such a primitive signal-response causal chain, like lifting a finger, as a role model for the complex internal reasoning processes that storm through our minds, often modifying and repeating considerations in tactical and strategic ways.

The scope of many neuro scientists is limited to the connectome of the brain on the level of psychology and they tend to ignore the content on the social level. But this view is incomplete on a systems level and therefore, they run the risk of explaining the whole system from a fragment. This is as useless as it is in vain to explain the functioning of a computer just from its BIOS or operating system and to ignore the running app and the data flowing through it.

But, the important part is in the content at the highest level which makes the meaning.

And it is here where progress takes place. We are continuously building up and improving our libraries and hand them over to others and to our children. We don't have to wait for millions of years until the random walk of evolution modifies our DNA to perform better here and there. And the result of this makes the difference to our predecessors and fellow mammals.

You and Freeman got it right: we establish and continuously refine goals as ideas in circular processes where logical subroutines or strategies make the difference and live lives of their own.

And this is the reason why human behavior is so hard to predict in the long run. We simply don't know what the next game changing idea is, like the wheel, iron, value-added tax, parliaments, TV, internet, or smartphones in the past. Maybe, in the next round it will be artificial intelligence accelerating the whole process of self-organization of living matter like a super-enzyme.

The difference to the slow process of evolution of self-organization which took billions of years is that now via circular causality, intentionality comes into play. Very much like another kind of intelligent design — this time without a god!

Randomness still plays a role, of course, because we simply don't understand the whole system in which we act nor do we know the various states of its elements which makes it impossible to calculate.

This non-calculability has nothing to do with the principle of uncertainty, which is postulated in the theory of quantum mechanics. Mathematicians and physicists know that there are deterministic but not calculable systems in plain Newtonian physics since Poincare dealt with the n-body problem 130 years ago and thus, laid the foundation for chaos theory. Therefore, there is no need to inject an extra dose of uncertainty based on quantum theory to explain the unpredictability of human actions or the free will phenomenon.

But many people, especially philosophers and philosophizing physicists, seem to be in love with this idea of fundamental, mystical indeterminacy. Although hardly anyone understands exactly what quantum mathematics is all about, it has become established that almost no work on this subject can do without a reference to Heisenberg and his principle of uncertainty. It is almost like a religious ritual or a perfunctory bow to a shrine in the hope that one will not be struck by the lightning of a holy member of the scientific inquisition. Sober engineers, like Carver Mead, however, can easily live without such mysterious quantum spice to the sauce because they know how reliable and predictable machines can be built on such shaky grounds and how successfully they can run useful algorithms.

From an engineer's point of view, our brains are very clumsy and unreliable devices for rational reasoning because evolution shaped them mainly to do other things, but they are quite helpful nevertheless, when it comes to bringing rational light to mystical darkness.

Your arguments on the free-will problem are a brilliant example! ☺

And bringing light to darkness works best when you combine many brains, a method called science. ☺

Right when I was dealing with your arguments about the future of machine translation I was surprised by a service which seems to supply a superior quality of machine translation. The name of the service is Deepl.com and it is even more surprising that it comes from a very small company in; I could hardly believe it, Cologne, Germany.

It came like a big bang to the tech community just this week with lots of coverage from the international media. Many journalists tested it and all that I have seen, so far, seems to confirm DeepL's bold claim to have reached a new level of quality in machine translation. Here you find a more thorough analysis made by TechCrunch:

https://techcrunch.com/2017/08/29/deepl-schools-other-online-translators-with-clever-machine-learning/

I also tested it in the German-English translation and found it very helpful indeed, and much more useful than any other machine translator I've used so far.

The system isn't just a fully automated translation machine but also a supportive tool for professional translators. So, when you don't like a specific term in the result of a translation you may click on it and will be offered a whole list of alternatives.

Really very nice and the only pity is that Mandarin will only be available next year. ☹[1]

I made a similar test as the one which you made when you translated my citation of Dennett's comment on the problem of translations from English to German and back to English.

As you can see here, the result is much better as what you got from Google's translator.

"Firstly, there is a culture of scientific writing in research journals that favours an impersonal, reduced representation of topics with a minimum of blossom, rhetoric and allusion. There is a good reason for the relentless sadness on the pages of the most reputable scientific journals. As one of my doctoral supervisors, the neuroanatomist J. Z. Young, wrote to me in 1965, when he turned against the somewhat imaginative prose in my dissertation in Oxford (philosophy, not neuroanatomy), English became the international scientific language, it is up to us English native speakers to write works that can be read by "a patient Chinese (sic) with a good dictionary". The results of this self-imposed discipline speak for themselves: whether you are a Chinese, German, Brazilian — or even French — scholar, you insist on publishing your most important work in English, naked English, translatable with minimal difficulty, relying as little as possible on cultural allusions, nuances, word games and even metaphor. The level of mutual understanding achieved by this international system is invaluable, but a price must be paid: some of the approaches that seem to need to be done require

---

[1] There has not been such service until the end of 2018.

informal metaphorisation and imagination, which attacks the barricades of closed minds with every trick, and if some of them are not easy to translate, then I only have to hope for virtuoso translators on the one hand and the growing eloquence of the world's scientists on the other."

Dennett's text is not an easy one and although the translation isn't perfect, the result is not bad in my eyes, especially when you consider that the text has been translated twice. The "simple" English-German translation DeepL performed with one single and less relevant flaw.

I'm not a translator anyway but I'm not sure whether I could have done it that way. English to German to English may be an easier job however then the English to Chinese to English test you have done.

The founder of DeepL, Gereon Frahling, used to work with Google for a while. He seems to have been inspired to use neural networks (convolutional networks in this case, originally designed to deal with visual objects) by the article of Bahdanau, Cho and Bengio "Neural Machine Translation By Jointly Learning to Align and Translate", 2015, which I have cited the other day.

The quality however doesn't only come from the CNNs. I guess it is mostly rooted in the huge amount of data DeepL has collected from many millions of documents for its online-dictionary Linguee over a couple of years. And they also have been creative in building their own super computer. It's ranking as the 23rd most powerful in the world as they claim and is located in Iceland, probably because energy is cheap there and you have to spend less on cooling the machines.

I don't know how long the brave young people in Cologne who dared to twist the dragon's tail will be in the lead and whether the big guys will strike back soon.[1]

In any event, this achievement of DeepL very much as the one of DeepArt, the company that allows you to create art from your photos, is a clear indication of how useful neural networks are and that AI-engineers are making progress not just in playing games.

---

[1] DeepL announced an investment made by US venture capital company Benchmark in December 2018.
https://slator.com/ma-and-funding/benchmark-capital-takes-13-6-stake-in-deepl-as-usage-explodes/

I get similar signals from very different sectors in the industry from robotics and machine tool controls, to autonomous driving, and from speech recognition, to predictive targeting in online marketing or high-speed stock-trading.

When Norbert Wiener started the cyber age with the idea of improving dynamical technical systems by applying closed loop controls on them it was already clear to him that the field of applications was almost unlimited. It took somewhat longer than he thought, but now 70 years later, it has become obvious that there is almost no process in our modern world that will not be influenced by automated machine learning and dynamic closed loop control systems. It's not clear yet what exactly the methods will be. Even to the experts who have created those new translation systems it is unclear. They even seem to be surprised how easy it was to make such progress. There is an interesting series of interviews Andrew Ng has made with the leading people in the business under the headline "Heroes of Deeplearning" on Youtube and the overall message is: "we have just started and we don't know what the next steps will be, but there is a lot to come". [1]

The progress we have seen in machine learning applications also came as a surprise to me. I wouldn't say yet that Kurzweil is right, and we will see his predicted singularity soon but I have to admit that progress came quicker than I thought and that I'm really impressed. I have a few young people around here who felt attracted to the AI field and they all report that the frameworks supplied by the big companies are easy to use and that they made quicker progress with useful solutions than they had expected.

This made me think whether neuro-scientists may overestimate the complexity of the problem. Maybe it's not that difficult to build much more powerful artificial intelligence system as they think, even in areas that were considered to be particularly difficult. Especially when we don't try to copy biology.

That's at least how my young sharks feel and instead of telling them that they are trying to achieve the impossible I rather encourage them and let them try to realize their dreams and maybe surprise me.

Well, at least I'm surprised and impressed by what happened over the last month, but I

---

[1] https://www.youtube.com/watch?v=-eyhCTvrEtE&list=PLfsVAYSMwsksjfpy8P2t_I52mugGeA5gR

don't seem to be the only one who's impressed.

At least your government also seems to be impressed because it has taken surprising action as you have noticed for sure. The state council of PR of China announced on July 8th 2017 a huge industrial AI offensive which should bring China in the lead by 2030.

This is very interesting indeed and the message created lots of responses all over the world. In a paper in the *New York Times* from July 20th [1] the tone is a bit too dramatic in my eyes because this AI-offensive is not about a declaration of an industrial war. But it is very characteristic of the sensitivity with which the growing technological competence of China is seen in the USA and also in Europe. Until lately, many people in the West still believed that their technological edge is something fundamentally guaranteed and that everyone else has to follow them. I have to admit that when I visited China in 1993, I also came with the perspective of seeing a developing country.

I had the typical prejudices of western culture in me, which feels superior and believes that modern technologies can only come from the West and all others have to learn from us.

I had seen a pattern of the drabness before, with low productivity and a shortage economy. I expected to see something of the same kind in China, but ten days in your country where enough to change my world view.

So far, I had been most impressed by California where I had met a kind of entrepreneurial spirit that made Europe look old and tired. In China to my great surprise, I met determined and energetic entrepreneurs with a speed that made look California slow. I was stunned to see something like this in a country called socialist. What I saw then was very puzzling to me and it still is. Although I tried, I'm far from understanding this amazing rise of a giant.

The article in the *NYT* is stressing the point that when AlphaGo defeated the Chinese Go Champion Ke Jie this was kind of a "Sputnik-shock" for China that sparked this AI-project. You may tell me whether this idea was born in the fantasy of a journalist or whether this was really the case. In any event, I'm very curious to learn what you think about the relevance and the chances of this huge project which makes other events like

---

[1] https://www.nytimes.com/2017/07/20/business/china-artificial-intelligence.html

the HBP or the US BRAIN Initiative look pretty tiny and negligible.

I'm sure you are also excited about this unexpected move and have an opinion on it.

Best to Shanghai.

Karl

PS: I had a longer phone call with Hans Braun and it turned out that he too is a great fan of Walter Freeman. He told me the story how he met him for the first time. It was at a congress at Kusadasi, Turkey, in 2004. Walter had the honor of giving the opening speech. In the following Q & A session, Hans asked him some critical questions and even insisted when he asked more questions after the answers. The organizers seemed to have found it inappropriate to attack such a famous man. But Walter seemed to like it because afterwards he came to Hans to tell him that his questions had been very interesting and that he would like to discuss them in more detail. So, it came that Hans was invited to sit beside Walter at the subsequent dinner. Very funny and very typical for the open mindedness of the two.

He sent me a photo of this event where you see the two of them beside Hans' friend, Herman Haken, and his wife. Hans has allowed me to share the photo with you which is a historic document of the inventors and protagonists of circular causality.

III-010 Fanji

19. 10. 2017

Dear Karl,

Thank you so much for your kind words. Thank you very much also for giving a deep analysis of the dilemma about free will vs. determinism, which I tried to give an explanation that is not as clear as yours. I am glad that your response is positive so that I can know that my idea may be on the right track, I am also glad that both of us think that we must be obliged to our friends, Herman Haken, Walter Freeman, and Hans Braun, who proposed and /or developed the concept of circular causality, which seems to be the key to solving the dilemma. I am regretful that I did not learn Hans' work on free will earlier but am happy that our thought is not far from his. I should read his article someday, after translating it into English with DeepL. ☺ Your arguments are really forceful, at least in my eyes, I could not say anything more.

The quality of the DeepL translation is far beyond my expectations! It is really amazing! I have to admit that the progress of machine translation is much faster than I expected. Your remarks sound correct that modern technology develops exponentially, while neurology and medicine only linearly. I am looking forward to the new service of DeepL to translate English into Chinese and the vice versa. As you suggested, I expected some Chinese domestically-developed translator may be better. However, up to now, I have tried Baidu and iFly translator, both are not satisfying just as Google and Bing translator. I don't doubt that these translators could reach the standard set by DeepL someday, I just hope that they can be improved quickly.

Your following remarks may give a summary after our long correspondence:

"*This made me think whether neuro-scientists may overestimate the complexity of the problem. Maybe it's not that difficult to build much more powerful artificial system as they think, even in areas that were thought to be particularly difficult. Especially when we don't try to copy biology."*

I might overestimate the difficulty which AI met, which you criticized in the above statement, and even mentioned the same idea at the beginning of our correspondence. I have changed some of my ideas based on your arguments and the rapid progress of AI during this period. Just as an old Chinese saying says: *"Facts speak louder than words"*, the speed of the progress and the way the progress has taken shows that your above statement is correct. As we emphasized at the beginning, the approaches used by Mother Nature and engineers are quite different. Engineers can be inspired by nature's approach but should not copy biology for every detail. I prefer the term "brain inspired" to "brain like", although people often use the latter, especially in China.

Thank you very much for telling me the *"New Generation of Artificial Intelligence Development Planning"* issued by State Council of my country, which I haven't paid enough attention to in time ⊗. Of course, people here are impressed by the rapid progress of AI, especially the victory of AlphaGo over Lee Sedol, the former world champion. As Go is a popular game in China, and Lee is also popular here, people were very interested in the competition. Just before the competition, several top Go players, including the senior master Nie Weiping, predicted that Lee would win. Therefore, AlphaGo's victory gave people a shock. As for the shock brought by AlphaGo's defeating the Chinese Go Champion Ke Jie, also the current world champion, was almost nothing, as most of the people now foresaw the result. So, it was only incidental that the state council announced the plan, almost at the same time when AlphaGo defeated Ke Jie.

In fact, the Chinese government has already been concerned about AI. In March 2016, the Chinese government proposed developing AI technology in the thirteenth five-year-plan. In March 2017, Premier Li Keqiang emphasized the importance of AI in his government report to the people's congress for the first time. Then the *New Generation of Artificial Intelligence Development Planning* was laid down on July 8th and announced on July 20th this year.[1]

The Plan elevates developing AI to the level of national strategy, and it defines the

---

[1] http://www.gov.cn/zhengce/content/2017-07/20/content_5211996.htm

strategic goal for the development of a new generation of artificial intelligence in China. By 2020, the overall technology and application of artificial intelligence will synchronize with the leading position in the world, and the artificial intelligence industry will become a new important economic growth point; By 2025, there will be a major breakthrough in the fundamental theory of AI, some of the technology and applications will reach an internationally advanced level, artificial intelligence will become a major impetus to drive the industrial upgrading and economic transition of China; by 2030, AI theories, technologies and applications will reach an internationally advanced level in general, become one of the world's major artificial intelligence innovation centers.

The Plan is ambitious. It puts forward key tasks and measures. The 6 key tasks proposed are: (1) Build an open and cooperative technological innovation system of artificial intelligence; (2) Cultivate a high-end and high-efficiency intelligent economy; (3) Build a safe and convenient intelligent society; (4) Strengthen the integration of military and civilian in the field of artificial intelligence; (5) Build a safe and efficient and intelligent infrastructure system; (6) Prospective arrangement of major scientific and technological projects about a new generation of artificial intelligence.

Government financial support and social capital will flow into the development of a new generation of artificial intelligence in a broad range of ways. Pilot projects will be explored, and successful experiences will be spread. Considering the scale of Chinese population, the data are huge. We also have top supercomputers.

The Plan also pointed out that there is still a gap in the overall development level of the artificial intelligence in China compared with that of the developed countries. The lack of significant original achievements may be the biggest problem, including those in basic theory, the core algorithm, the key equipment, high-end chip, etc. ; Qualified scientists and technicians are anxiously needed; Infrastructure, policies and regulations, and the standard system need to be improved.

I am optimistic with that plan. China has a tradition of focusing her resources and energy to developing some key businesses. The fields the plan is involved with seem broad enough. It is expected that China will have a big change when the plan is fulfilled. It seems to me; the key point of the plan is to use a variety of AI approaches to solve practical problems in different fields. It's mainly a practical engineering plan, although basic theoretical research, exploring new technologies, and interdisciplinary studies are also emphasized. I think this is proper for China as a developing country.

The photo you attached with your last letter is marvelous! Just as you described that it "*is a historic document of the inventors and protagonists of circular causality*". I have a similar one taken in 2000 when Freeman and Haken were invited to attend a workshop held in Hangzhou, China, organized by Prof. Qinye Tong and me. It is obvious that the two giants appreciate each other, and they cited the other in their talks. It is lucky for me to know them by personal contact and to get their advice many times.

Time flies so fast, almost 5 years has passed since we started our correspondence. We have discussed many open problems both in brain science and artificial intelligence, for some we have reached a consensus, for some we still dispute, and for some others we just don't know. The two fields develop quickly, although maybe one grows linearly and the other exponentially, which must influence every side of our society and daily life deeply, which our ancestors could not even imagine. Some progress in the field of AI is even beyond my expectations. I have to admit that Kurzweil's "Acceleration Law" may have some sense, although I am still heavily skeptical about his prophecy that "Singularity is near". We are lucky to witness such earth-shaking change in our society. I do not scare so much for the results of the development of AI. Up to now, AI products are still tools to help people solve their problems, they don't have their own will, or even have their own inner world, and I don't think that they will in the foreseeable future. I just think that why should people create a machine with its own will? Such work should be banned, although it is not an urgent matter at the present moment. I am just wondering what will happen in the next 5 years. What is next after the deep learning approach is widely applied to a variety of fields.

Will there be any other breakthroughs in AI in the near future?

Fanji

P.S. I suppose that you have noticed that just yesterday, a news about AlphaGo was announced in the magazine Nature, it is announced that a new version of AlphaGo — AlphaGo Zero beat its old versions, and beat the one which beat Lee Sedol with a score of 100 : 0. In addition, it is said that the new version does not need any human player's data, it just learned to play by self-competition from scratch. All the knowledge it needs is only the basic rules of playing and the criterion to judge who is the winner. Only after three days training, it beat AlphaGo Lee! The result is amazing. It is said that AlphaGo Zero could be used to solve problems in other fields, such as diagnosis of medical images. Is AlphaGo Zero an artificial general intelligence or just a collection of weak AI products? Will it forget all the old skills it has learned when it learns a new skill? How do you evaluate the meaning of AlphaGo Zero in the development of AI?

15. 12. 2017

Dear Fanji,

Thank you for your kind and inspiring letter!
I very much liked the Chinese saying you cited *"Facts speak louder than words"*.

It describes best my own pragmatic attitude towards science and technology and it reminds me of a story Karl Popper tells in his autobiography. Before he became an academic he completed an apprenticeship as a cabinetmaker with Master Adalbert Pösch in Vienna. And this all-round interested and self-educated craftsman taught the young man his quite sophisticated but pragmatic world-view while they were jointly working on furniture in the workshop. It must have been an interesting time because Popper claims that he learned more from this master cabinetmaker than from many famous philosophers. Pösch questioned everything, even the most fundamental things in physics like the "law of thermodynamics". Thus Popper quotes him right at the beginning of his autobiography with reference to a perpetual motion machine:

"They say you can't make it; but once it's been made they'll talk different!" [1]

Well, Master Pösch didn't live long enough to experience the triumph that Ilya Prigogine won the 1977 Nobel Prize in Chemistry for his work on dissipative structures. He had showed that in open chemical systems, with their importation and dissipation of energy constellations, can occur far away from equilibrium where entropy doesn't maximize and the second law of thermodynamics isn't applicable.

Many people who cite the second law as the most reliable physical law of all, indeed overlook that it is only defined for closed systems. And the problem is that truly closed

---

[1] Popper K. Unended Quest: An Intellectual Autobiography[M]. London and New York: Routledge, 1976, p.1.

systems only exist in books or our phantasies. There is no fully closed system in the whole universe however. I'm not sure whether this was the point which made Master Pösch so skeptical about the "laws" of thermodynamics or whether it was his general skepticism against any kind of postulated truth in the sense of eternally valid dogmas.

I have to admit that I also believed in the second law in a naive way and that it was Herman Haken himself who gave me a lesson that cured the problem. And so, it happened: Adi and I were invited to a party at Hans Braun's house where we also met Herman Haken and I complained that although I have the most wonderful spouse of all, she does not believe in the second law. Actually, she and I had a debate about this over many years where I failed to make Adi a believer in thermodynamics. And I have to admit that Adi was about as creative as Einstein was when it came to thought experiments against the Copenhagen interpretation of quantum mechanics. So, I asked Haken as one of the most eminent living physicists to take sides with rational science and explain to this stubborn woman why it is foolish to doubt the second law. To my great surprise he took sides, but not with me but with Adi. He told me that I should be happy to have such a smart spouse because the second law is only applicable to closed systems, which means almost never.

That's about 20 years now and I'm still grateful for this lesson. ☺

At the time I didn't know much about biology, self-organized systems, phase-transitions, emergent behavior or circular causality and the little I knew about physics I believed firmly. Very much as we all tend to do in matters we only know superficially. Actually, our brain seems to be a device with the core competence to act (and feel well) based on incomplete, inconsistent, or even wrong information and insight. And as long as there are rails our mind's engine is happy to run on them, even if they are only castles in the air.

Sometimes it takes a long time to spread new insights. And although Prigogine's and Haken's similar work became quite influential for further research about self-organizing systems they get little credit in the mainstream neurosciences. And as with Walter Freeman you don't find their names in Kandel's bible.

Thank you for the insight you gave me about the background of China's "*New Generation of Artificial Intelligence Development Planning*".

It's interesting that your analysis reveals that the project didn't result from a "shock" caused by AlphaGo's victory over Ke Jie, as cited by the NYT journalist, but was part of a master plan that existed for quite some time.

The plan may be ambitious, especially considering the tight timeline, but to me, it makes much more sense and is more realistic than installing another brain research project in the style of the HBP or US BRAIN Initiative. When the project didn't result from a shock in your country it for sure created some shockwaves in the West, especially in the US.

Numerous studies and articles have been published since, and the reactions in the West ranged from surprise to being impressed. Some people seemed almost insulted that a former developing country has learned to play the industrial game and is now meeting the West on an equal footing with cutting-edge technologies.

An interesting example is this brand new statement "Chinese Technology Development and Acquisition Strategy and the U.S. Response" given by Adam Segal before the "House Committee of Financial Services, Monetary Policy and Trade Subcommittee" of the Unites States House of Representatives on December 12 2017[1].

There have been other comments as well that dealt with the shock in a way which is more in the tradition of Americans being good sports and who like competition.

Will Knight published an article in Technology Review on October 10th 2017 under an unusual double headline in English and in Chinese:

China's AI Awakening[2]

中国人工智能的崛起

And the sub headline is:

---

[1] https://financialservices.house.gov/uploadedfiles/hhrg-115-ba19-wstate-asegal-20171214.pdf
[2] https://www.technologyreview.com/s/609038/chinas-ai-awakening/

"The West shouldn't fear China's artificial-intelligence revolution. It should copy it."

And on it goes with a sober analysis of the situation and estimation whether Chinas plan is realistic:

"*one thing is clear: artificial intelligence may have been invented in the West, but you can see its future taking shape on the other side of the world.*"

"*China's AI push includes an extraordinary commitment from the government, which recently announced a sweeping vision for AI ascendancy. The plan calls for homegrown AI to match that developed in the West within three years, for China's researchers to be making ' major breakthroughs' by 2025, and for Chinese AI to be the envy of the world by 2030.*

There are good reasons to believe the country can make this vision real. In the early 2000s, the government said it wanted to build a high-speed rail network that would spur technological development and improve the country's transportation system. This train network is now the most advanced in the world."

What Knight says may not flatter the national pride of his countrymen, but it is a realistic analysis of the situation in my eyes. The days when one could call China a copycat are gone. The increasing amount of patents granted to Chinese companies or IP acquired by company takeovers, especially around AI, documents that China has learned to play this part of the game as well.

I also agree with what Knight says about the industrial relevance of this move. And it fits pretty well that yesterday Andrew Ng published that he decided to bring AI to the production industry with his new company landing. ai, beginning with the partner Foxconn.

Ng who brought up the Google Brain Deep Learning project and was later chief scientist of Baidu is, despite his young age, rightfully called a veteran of AI, for sure a no nonsense researcher and manager who knows what he's doing. I've been following his activities and those of his students for a while and I think he's right on the spot when he now turns on what we call Industrie 4.0 in Germany.

When Chris Anderson in his book " Makers — The New Industrial Revolution " proclaimed a new industrial age 5 years ago he was a bit premature. We talked about it then and his vision seemed to make sense to me. However, his assumption was wrong that the revolution would be triggered by 3D printers. But he was quite correct in terms of the general trend towards automation, and also in terms of the importance of industrial robots.

His own attempt to demonstrate that lost production territory can be brought back from China to the US was a complete flop however. His company 3D Robotics once started to become the number one drone manufacturer, burned 100 m dollars and in the end was defeated by DJI, China's dominating drone-giant.[1]

I'm not sure whether this was caused by "*classic Silicon Valley hubris*," as a former employee is cited in the article or whether it was more unlucky timing or the romantic and naive approach an unexperienced former journalist took which caused the failure.

Tesla was presented as a role model for this new industrial age in Anderson's book and here Silicon Valley hubris, if not to say megalomania, quite likely, come into play. Looking at the many problems Elon Musk is facing in his attempt to meet the necessary quality standards in the mass production of cars, some people already start to bet on his failure.

I hope this won't happen because my sympathy is always more for those who take a risk and take it forward than for the scaremongers who mainly want to preserve the existing conditions. In any event we Germans must be particularly grateful to Musk, because he may have awakened our car industry, which is still the basis of our national prosperity, just in time.

And I hope the same awakening effect will happen now when China is putting so much energy and effort into developing AI and bringing its industry to the next level. And as you said the other day, China will have to do this because it may be the only way to keep 1.3 billion people on track. When the goal is that everybody should live a good life, the question is how to increase productivity. A problem the Soviet Union never could overcome and which in the end caused its collapse. I'm far from understanding the political and economic situation in China and that's why I find it hard to give an

---

[1] https://www.forbes.com/sites/ryanmac/2016/10/05/3d-robotics-solo-crash-chris-anderson/#65afe4073ff5

estimation which is not only made up of the usual clichés. I told you that I was puzzled to meet in China, a socialist country, the most creative and ambitious entrepreneurs I hadn't even seen in California. On the other hand, I see in the West academically educated elites that feel uncomfortable with the capitalist system and its industrial culture. At the same time, however, this system allows them to live the pleasant and free life that they enjoy so much. I'm surprised to see when I talk to young people here that even the bright ones seem to take economical welfare for granted. Their focus is often on how to share prosperity in a more fair way, and not so much on how to create it.

When I talked to young people in China however, my impression was that they primarily wanted to make it in life and that it was clear to them that the best way to achieve it was hard work and discipline. Virtues that have somehow gone out of fashion here and have been replaced by higher principles of political correctness and by the optimization of a function called "work-life-balance".

My friend Paul Wahl, former general manager of SAP – USA and later president of Siebel, made a similar observation, already back in the late 1990s, after doing business in China. He summed up his opinion in a sentence, which always comes to my mind when I look at the actual development:

"If you look at the younger generation in China and compare it to ours, you can only hope for the benefit of our young people that they will never meet on the world market".

Some people believe that China will learn from the West and also adapt to the work-life-balance mode. I wouldn't bet on it, and for the sake of Europe rather follow Knight's advice and hope that we will learn from China at least when it comes to using AI as the next booster to improve global prosperity.

And we may also learn from your investors, and it's not just the industrial heavyweights Baidu, Alibaba, Tencent, or Xiaomi who display very smart strategies from incubators and startup investments all the way up to strategic acquisitions. Midea's acquisition of KUKA is a good but also alarming signal because no German player had seen the strategic value of this gem of the automation sector.

All the questions you ask about the near future of development in AI I'm asking myself. And after what I have seen over the last month I got very cautious in making predictions.

I also don't think that Kurzweil's singularity is near and even don't believe that we have entered into general AI. We have just started to scratch at the surface of simple AI but there is tremendous progress. What you have described in your PS with Alpha-Zero beating its own predecessor AlphaGo is a good example. But machine learning is much more than deep learning and neural networks. I also believe that we'll see a renaissance of GOFAI and all kinds of hybrid-systems and also lots of AI-dedicated hardware solutions.

My opinion that we have merely seen a start and that the really important things are just coming, corresponds with the outlook Karpathy and Goodfellow made in their interviews with Andrew Ng and also with what Hassabis says in the cited Wired interview. Unsupervised learning and new variants of GANs (Generative Adversarial Networks) enable a new kind of machine learning where the machines do better when they explore the territory in their own way not following the guidance of humans.

I'm pretty sure that we will see amazing things, but please don't buy stock of AI-related companies based on my outlook because in economy everything is possible and nothing is granted. ☺ It's all vortices in the sense of Navier-Stokes. The next financial crises may be just around the corner and maybe a collapse of cryptocurrencies like bitcoin, the next gamble so fascinating and attractive to many (because they don't understand it) may cause a drawback.

But in the long run, Machine Learning (ML) will work. The longer I think about it the less applications come to my mind where ML is not applicable. Actually, it is hard to find domains that will not be influenced by ML. This lesson I have learned from ML-applications I would never have thought of.

One is dynamically modulating the magnetic field that holds the hot plasma in place in a Tokamak hydrogen fusion reactor![1]

---

[1] http://www.digitaljournal.com/tech-and-science/technology/google-optometrist-algorithm-brings-us-closer-to-fusion-power/article/498508

When you think about it, you will find that there is no physical or chemical process where it comes to quick control that couldn't benefit from similar applications.

The other was recently published by Andrej Karpathy under the name software 2.0[1].

The idea is as surprising as powerful: Let the machine read your code and improve it!

This is another hint to a similar phenomenon as described in Moravec's paradox.

The more complicated the task is the better AI will do! There is great debate about people losing their jobs, and in the media, you always see people at the lower skill levels replaced by robots. My guess is that the major effects we will see at the high end as demonstrated in Karpathy's software example. But it won't be programmer's losing their jobs to machines.

We don't have enough qualified people in this sector anyway but we have huge problems with the safety and robustness of large computer programs. The majority of programs on which our world relies, from banking to flight control systems is about as old and in need of renewal as highway bridges or wastewater canals in some old cities.

On the other hand, jobs of nurses, cooks, and strawberry pickers won't be threatened by AI for a long time. It would be naive to say that all this innovation coming with AI will be no problem at all. There will be tremendous change across all domains from engineering to the legal system. The debate in the US about fake news, filter bubbles, and the control of social networks are all signs of what we are facing.

Some people may be scared by it while others may see it as a tremendous opportunity to give this world a new shape.

In this respect, I recently had a very encouraging and refreshing experience. We talked about this German Cyber-Valley project around AI the other day and also this amazing CNN (Convolutional Neural Network) application called DeepArt. It happened just by chance that my daughter Jelka at a conference met one of the three founders, Matthias Bethge, who is also one of the fathers of DeepArt.

She made a contact and so it came that I paid him and his team a visit at nearby Tübingen where I also met the two other Cyber-Valley founders Michael Black and

---

[1] https://medium.com/@karpathy/software−2−0−a64152b37c35

Bernhard Schölkopf. Well, as you can imagine, this was a similar pleasure to me like a visit to the chocolate factory for a six year old. The visit was worthwhile and interesting in several respects.

First, I was pleased to see that all three are very modest, down-to-earth and likeable guys, although they play in the world league of AI researchers. Or maybe I should say *because* they play in the world league. In my experience, the really excellent people are the most pleasant, while the mediocre people often tend to perform like the conceited viceroy.

The other good thing was that we got into a discussion very quickly about what is going to happen next in the AI and from whom we have something to expect. The global community of the leading people in this sector seems to be still pretty small and they all know each other. This seems to create a friendly and nevertheless competitive atmosphere which reminded me of the situation Wilhelm Ostwald describes when he and his international friends jointly laid the ground of modern physical-chemistry on which later whole industries were built.

They knew back then that as pioneers they would break new ground and make incredible discoveries and inventions. And this consciousness connects them today all around the world. You can almost feel it, when you enter the building, like a vibration — very refreshing and very good.

We discussed a few things you and I have already discussed and I wished you had been with me because you would have heard opinions on familiar names and concepts and also liked to see that we aren't that much off the track with what we have discussed over the years.

So, I highly recommend a trip to such a place and while I know that you don't like long distance flights there is good news for you because there are similar AI-chocolate factories around you in plenty. Ingo Hoffmann, a former hipo and good friend of mine just came back from a trip to China. He's specialized in AI and runs one of the best blogs on the subject I know. [1]

Ingo is well connected in China and had prepared himself very carefully for this visit

---

[1] https://ingo-hoffmann.com/

where he attended the "World Intelligent Manufacturing Conference" in Nanjing also as a speaker.

Through him I got lots of interesting information about what is going on in the industrial AI business and I recommend you to visit his site where you will find links to articles like this one by Dr. Kai-Fu Lee who deals with the AI-related subjects I have mentioned and their relevance for the Chinese economy in a much more qualified way than I can do. [1]

What I also found is a very interesting AI-chocolate factory just around the corner in Shanghai. Its name is Westwell Lab and they have a neuromorphic Chip and claim to have overcome the handicaps of the von Neumann architecture. [2]

I can't tell from a distance how good they are, but my guess is they are and ambitious they are in any event, and also seem to be well funded. [3]

And of course they are not the only ones. Actually, it looks like AI-chips seem to be a hot spot in China, with Cambricon being the sector's famous billion dollar unicorn. [4]

When I saw all this and ran through Ingo's China-AI-material I could very well understand what you meant because there is quite obviously a Chinese AI-industry in full swing and nothing has been created hastily after a possible AlphaGo shock.

So, my impression is that we both, and the rest of the world, can really be curious about what wonders the young generation will surprise us with.

Good night my dear friend and best regards.

Karl

---

[1] https://www.eurasiagroup.net/files/upload/China_Embraces_AI.pdf   Meanwhile Kai-Fu Lee has published a very interesting book dealing with many questions we have discussed here. Kai-Fu Lee. AI Superpowers: China, Silicon Valley, and the New World Order. Boston, Massachusetts: Houghton Mifflin, 2018.
[2] http://www.westwell-lab.com/about_en.html
[3] https://www.chinamoneynetwork.com/2017/06/23/fosun-unit-leads-series-a-round-in-chinese-ai-chip-maker-westwell-lab
[4] https://www.chinamoneynetwork.com/2017/08/18/chinas-state-development-investment-corp-leads-100m-round-in-ai-chip-maker-cambricon

Epilogue
August 2018

When our unexpected correspondence began almost 6 years ago, there was no plan and no goal for our journey. Fanji had stumbled across some inconsistencies in the concept of the prestigious and ambitious HBP but he wasn't sure whether his apprehension was justified or not. In the beginning, we were uncertain and simply did what car mechanics do: with much curiosity, we looked under the hood and tried to understand what was going on.

Looking back, we ourselves are surprised at what we discovered and what path our discussion about the brain, the mind, AI and big science projects took.

As we have pointed out in the preface it was more like a random walk; wandering from field to field, stopping to study something in depth whenever we wished. We were guided only by our curiosity, and simply put in more effort when we had the desire to understand things a little more precisely or when we felt the need to fill gaps in our knowledge. And often, we enjoyed following the streams of knowledge back to their sources, including some excursions into the history of our very different cultures. But as chaotic as our journey was, we feel that through our continuous and sometimes controversial debate, we gained insights we would have not acquired had we chosen a more systematic approach.

In retrospect, we are quite surprised at what we thought about certain things, theories and concepts, and in what different light they appear to us today. We have noticed that we have experienced a gradual change in perspective, the extent of which only becomes clear to us in hindsight.

Karl, for example, learned that there are more promising perspectives with which to look at the brain as opposed to the conventional approach, which is to think of it as an information processing machine.

At the beginning of our correspondence, Karl was (and still is) an amateur in neuro science and was simply happy that a seasoned expert like Fanji was willing to discuss such delicate questions relating to the latest state of research with him; while Fanji was (and still is) a layman of information technology (IT), although he is much interested in the rapid progress in recent years in this field. It was difficult for him to find an IT expert who knows brain science as well just like Karl to ask. Experts rarely enjoy discussing their business with outsiders, especially when their opinions may be questioned. Not many of them like to be bothered with the kinds of basic questions children ask, or to express their ideas in a way that children can understand.

We both come from very different fields and our set of professional and cultural experiences couldn't be more diverse. However, we share a passion for curiosity and also for rational reasoning and have a weakness for questioning everything, especially when it comes to established academic insight and the claim for scientific or technological progress.

Although we have already said the following in the preface we would like to repeat it now because of its importance:

We both like to take the perspective of a child who asks simple questions in order to grasp what is happening. Sometimes the child can see that the emperor's new clothes are not as brilliant as they are presented. But we don't want to overdo it with this perspective because it would be presumptuous to say that we are the child in the famous fairy tale "The Emperor's New Clothes" who can see or can't see what others see.

However in the case of HBP, Karl insists that very early on, while others were still praising it, Fanji was able to recognize that there were flaws in this impressively presented project.

We spent a great deal of effort in our early letters demonstrating and assuring ourselves that more than one thing was wrong in the concept of the HBP, and that we shouldn't expect too much from it. It initiated our correspondence, and it became a good model to explore many of the basic elements of the brain and the mind and a possible link with artificial intelligence and computer-technology.

Today, after the facade of this project has been seriously damaged in public such criticism is common and our verve in the old days may seem to some as an attempt to kick a dead dog. Maybe the meanwhile common critique is even too much because in our eyes there are interesting parts in the HBP-concept that have deserved a second try.

Fierce, content-focused debates were once the usual style of scientific disputes, similar to those in sporting competitions. But today this very successful method seems to be going out of fashion, so that one almost has to apologize for using it.

It may also be that this method is accepted more in Europe and may be seen as rude and inappropriate in China.

But whatever the feelings we've triggered in our readers, it's important to remember that we probably owe this harsh method a great deal of the astounding progress made in the golden era of the natural sciences. A glance at publications shows that the debates among scientists at the turn of the 19th and 20th centuries weren't based on polite consensus and harmony, but were often violent disputes. It was often friends who fought as hard as they could for truth and knowledge, but only with the aim of destroying the opponent's theory and not his character. Of course, such fights could hurt and leave wounds, but they also released the energy to come back with better solutions and to fight for ones cause.

In any event, we, and especially Karl, want to make clear that if we sometimes used strong words or metaphors when criticizing a theory or perspective, we did so in this traditional, sportive sense and aimed it only at the argument itself and never with the intention of discrediting an individual.

※　※　※

When Karl received Fanji's first e-mail it was full of questions he couldn't answer and some which he hardly understood. But he was fascinated by the subject and also by the seriousness of the mind he encountered in Fanji, a mind which was trying to understand and fully penetrate the concepts. He was also impressed by the sharp-witted arguments of his counterpart, and his braveness to ask critical questions even when they challenged what great authorities were doing. Even today, Karl wonders whether this unusually determined focus and persistence, not to give way until everything is as clear and transparent as possible, is an original part of Fanji's personality, or a skill he has acquired while translating many of today's most relevant books about neuro science into Chinese.

In any event, it seems to Karl that Fanji has honed his skills in his discussions with many first-class authors to a point that can hurt. Karl got to experience how good Fanji is at forcing an author to make clear what he wanted to say when the time came to translate our English correspondence to Chinese; Fanji came back with questions again and again to make sure that he had caught the true meaning of a statement or sometimes even a single term. This process wasn't always easy and pleasant because sometimes it revealed that the problem was not in the translator's ability to understand, but maybe in the formulation of the idea or, even worse, in the idea itself.

Fanji appreciates Karl's erudition, earnestness and answering every question he raised. He even subscribed to books in order to answer questions. It is difficult to find a friend like Karl who is willing to think, answer and discuss with him the questions that confuse him. Fanji knows that he has found a good friend and teacher.

This continuous process of clarifying what we wanted to say and truly understanding each other's points of view, as well as taking into account subjective and cultural differences, wasn't only a pleasure in its own right, but also had a lot to do with the subject of our mutual interest.

The question of whether and how it is possible to transfer the content of a statement (in the sense of meaning) correctly and completely from one language into another is a recurring theme in our conversation. We came back to it time and again, not only because Fanji is an experienced translator who has often struggled with this problem, but also because it is a well-suited empirical testbed. This translation-problem has a correlation to the general difficulty of transferring content from one brain to another and also to machines. The subjectivity and privacy of the way in which consciousness is composed, as well as how it builds meaning, occupied us a lot. And the question of whether such content can be completely reduced to physical processes and be accessible through objective, rational research is another recurring topic in our correspondence. This of course is relevant to the popular discussion of whether or not, with the help of modern AI, we can build machines that have a consciousness similar to a human's. This fundamental question is the natural link between the two enigmas we are dealing with in our letters. The first is the natural intelligence of the brain, which we understand so little about even after generations of the smartest researchers having spent so much effort to break its secret. The second is the artificial intelligence of man-made algorithms running on computers which are totally different kinds of machines, totally unlike the brain, and that improve their competence with astounding speed.

Actually AI is already superior to human abilities in many respects, something people tend to overlook because the discussion is focused on what machines still can't do.

The key questions we discussed were whether autonomously learning machines will equal and outperform humans in all domains of intelligence, including creativity and other abilities, which many consider to be genuinely human, like emotions and humor. And if so, will they ever be able to develop consciousness and subjectivity at all?

These are issues where we still have different opinions, although our views have altered under the impact of each other's arguments. Before we began our exchange, Karl had thought little about the problem of subjectivity, and under what conditions and how consciousness may emerge. To him, it seemed more like a pseudo-problem that could vanish like the old "élan vital" problem in physics has vanished. Many engineers believe that consciousness may somehow emerge by itself when AI has piled up enough algorithms in high-speed computers to finally perform all intellectual functions our mind is providing. When Fanji insisted that this was not a trivial problem, Karl at first was suspicious whether his opponent may retire to a metaphysical position, and thereby leave the rational turf of the discussion. And Fanji wasn't quite sure whether Karl had grasped the core of the problem that had kept him from sleeping for many years. On the other hand, Karl felt that Fanji may have underestimated the pace with which hardware was still developing, and may have shared the popular belief that Moore's law has to come to an end soon.

We explored these themes together and tried to get an overview of the developments in the various related disciplines in science and technology. And by exchanging arguments back and forth, we finally managed to better understand the issues and what our counterpart's position was. With the difficult case of consciousness, Karl was impressed that Fanji did not escape into metaphysics but presented strong arguments supporting his position.

We finally agreed that a 1 : 1 to copy of a body and brain could contain internal states of subjectivity, and even the mysterious qualia. But Fanji was successful in demonstrating that it's close to impossible to create such a copy. Karl, on the other hand, had to realize that it may be much more difficult than he had hoped to build strong, general AI systems that can develop consciousness and subjectivity. And Fanji had to admit that he had underestimated the speed of AI's development.

So when asked today, Fanji may still be skeptical of whether strong AI is possible at all, while Karl would still opt for a positive answer. But although we never expressed this explicitly in our letters, we both have a feeling that we have managed to plant a seed of doubt in the mind of the other.

In any event, we both agree that mind-uploading is close to impossible, and that it will take much longer to build even mere artificial zombies than the Kurzweil flock hopes. Not to mention, an allegedly near singularity. And we also agree that establishing neural links between machines, the body and the brain is much more difficult than many believe.

Another question that occupied us was how engineers could be so successful with their concepts of computers and AI software, even though both were based on obviously quite inaccurate models of the brain?

And would it help engineers to reach their goals if they would try to make their machines more brain-like? And would neuro scientists be inspired to better understand the brain and the mind by knowledge of AI concepts and getting more help from engineers, mathematicians and logicians? And would bigger and faster computers help in simulating, and thereby improving, neurological brain models?

We both agreed that a computer-simulation of a highly dynamic functional system like the brain which we don't even understand on the physical level, not to mention the higher functional level, doesn't make much sense. However, trying to create one may help in improving the tools needed to achieve understanding.

We also finally agreed that trying to reengineer the biological brain on silicon isn't a promising approach. Karl proposed early on that engineers wouldn't benefit much from the results of brain research, and would be better off going their own way and disregarding the biological model. An idea that seemed strange to Fanji in the beginning, but which he finds less absurd now.

We still can't answer many of the questions that puzzle us, nor can we be sure that what we have agreed on is valid, and of course, we are not claiming that we have gained any new scientific insights.

※ ※ ※

The point is, that the two of us have made progress in getting a better view of the field in order to gain relevant insight for ourselves. The journey we have made was very rewarding for us, and it may inspire others to explore the field in a similar way. Readers can decide which position suits them more, and some may disagree with both of us.

By publishing our correspondence, it is not our goal to proclaim theories, but rather to propose a method of acquiring knowledge and exploring areas previously unknown to us as individuals.

As an inspiration for those who are interested in this kind of subjective exploration, we would like to reiterate once more what we considered to be the most helpful and inspiring perspective.

There was one man who helped us more than anyone else in discussing all of this, and who gave us a theoretical framework of reference which we found to be very helpful. This was Walter Freeman, Fanji's friend, who sadly passed away during our correspondence. We acknowledge him as one of the most eminent thinkers and researchers in the field. It's a pity that although he's well-known, he hasn't received the recognition in mainstream neuro science to the extent he has deserved in our eyes.

For Karl, being introduced to the work of Walter Freeman by Fanji, which isn't easily accessible, was the most challenging and rewarding experience in their correspondence. Some years prior, when Karl was confronted with the neural system for the first time, he had looked at the brain as an information processing machine, and had believed in the possibility of mind-uploading, and that the paradigms built around logic and mathematics by von Neumann, Shannon and Turing seemed to be the right approach to crack the enigma. However the more he looked into the problem, the more he grew uneasy about the digital nature of this approach and the idea of the brain as a device doing calculations.

Freeman was the first person Karl encountered who understood as much about the brain as one possibly could, and who not only shared this unease, but went even further by calling the principle of "information processing" as a basic misconception of brain function. And via our discussion, even Fanji gained new insight into how helpful Freeman's framework could be, although he had already translated Freeman's work to Chinese.

Trying to understand Freeman's perspective, and especially his point that our brain doesn't just calculate information but is trying to filter meaning in a continuous process of interaction and loops of "circular causality", had a pivotal effect on both of us. Although this perspective may be imperfect, or not applicable for all brain regions and functions, and without a doubt will not be the last word spoken in this debate.

We still do not agree on other things, such as the essence of subjectivity and whether and how consciousness may emerge from matter and be acquired by machines. We also still have a diverging estimation of some authors like Chalmers and Dennett. But our agreement on the value of Freeman's work is much more important than what we still do not agree about.

To us, Freeman's perspective, which he had already presented in "*Consciousness, Intentionality, and Causality*" (1999), might also make a promising theoretical framework that could be applied and tested in research projects like HBP and the BRAIN Initiative.

The idea that our brain is not a digital information processing machine built around logic doing numerical calculations, but is instead, an analogue device that filters meaning from noise, may not sound attractive to mathematicians and programmers eager to solve partial differential equations on ever quicker computers. Nevertheless, we hope that it could be an appealing and inspiring starting point for young, unbiased explorers looking for a different entrance to the magic castle.

If we manage to inspire even just a couple of young students or one or two experienced researchers to try out our idea, our work will have been worth the effort.

In addition to questioning the miracles of the brain, the mind and AI, we wrote a lot about the methods of doing science and how best to make progress in knowledge. Naturally, we don't claim the right to tell others how to do their science, although we have widely criticized how some research is conducted.

The point is, that science itself is subject to rational reasoning, and it is apparent that some methods and research arrangements are more promising than others in the quest for cracking scientific enigmas. First of all, it is noticeable that the treasure of substantial knowledge that we have at our disposal in the natural sciences has been accumulated to a large extent by former generations. And when we look at the curve of knowledge growth, it is amazing how steep the increase was in the phase around the turn of the 19th and 20th centuries. And it is all the more astonishing how small the group of people who prepared the ground for our scientific worldview was, and how little it cost.

If we look at it as a building, we realize that the basement and the first few floors were built by just a few craftsmen in a short time. Today, a whole industry of craftsmen is working at the construction site, yet progress seems to have slowed down and even minor decorations are costing more than those first couple of floors.

Some people argue that the modern science industry has cranked out more knowledge over the last couple of years than has been accumulated in all of history. This may be true if we judge by the amount of papers published. But what if we judge by "creative competence" instead? Some measurement that would indicate what we can do today that our ancestors couldn't. This is quite a trivial and pragmatic perspective, but may be a very helpful one.

We have stated that physicists, and especially engineers, have performed better than life scientists in this discipline, and that engineers seem to be progressing at an exponential rate while the advancement in neuro science and medicine is developing only linearly.

We have been especially critical and have expressed dissatisfaction with the output of many research projects and results published as "interesting findings" where patients are waiting for cures. Maybe we have done this to an extent which was too much for many who are working hard in these fields and may find our critique unfair and depressing. This may especially be the case when they are involved in big science projects that leave them little freedom of choice. But again, we are not aiming at any specific scientists but are simply discussing what caught our attention, and fully understand how difficult the remaining problems are, after low-hanging fruit have already been harvested by our predecessors. Fanji thinks that a big scientific project may be important and feasible, if it is technical in its essence, has a solid foundation of scientific theory and aims at solving a well-defined important task. The multidisciplinary development of new research tools and the large-scale collection of basic brain data may be important for brain research, and need huge investment and team tackling. But this does not solve all the problems in brain research, especially those in which creativity and insight plays key role.

※ ※ ※

So our point is not that we don't see enough wonders, but that some are promising more than they can deliver which seems to push the whole sector into a mode of overselling. This is intensified by the established academic reward-system where publications and citations are the currency. This favors a tendency in science and research to sell mere leafage dressed up as substantial fruits simply because many published "interesting findings" around a few enigmas may be a safer way to make a career than the risky way of trying to crack a hard enigma.

We don't have an easy solution to this problem but it wouldn't be right not to mention it.

We are well aware that criticizing is easy and doing better is difficult. And although we don't believe that we have a miracle-solution at hand, we don't want to avoid trying to give an answer to the question: "what should we do?"

When we summarized our conversation and prepared for this epilogue, Fanji asked the question below with reference to a comment Karl had made in one of his early e-mails when he, half-jokingly, said that he would rather give one million to a thousand smart students than pour a billion into a research monster like HBP:

"I agree with what you said about the state of the art in medicine, but there is always a question in my mind, after our discussion, i.e. what should we do in medicine? Cut financial support to these researches? I don't think so, although your questioning is correct, it seems that the way in the past did not work, 'big science' seems also not working, then what should we do? Follow the example of information technologies? Owing to the different natures, this may be difficult. Then what else we can do? Give the money to many post-doctors? Maybe, but what should they do?"

The questions Fanji had raised made us think and discuss this problem once again.

Karl still believes that a seemingly crazy idea like betting on the creativity of a few young, wild and obsessed students may be a valid alternative in some cases. But of course it's not a realistic alternative for all big science, and it may only be an option for small experiments simply because big science is an established, multi-billion dollar industry that can't be revolutionized so dramatically over night. This is even more so in medicine, which is a highly regulated sector with many legal restrictions and ethical boundaries that have been established for good reasons. All this adds to the inertia that has been building up in science for years now, as we can see from Max Planck's complains from over a hundred years ago.

So when we try to give a pragmatic answer to this "what should we do?" question, we don't have the battleship captains or the fleet commanders of big science in mind because they already know all about the calamities we are describing. What we propose may only increase their frustration when they look at how little room there is for maneuverability when it comes to even small course changes in such large structures.

So the most we can expect from them is to hope that they don't take us for naive romantics. Our correspondence suggests that, indeed, we are both a little romantic for determined and self-reliant thinkers who are permanently concerned with a problem and won't let it go until it's solved.

But we are of course aware that modern science is a team sport and that it needs different kinds of people to build a successful team. Besides the individual skills and ambitions, it is the team spirit that is decisive, and it is meaningful that in the history of science, this team spirit, in the sense of a passion focused on a given problem, has been very helpful.

So when we make recommendations we have passionate individuals and small teams in mind, like the captains of small research-speedboats and their crews, and especially young students who already feel that they have a high potential, and are looking for a real challenge and who might profit from a little encouragement.

To them, the very simple and not surprising advice is this:

• Work hard and wrestle intensely with a matter/problem that fascinates you (as Walter

Freeman did), very much as a sportsperson preparing for the Olympic Games.

- Try to become a passionate expert in a field early on, and truly master it in depth.
- Don't mistake good grades and winning prizes for real insight or true mastery (which means avoiding what Feynman describes as the scholastic mistake of the Brazilian university).
- Find/produce real fruit and don't fool yourself into mistaking leaves for fruit.
- Don't dress up negligible results into "interesting findings" just to publish them.
- Don't impress just your peers, but also others in the community who will realize your caliber and cooperate with you to crack a well-defined enigma.

Truly, we don't think that most of those whom we have in mind really need this advice, because they are already operating this way, and they are so busy and obsessed by what they are doing that they wouldn't care about our advice anyway.

But there may be some who are still undecided which way they should go and may get some benefit from our outlook and some encouragement if they are already leaning to the side we prefer.

Karl got an interesting response when he showed his daughter and some of his young friends the draft of this epilogue to see how our advice sounds to the audience in question. They argued that we should not only focus our attention on the genius of exceptional inventors and discoverers, but also on the slightly less gifted that can also make important contributions. Most of us do not fall into the genius category and it doesn't help us to behave as if we do. However less exceptionally gifted people can make very helpful contributions in teams, often by compensating for the typical deficits of the genius-type people.

This is a valid argument and it's especially true when it comes to building teams and companies where scientific inventions have to be transformed into products. And here it is often the case that social skills can carry far greater weight than scientific excellence. We have mentioned previously that much of science is a team-sport where the right orchestration of resources is crucial for success and we should not forget this lesson!

※ ※ ※

There is another, more general, point that stands out when we look back to the golden age of the natural sciences, which is true for everybody participating in the game. It is the average age of a student, who is finally starting out with his or her own research. Typically, they are much older today than a hundred years ago, simply because it takes so long for academic rituals to be completed. The question is whether it really has to take so long in all cases?

There are encouraging examples that this is unnecessary. A hundred years ago, a couple of 19-year-olds were encouraged by their academic peers to redefine the world of physics and they did so quite successfully, leaving their professors utterly baffled. The driving force behind their accomplishment was curiosity, ambition and a burning desire to understand and hopefully crack a fascinating enigma. When Wolfgang Pauli and Werner Heisenberg left gymnasium and arrived at university, they already knew almost everything that there was to know about the theory of relativity and quantum theory at the time. Not because they were super intelligent child-prodigies, but simply because they had been fascinated by a very specific puzzle for a long time. The two were lucky that their passion and skill was recognized by their professor, Arnold Sommerfeld, who accelerated the fulfillment of their academic perquisites. Not everybody in the faculty liked this, but Sommerfeld's brave move was rewarded by the fact that both youngsters did their major work at age 24 which later secured each of them a Noble Prize.

Today, we also have such passionate youngsters in the world. Whatever their fascination is, it keeps them up at night, and the more obsessed they are the more they tend to neglect everything else around them because if it has nothing to do with their primary interest, they find it boring. Traditional academic entry criteria, based on superior school grades or the skill to excel in tests, tends to filter such people out.

So the question is, how do we attract these people and bring them to the playground?

Of course 19-year-olds can't crack all the enigmas in science. But science isn't so different from other kinds of high-performance sports where we often see a decline in performance after the age of 30.

When we look at history, we see how many breakthroughs we owe to extremely young researchers, especially in math, where we have always had a high fruits to leaves ratio. A basic principle of mathematic statistics is the method of least squares. It was developed along with what we call today "curve fitting" by an 18-year-old, Carl Friedrich Gauss, in 1795, and is the core principle around which modern AI deep learning algorithms are built.

Of course we are not suggesting 19-year-old hacker-neurologists should be doing brain surgery on their neighbors in the garage. Not least because there are good reasons for protecting patients by established professional rules and regulatory bodies.

The point is, that other very bright young people have done similarly crazy things in their garages with electronics and software and created entire new industries over the last few decades. Many of them were university drop-outs who found it more compelling to do wonders in the real world than conform to the rituals of academia.

And this brings us to the point of the technological earthquake we are experiencing right now.

We didn't use the term "trade war" in our correspondence but towards the end it could be read between the lines. Yet similar terms made news headlines at the time and everybody talked about it after the strong US reaction to China's announcement of becoming a main world center of AI innovation by 2030.

We were quite surprised to find the technical part of our field of interest in the middle of this debate. It seemed to have come out of the blue after China had made this announcement.

When we started our deliberations over the understanding of the brain, intelligence and AI this was a subject for initiated people and a small group of experts. Today AI is a dominating topic in the media, and is discussed everywhere from science to economics and politics. In 2016, some had predicted the AI hype-bubble and that it would see its next winter soon. But the opposite happened: AI became the most hyped-up technology ever. And now after China's declaration of her AI project and US' strong reaction, interest is ever increasing as the world follows with interest as this showdown develops.

We felt as though a tsunami had hit the small river on which we had been calmly rowing whilst contemplating the topic in our little boat.

Everybody seems to have an opinion on AI now, and to understand what is going on. Of course we don't claim that we have a better understanding of what's going on in the field, or what will happen next, only because we have been dealing with this subject a little longer than some of those who talk and write about it today.

We are even impressed by how well some journalists and analysts have managed to wrap up the essentials of the field and to highlight the crucial points and challenges. That's another example of the enormous help which older versions of AI, like the good old internet with its archaic search engines, provide when it comes to gaining insight into new fields and complicated matters.

And we, too, profited from this increased popularity of the field because many clever people have shed more light on the topic which helped us gain new insights.

So, readers may hear us talk about things that were little known until just recently and have become popular almost overnight. We don't think that we were much off-track with our reasoning, and hope that it may encourage some to have a closer and sober look at the AI-issue.

The major, and maybe game-changing, effect of this AI-hype is that even more money is now flowing into the field. At the time of the big brain research projects one could get the impression that no government of an industrialized country could afford not to have such a project. This is even more true for AI, especially after China revealed its ambitious plan last year. And the price tickets are much higher. A billion dollars seemed to be an breathtaking sum for a research project five years ago, but now, Cambricon in China, a single start-up company working on a hardware-chip to run AI-applications, is valued at a billion dollars (at the time we checked this we found that it had raised to 2.5 billion in June 2018). And there are an incredible number of such start-ups popping up like mushrooms everywhere.

A side effect of this rush is that people who understand how to build AI systems, and especially those who are also competent in the neuro sciences, are very much in demand. Industrial leaders all over the world are trying to recruit these rare experts from where ever they can, especially from universities and research institutions. This is good for skilled people, but a problem for these institutions because they often can't compete financially. At the same time, however, it also opens up opportunities for new forms of cooperation between academia, industry and investors to build start-ups and all kinds of research arrangements and new forms of research institutes. It is amazing to see how many creative constellations are currently emerging and how quickly bureaucratic obstacles are removed. It seems that China is particularly creative and fast in this respect.

It is therefore possible that some of the problems described above, which are related to purely academic research, will also be solved this way. We would not be surprised if part of the research will shift from universities to industry and that we'll experience growth and development similar to the IT industry over the last few decades, where relevant basic research migrated to the laboratories of big companies like Bell-Labs or IBM, which also produced a number of Nobel Prize winners.

Because passion and team spirit is so compelling in such research projects, we see an advantage for our favorites, the small speed boats, over big science battleships, at least in the early phase.

But this is already speculative, and we cannot make guesses to predict the future. Right now we don't see much more than what we have discussed in our letters, the major difference being that we may see change faster.

We left out many things or only mentioned them briefly. For example, the vast field covering the possible social effects of AI, such as people losing their jobs to robots or concerns over the privacy of personal data.

We didn't go deeper into this because we didn't see it. We simply didn't do it because there is already so much speculation and noise in this discussion that we didn't know what we could substantially add to it, and, we also didn't want to trump others with even wilder negative speculations.

However there is one more observation we want to add.

We often cited John von Neumann who was certainly one of the true geniuses of the 20th century and made more contributions to modern science than most. One of these was his work on game theories which he published in 1944 together with Oskar Morgenstern. It was almost as successful as the computer-architecture carrying his name. But as impressive as it was, it also came with some drawbacks. The major one being that it focused on fixed-sum games where the victory of one player was the loss of the other. This perspective became a very popular paradigm, especially among generals of the Cold War era.

It took another scientific maverick and mathematical genius, John Nash, to demonstrate that in economics there are more promising cooperative strategies for improving welfare than trying to defeat a competitor.

And those cooperative strategies surely apply in the process of gaining knowledge in science and technology, and especially when it comes to a war worth fighting, such as the one on Alzheimer's, Parkinson's and similar scourges of mankind.

And thus, the Chinese scientist and the German engineer agree that we better off doing this in friendly competition rather than under the false assumption that we are playing a zero-sum game. And if this idea does not spread easily in people's minds, there is still hope that the next generation of artificial neural networks will find out for themselves.

Scan the QR code for
more information about
the brain research and AI.

Translation Postscript

This is the first time I have translated a book coauthored by myself. Although translating the part written by myself is easier, as I know what I wanted to say, almost two thirds of the book were written by Karl, which is not so easy for me to translate. This is due to his deep thinking and broad knowledge, especially when the topic involves advanced technology, western philosophy and science history which goes beyond neuroscience. Fortunately, Karl is kind and patient enough to answer my questions in detail, or even re-write paragraphs so that I can better understand them. This makes the translation be true as possible to the original. Therefore, I would like to express my sincere gratitude to my dear friend for his kindness and patience. Without his help, I really could not have translated the whole book in its present form. This reminds me of a problem with current machine translation, which doesn't understand the meaning of the text! Even a poor human translator can know something of the original text that he or she doesn't understand, and if he or she is serious, then he or she may consult dictionaries, encyclopedia, background materials or even ask the original author or other experts, until he understands. However, the present machine translator never asks a question, it even translate a typo as if it were not a typo!

As we wrote in our preface, this book is mainly to inspire readers' thinking rather than tell them what conclusions we came to. The authors do not pretend to know everything and have all the correct answers, and neither do they want to pretend to be prophets and predict the future. They just want to tell readers what they thought based on rational reasoning and hope readers take a similar approach. I have written several popular science books on the brain and mind, and one of my friends once asked me how I could guarantee all that I said was correct. My answer was "No, I could not guarantee that, and I don't think any author could, what I could do is make sure all the material I talked about has a strong foundation and can be justified or rejected empirically." This is also true for this book, where to take things even further, the two authors still have different opinions on some topics, just as Karl summarized in our epilogue. Each of us had their own reasons, but could not persuade the other to believe, while for some topics we may agree but to a different degree. For example, we both think that the simulation of the whole human brain is unrealistic and perhaps even meaningless, but to simulate neural circuits and the like with most parameters within the physiological range is meaningful. To understand the brain one cannot depend purely on big science projects. However, when the subject involves very expensive installations, huge data collections with sound theoretical foundations and requires broad cooperation between experts from different fields, big science project may still be necessary. The problem is the content of the project and how to organize it. And how to have a proper balance between the battleship and the small speedboats. We both agree that technology develops exponentially, while neurology and medicine develop only linearly, but is this difference due to the dominance of "Interest-performers" in the former, and "System-performers" in the latter? Furthermore, in the translator's eye, there is also struggle between "Interest-performers" and "System-performers" in the latter, although the degree is different. Maybe the difference stems from the nature of the two fields, although the different degree of the struggle between the two types of performers may also play role. These are still open problems, and deserve the readers' careful pondering.

Anyway, writing this book and the discussion with Karl has changed many of my ideas, and even the way I think. I hope readers with similar ideas as mine may have the same feeling. Translating gives me a chance to review the whole process and to tidy up my ideas. Now I hope that as I am at a new starting point and can see these open problems from a new perspective, I may see a little clearer and am more sensitive to problems and the rapid progress of these fields. And although these problems haven't been solved yet, and are not expected to be solved in the foreseeable future, it is exactly this point which makes study of these fields so charming.

Fanji Gu

Acknowledgements

The most relevant currency in the world isn't money but time. Our family, friends and colleagues devote to us with their attention and cooperation. We are very grateful for the enormous amount of time others have devoted to us until this book could be presented.

We both enjoy the rare and unlikely happiness of being married to a woman who is an intellectual partner who has been tirelessly reducing entropy in our lives for more than 5 decades. And we both enjoy the good fortune of having a daughter who is also a friend and valuable advisor. Yuemei Ma, Adi, Yimeng Gu, Jelka and Jürgen Seitz deserve our thanks first and foremost. Special thanks to Adi for arranging and editing our drafts in endless loops. And we owe Jelka and her husband Jürgen exciting and controversial discussions about the conclusions to be drawn from our investigations and how they could be implemented. As a now 4-year-old mentor and teacher of his grandfather, Alexander Seitz may not have noticed how much he helped him with the question of how mind and consciousness develop.

With regard to causality, our thanks go to our mutual friend Prof. Hans Braun ("German Hans"). He initiated our contact, our friendship and thus this book. Some chaos-experts say that a butterfly's wing beat in Shanghai can't cause a tornado in the Caribbean. But Hans, a distinguished chaos-expert, has demonstrated that it is possible to make two people work for six years by just redirecting a Shanghai-e-mail from Marburg to Karlsruhe. A performance that should make butterflies, tornados and chaos-experts think!

There is a dear friend and colleague of "German Hans", to whom we also owe a special debt of gratitude. It is Prof. Hans Lilienström ("Swedish Hans") a longtime friend and valuable advisor of Fanji.

The support and advice we received from Fanji's friend Prof. Nelson Y. S. Kiang was especially helpful and we would also like to thank him for his kindness, which allows us to quote from his personal communication.

There were many other friends and colleagues who helped us answer specific questions or expand our research in the course of our conversation. We cannot name them all and would like to mention some of those whose support has been particularly encouraging.

In China it was Yizhang Chen, Tiande Shou, Peiji Liang, Pei Wang, Xiaowei Tang, Aike Guo, Xiongli Yang, Bo Hong, Si Wu, Jintao Gu, Hongbo Yu, Longnian Lin, Ying Pan and Amenda M. Song.

In the West it was Dietmar Harhoff, Gert Hauske, Ehrenfried Zschech, Matthias Bethge, Dirk Kanngiesser, Susanne and Rafael Laguna, Ingo Hoffmann, Serdar Dogan, Ben Hansen, Andreas Bogk, Bernd Ulmann, Hendrik Höfer, Mirko Holzer, Miro Taphanel, Matthäus Paletta, James Wright and Leslie Kay.

Y. Chen, P. Liang, P. Wang, X. Tang, J. Gu, A. M. Song, H. Braun, D. Harhoff, G. Hauske, E. Zschech, and M. Bethge were so kind to read the full manuscript. They made valuable suggestions and comments which helped us to reduce errors and unclear passages before printing.

We are infinitely grateful that they have also taken on the great effort of writing commentaries and reviews, which you find as introductions to some sections of our letters. While we are ashamed that they are generally much too friendly and generous with us, we believe that they perhaps convey the essence of our discourse better than we could have done ourselves.

Thank you to Patrick Hartmann, Agnieszka Paletta and John Broomfield for fixing some of the worst mistakes in our Chinglish-Denglish gibberish without losing the original tonality of our e-mails.

Very special thanks go to our editors: Mr. Wei Huang and Ms. Mingyue Shen from Shanghai Educational Publishing House for their great support and many helpful ideas on how to make the book better, such as the suggestion to have several tables of contents, and the wonderful form to make the book both in Chinese and in English. They have also invited an artist, Mr. Chuqiao Chen, to draw drawings for the Chinese content and suggested what should be drawn. Of course, we would also like to express our gratitude to Mr. Chen here.

We don't know if Fanji's friend Walter Freeman would have liked what we have to say, but we both feel the need to express our gratitude, admiration and respect by dedicating this book to the memory of this great researcher whom we deserve so much inspiration and insight.

还有许多其他朋友和同事在和我们的讨论过程中帮助我们解答了一些具体问题，或扩展了我们的研究。我们不能一一列举他们的名字，而只能提到一些其支持对我们特别有鼓舞的人。

在中国方面我们要感谢陈宜张教授、寿天德教授、梁培基教授、王培教授、唐孝威教授、郭爱克教授、杨雄里教授、洪波教授、吴思教授、俞洪波教授、林龙年教授、童勤业教授、曹建庭教授、孙哲博士、顾金涛博士、潘颖女士和宋蔓女士。

在欧美国家，我们要感谢哈霍夫（Dietmar Harhoff）、郝斯克（Gert Hauske）、恰伊赫（Ehrenfried Zschech）、贝特格（Matthias Bethge）、吉塞尔（Dirk Kanngiesser）、拉古纳夫妇（Susanne and Rafael Laguna）、霍夫曼（Ingo Hofmann）、多甘（Serdar Dogan）、汉森（Ben Hansen）、博克（Andreas Bogk）、乌尔曼（Bernd Ulmann）、赫费尔（Hendrik Höfer）、霍尔策（Mirko Holzer）、M.帕莱塔（Matthäus Paletta）、赖特（James Wright）和凯（Leslie Kay）。

陈宜张、梁培基、王培、唐孝威、顾金涛、宋蔓、布劳恩、哈霍夫、郝斯克、恰伊赫、贝特格审阅了全稿，并提出了许多宝贵的意见，这使我们得以在付印之前减少错误和写得不清楚之处。我们非常感激他们付出巨大的努力为本书撰写评论和书评，这些评论是对我们信件的某些部分的介绍。使我们感到不好意思的是他们通常对我们过于宽容和慷慨，我们相信他们或许能够把我们自己想表达的意思讲得更为清楚。

我们要感谢哈特曼（Patrick Hartmann）、A.帕莱塔（Agnieszka Paletta）和布鲁姆菲尔德（John Broomfield）纠正了我们的中式英语/德式英语中的一些主要错误，同时又不失去我们电子邮件中原来的语气。

特别感谢上海教育出版社的编辑黄伟先生和沈明玥女士，感谢他们巨大的支持和对改进本书的许多有益的想法，例如列出多种目录的建议，以及使本书同时兼具中英文版的奇妙形式。他们还邀请了一位画家陈楚桥先生为中文部分绘制插图，并对插图的内容提出建议。当然，我们也要在此向陈先生表示感谢。

我们不知道凡及的朋友弗里曼（Walter Freeman）是否会喜欢我们要说的话，但是我们都觉得有必要通过将本书献给这位杰出的研究者来表达我们的感激、钦佩和敬仰，从他那里我们得到了如此多的启发和洞见。

致 谢

致　谢

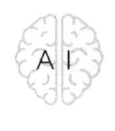

　　世界上最可宝贵的并非金钱，而是时间。我们的家人、朋友和同事以其关心和合作帮助了我们。我们非常感谢他们投入的大量时间，使本书得以最后出版。

　　我们都以有一位睿智的妻子为傲，这是我们的福气。我们的妻子在50多年来一直在不知疲倦地减少我们生活中的"熵"。我们也都有幸有一个女儿，她们既是我们的朋友，也是非常有价值的顾问。因此我们首先并且永远要感谢马月美，阿迪（Adi），顾以蒙，杰尔卡·塞茨（Jelka Seitz）和于尔根·塞茨（Jürgen Seitz）。特别要感谢阿迪一遍又一遍地整理和编辑我们的草稿。我们也要感谢杰尔卡和她的丈夫于尔根。关于由我们的调研中可以得出些什么结论以及如何实现的问题，他们和我们进行了令人兴奋的讨论和争辩。亚历山大·塞茨（Alexander Seitz），作为他外祖父的一位现年4岁的老师，可能并没有注意到他在心智和意识如何发育的问题上给了他外祖父多大的帮助。

　　关于因果关系，我们要感谢我们共同的朋友汉斯·布劳恩教授（"德国汉斯"）。正是他启动了我们之间的联系和友谊，因此也就有了这本书。有些混沌学专家说，上海蝴蝶翅膀的扇动不会导致加勒比地区的龙卷风。但是杰出的混沌学专家汉斯已经表明，通过只是将一封来自上海的电子邮件从马尔堡转发到卡尔斯鲁厄，可以让两个人工作六年。这件事值得蝴蝶、龙卷风和混沌学专家思考！

　　我们也要特别感谢"德国汉斯"的一位亲密朋友和同事。这就是凌瀚思（Hans Lilienström）教授（"瑞典汉斯"），他也是凡及的老朋友和极有价值的顾问。

　　得自凡及的朋友江渊声（Nelson Y. S. Kiang）教授的支持和建议特别有帮助，我们也要感谢他慨允我们引用他在私人通信中所说的话。

甚至可能是毫无意义的，然而，如果大多数参数都在生理范围内，那么对神经回路等的仿真还是有意义的。认识脑不能光依靠大科学工程，但是当问题涉及昂贵的设备、有坚实理论基础的大规模数据收集、不同领域的专家之间的广泛合作，大科学工程可能仍然是必要的。问题是这种项目的内容、组织形式以及如何在战舰和小型快艇之间取得适当的平衡。我们两人都同意目前技术正呈指数式增长，而脑科学和医学则呈线性增长。但是其原因是否就是因为前者兴趣派占了主导地位，而后者则是规矩派一统天下？在译者看来在这两个领域中都存在着兴趣派和规矩派的斗争，虽然他们在这两个领域中的比重不同。他们发展速度之差主要是由其学科性质决定的，虽然这两派的斗争也起作用。这些都仍然是没有解决的问题，值得读者思考。

不管怎么说，写这本书和与卡尔的讨论改变了我的许多想法，甚至改变了思考问题的方式。我希望有类似想法的读者在读了本书之后也会有同样的感受。翻译让我有机会重新审查整个过程并整理我的想法。现在我希望以新的视角看待这些尚无定论的问题，我可能可以看得更清楚一点，对这些领域快速发展中出现的问题更敏感一点。尽管这些问题最终还是没有解决，甚至也不能在可预见的将来得到解决，这也正是这一领域之所以如此迷人之处。

仔细的读者也许会发现，本书的中文部分比英文部分多了一些内容。其实这些内容只是为了帮助国内读者更好地理解书中的某些内容所提供的背景材料，其中包括译注、专栏(其实专栏只是一种扩大型的译注)和插图。对于熟悉这些背景材料的读者来说，当然无需去阅读它们。由于这些材料并未译成英文让卡尔过目，因此在这些材料中如果有什么错误的话，那么责任应该完全由译者来负。

顾凡及记于复旦大学

2019 年 6 月 30 日

译后记

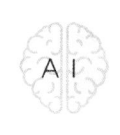

　　本书是我第一次翻译一本自己合著的书。尽管翻译自己写的部分比较容易，我当然知道自己想说些什么，但是，本书的近三分之二是由卡尔写的，由于他的思想深刻、知识面广，这使我在翻译时很不容易，尤其是当话题涉及先进技术、西方哲学和除神经科学之外的科学史时更是如此。幸运的是，卡尔非常耐心地回答我的问题，甚至重新撰写某个段落，以便我能理解。这使翻译得以尽可能地忠实于原文。因此，我想对我亲爱的朋友表示衷心的感谢。没有他的帮助，我就无法把本书翻译成现在的样子。这让我想起目前机器翻译的一个问题，它不理解文本的含义！即使是一个很差的翻译人员也总可以知道自己在原文中哪些地方没有看懂。如果他或她是认真的话，就会去查阅字典、百科全书、背景资料，甚至询问原作者或其他专家，直到弄懂为止。然而，现在的机器翻译从来都不会提出任何问题，它甚至把错字也照翻不误，就好像它不是一个错字一样！

　　正如在序言中所说的，本书主要是为了引起读者的思考，而不是告诉他们结论。作者不想假装知道一切，并给读者正确的答案，他们也不想假装成为预言未来的先知。他们只是想通过理性思考告诉读者他们的想法，并希望读者也能类似地进行思考。我写过几本关于脑和心智的科普书籍，我的一位朋友曾问我如何能保证我所说的所有内容都是正确的。我的回答是："不，我不能保证，我也不认为任何作者能作这样的保证，我所能做的只是保证我所谈论的所有材料都有一定的根据，你可以实证地加以检验或否定。"本书也是如此。更进一步说，两位作者对一些话题都仍然有不同的看法，就像卡尔在跋中所总结的一样。每个人都有自己的理由，但无法说服对方。对于某些话题，我们可能会在认同的程度上有所不同。例如，两位作者都认为对整个人脑进行仿真是不现实的，

168　　　　　　人工智能的第三个春天　　　一位德国工程师与一位中国科学家之间的对话

我们经常提到冯·诺伊曼，他肯定是 20 世纪真正的天才之一，并且对现代科学的贡献比大多数人都多。其中之一是他在 1944 年与摩根斯坦（Oskar Morgenstern）一起发表的博弈论。这一理论几乎与以他的名字命名的计算机架构一样成功。尽管这令人印象深刻，但也有一些缺点。其中最主要的一点是它只研究总额固定的博弈，其中一个玩家的胜利就是另一个玩家的损失。这种观点不仅在冷战时代的将军中非常流行，而且还成了一种范式。

另一位科学特立独行者和数学天才纳什（John Nash）证明，在经济学中，为了改善福祉，合作策略比尝试击败竞争对手更有前途。

而这些合作策略肯定适用于获得科技知识的过程中，特别是当涉及一场值得为之一战的斗争，那就是对类似阿尔茨海默病、帕金森病等危害人类健康的疾病的攻关。

因此，中国科学家和德国工程师一致认为，我们最好在友好竞争中做到这一点，而不是错误地认为我们正在玩零和游戏。如果这个想法不容易在人们的头脑中传播，那么希望下一代人工神经网络能够为自己得出同样的结论。

扫一扫，获取更多有关
脑研究与 AI 的信息

2018 年 6 月我们发现投入已经上升到了 25 亿美元）。而且还有不计其数的这类创业公司正像雨后春笋一样到处出现。

这种冲击的副作用是，懂得如何建立人工智能系统的人，尤其是那些同时也懂神经科学的人非常抢手。世界各地的工业领袖都试图到处挖掘这样为数不多的专家，特别是从大学和研究机构挖人。这对专家来说固然是件好事，但对这些机构来说却是一个问题，因为他们往往无法在财务上进行竞争。但与此同时，它也为学术界、工业界和建立初创企业的投资者之间的新型合作方式以及各种研究合作和研究机构的新形式提供了机会。看到众多有创造性的新星正在涌现以及官僚主义的障碍正在快速被克服，这真是令人惊叹。中国似乎在这方面特别有创意和行动力。

因此，上面讲到的一些与纯学术研究有关的问题，也可能以这种方式得到解决。如果有一部分研究将从大学转向产业，我们丝毫不会感到惊讶，我们将看到过去几十年中 IT 行业那样的成长和发展，相关的基础研究将转移到像贝尔实验室或 IBM 那样的大型公司的实验室中进行，其中也产生出诺贝尔奖获得者。

由于激情和团队精神在这种研究项目中所起的重要作用，我们认为至少在早期阶段，我们所钟爱的小型快艇比大科学战舰更有优势。

但是这就已经是猜测性的了，要预测接下来会看到什么就更纯属猜测了。现在除了变化更快之外，我们还没有看到有比我们在信中讨论过的更多的问题。

有很多东西我们没有谈到，或者只是简单地提了一下。其中之一是人工智能可能产生的社会影响这样的大问题，从可能引起的失业潮到个人数据的隐私权问题。

我们没有深入讨论这些问题，这并不是因为我们没有看到这些问题。我们没有这样做，只是因为在这些讨论中对这些危险有太多的猜测和喧嚣，以至于我们不知道该再讲些什么真知灼见，我们也不想比别人提出更为阴郁的负面猜测。

然而，有一个方面我们想再讲几句。

许多变化。

当我们开始分析理解脑、智能和人工智能的作用时，这还只是一个有一定知识的人以及一小群专家关注的主题。今天其中有关人工智能的部分已经成了媒体中的一个主要话题，从科学界到经济界和政界，到处都在讨论。2016年有人曾预测人工智能是一种炒作泡沫，很快就会进入下一个冬天。但事实却恰恰相反。人工智能成为有史以来炒得最热的技术，而且在中国宣布其人工智能计划和美国做出激烈反应之后，人们对人工智能的兴趣与日俱增。

对我们来说，感觉上就像是一场海啸，涌入了我们小心翼翼划着的友谊小船的小河中。

现在每个人似乎都对 AI 有看法，似乎也都明白是怎么回事。当然，我们并不因为我们比某些在今天谈论这个问题和写文章的人可能考虑得稍长远一些，就声称我们对该领域正在发生什么以及接下来会发生什么有更好的认识。特别是凡及，他依然只是一个对此领域抱有强烈兴趣而好质疑的"门外汉"。

一些记者和分析师对该领域的要点做了很好的总结，突出了关键点和所面临的挑战，我们对此印象深刻。这再次说明，即使是借助像老式的互联网及其陈旧的搜索引擎那样的老版本 AI，在获得有关新领域和复杂事物的见解方面也大有帮助。

而且我们也从这个领域的日益普及中获益，因为许多聪明人已经对这个话题作了进一步的解释，这帮助我们获得了新的见解。

所以读者可能会听到我们谈论最近才知道的事情，现在这些事情几乎在一夜之间就变得广为人知了。我们认为，我们的思考并没有太偏离正轨，并希望它可能会让一些人对人工智能问题有一个更加深入和清醒的认识。

这场人工智能热的主要后果，或许也是会引起巨变的后果，是现在有更多的钱涌入这一领域。早在脑研究计划时，人们可能就有这样一种印象，即没有任何一个工业化国家的政府可以不启动这样一个计划。在 AI 这一领域就更是如此，并且奖券更高。在五年以前，10 亿美元对一个研究计划来说似乎已经是一笔惊人的数目，而现在，单是一家中国的初创公司寒武纪投在运行 AI 应用程序的硬件芯片上的资金就达到 10 亿美元（这还是我们当时得到的数字，到了

Sommerfeld)赏识他们的热情和才能,为他们作了特别安排以满足他们的学业要求。并不是每位同事都喜欢这样做,但是索末菲的勇敢举措得到了回报,两位年轻人都在 24 岁时完成了主要工作,并且都获得了诺贝尔奖。

我们今天在世界各地都有这样充满激情的年轻人。他们醉心一件事到了废寝忘食的地步。他们越是痴迷于某件事,就越容易忽视其他一切,因为这些事对他们的主要兴趣没多大帮助,使他们觉得枯燥乏味。传统的基于高分或应试技能的入学标准让他们名落孙山。所以问题是:我们应该如何吸引这些人并将他们引入"竞技场"?

当然,19 岁的年轻人不可能破解所有的科学谜团。但科学与其他类型高难度运动没有太大的区别,我们经常发现运动员在 30 岁时的表现就已经下降了。

当我们回顾历史时,我们会看到有很多突破是由极其年轻的研究者作出的,特别是在数学方面,那里的果实-树叶比总是很高。数学统计的一条基本原理是"最小二乘法"。它和现在我们所称的"曲线拟合"都是由 18 岁的高斯(Carl Friedrich Gauss)于 1795 年发明的。它是构建现代 AI 深度学习算法的核心原理。

当然,我们并不是说要一位 19 岁的黑客神经病学家在车库里为他们的邻居做脑部手术,对于这样的医疗手术,有充分的理由需要建立专业规则和管理部门来保护患者。

但我们要说的是,其他一些非常聪明的年轻人类似地在车库里就电子学和软件做出疯狂之举,并在过去几十年中创造出全新的行业。他们中的许多人都是辍学的人,他们觉得在现实世界中创造奇迹比完成学业更有吸引力。

正是这种情况才使我们有了现在所经历的技术巨变。

在我们的通信中,卡尔没有使用过"贸易战"这个词,但最后你可以在字里行间感觉到这一点。与此同时,在美国对中国宣布将在 2030 年成为世界主要人工智能创新中心的雄心作出激烈反应之后,类似的用语在西方成了头条新闻,大家也都在这样谈论。

我们很惊讶地看到我们感兴趣的这一领域的技术方面成了有关全球工业领导地位辩论的中心议题。在中国宣布其人工智能计划之后,世界上又发生了的能力,并与你合作以破解一个问题明确的谜团。

好吧,我们并不认为我们心目中的大多数人确实需要这样的建议,因为他们早就已经这样做了,而且他们也太忙并全神贯注于他们所做的事情,没有时间来理会我们的建议。

但是也可能还有人仍然没有拿定主意该走什么道路,如果他们倾向于我们所主张的道路,那么我们的看法对他们也许有点好处,并从中受到鼓舞。

卡尔把他起草的跋给他女儿和一些年轻朋友看,想知道我们的建议会让读者们怎么看,结果得到了一些很有意思的反响。他们认为我们不应该只关注优秀发明家和发现者那样的天才,而且也要关注那些天赋上稍差一点,但是也能作出重要贡献的人。我们大多数人都不是天才,如果想让我们像天才那样行事也没有用。然而,天赋不太高的人也可以通过很好地补偿天才型人才的缺陷,而在团队中作出非常有益的贡献。这是一个很有道理的论点,为把科学发明转化为产品而创立团队和公司时尤其是如此,社交能力往往比科学卓越更为重要。我们提到过有许多科学是团队运动,此时合理调配资源对成功起到决定性的作用,我们应该汲取这一教训。

※　　※　　※

当我们回顾自然科学的黄金时代时,还有另一个更普遍的观点值得注意。这就是学者开始进行研究的年龄。一般说来,比起100年前,现在开始研究的年龄要大得多,只是因为完成学业需要花费的时间很长。问题是在所有情况下是否真的有必要花这么长的时间。

有一些令人鼓舞的例子,说明并不一定需要如此。100年前,两位19岁的年轻人受到学术界同行的鼓励去重塑物理学世界,他们很快就取得了成功,并使他们的教授不知所措。在他们成就背后的驱动力是好奇心和雄心壮志,以及对理解并破解迷人谜团的渴望。当泡利(Wolfgang Pauli)和海森堡(Werner Heisenberg)中学毕业进入大学时,他们已经知道了当时所有有关相对论和量子理论的知识。这并非因为他们是超级聪明的神童,而仅仅是因为他们长久以来一直为一个特定的问题着迷。两人是幸运的,他们的教授索末菲(Arnold

跋
163

只是对某个小实验的选项。这是因为大科学是一个已经建立的价值数十亿美元的行业,不可能在一夜间就发生巨大的革命性变化。在医药行业就更是如此,无论如何,这个行业有许许多多规定,有很多法律限制和伦理界线,这些都是有充分理由建立起来的。正如我们从普朗克100多年前的抱怨中所能看到的那样,所有这一切还要加上科学长期以来形成的惯性。

所以当我们试图对"我们应该做什么?"这样一个问题给出一个实用的答案时,我们不用去考虑战舰舰长或者大科学舰队司令官的意见,因为他们早就知道我们所描述的所有问题。我们提出的建议可能只会增加他们的沮丧,因为他们看到这样大的结构即使要在路线上稍做改变,也没有多少回旋余地。

所以从他们那里,我们至多只能期望他们不要把我们当成天真的浪漫主义者。我们的对话表明,对于那些充满决心和自信心的思想家们来说,我们确实都有点浪漫,他们永远关心某个问题,并且在问题得到解决之前不会放弃。

但我们当然也知道现代科学是一项团队运动,需要不同类型的人来建立一个成功的团队。除了个人的技能和雄心之外,团队精神也是决定性的。在科学史上,表现为把激情集中在某个特定问题上的团队精神非常有帮助,这一点也很重要。

因此,当我们提出建议时,我们想到的是充满热情的个人和小团队,就像小型研究快艇的船长及其船员,尤其是那些年轻的学生,他们早就感觉到自己有很大潜力,并且正在寻找真正的挑战,这些人可能从一点鼓励中获益。

对他们来说,非常简单而不足为奇的建议是:

● 努力工作,坚持钻研令你着迷的问题(就像弗里曼所做的那样),这和备战奥运的运动员很相像。

● 努力尽早成为某一领域中充满热情的专家,并真正精通。

● 不要错把高分和得奖当作真正的洞察力或实际掌握(这意味着要避免费恩曼所形容的巴西大学中的那种学究式错误)。

● 寻找/生产真正的果实,不要欺骗自己,误把树叶当果实。

● 不要只是为了发表而将无足轻重的结果当成"有趣的发现"。

● 让你的同行了解你,而且还要让其他有关的人也理解你,让他们知道你攻关,这可能是大科学最适合的场所。但这并不能解决脑研究中的所有问题,特别是那些极度需要创造性和洞见的问题。

※　※　※

我们的意思并不是说我们没有看到很多成就,而是说有些东西言过其实了,这似乎助长了整个领域的某种浮夸风。现有的学术奖励制度只看出版物和引用数据,让这种情况更加恶化。这助长了科学和研究中普遍将叶子装扮成果实的倾向,因为围绕几个谜题发表一些"有趣的发现"可能比冒着风险尝试破解困难的谜题对前程来说要更保险一点。

对这个问题我们并没有一个简单的解决方案,但是因此就连提都不敢提也是不对的。

我们很清楚批评容易做起来难。虽然我们手里并没有什么神奇的解决方案,但我们不想避而不答"我们该怎么办?"的问题。

当我们总结这场对话并为写跋做准备时,凡及提到卡尔在早先的信中半开玩笑的提议,他宁愿给一千个聪明的学生每人一百万,而不愿把10亿欧元投入到像HBP这样的"研究怪兽"中去,并提出了以下问题:

我同意你所说的关于医学的现状,但是在我们讨论之后,我的脑海里始终存在一个问题,那就是:我们应该在医学上做什么?削减对这些研究的财政支持?我并不认为应该这样做。所以,虽然你的质疑是正确的,过去的方式似乎并不奏效,"大科学"似乎也没有解决太多问题,那么我们该怎么做呢?照技术界那样行事?由于性质不同,那也不太可能,那么我们还有什么途径可以选择呢?把钱分给许多博士后?也许吧,但他们又应该怎么做呢?

凡及提出的问题让我们再次陷入思考和讨论。

卡尔仍然认为,即使是这样一种看似疯狂的想法,就像把希望寄托在一些年轻的不受拘束和心无旁骛的学生的创造力,在某些情况下可能是一种有效的选择。但是,对于所有的大科学来说,这当然不是一种现实的替代方案,而可能

跋

些研究的做法,但我们当然不会声称我们有权告诉别人如何去做科学研究。

关键在于科学本身应该服从理性思维,很明显,在寻求破解科学谜团方面有些方法和研究做法对揭开科学之谜比起其他的来说更有希望。首先,我们在自然科学中拥有的实质性知识宝藏在很大程度上是由前几代人积累起来的。当我们观察知识增长曲线时,令人惊讶的是,在 19 世纪和 20 世纪之交的这个阶段,这个增长曲线是多么的陡峭。而且,更令人惊讶的是,为我们的科学世界观做好准备的群体是多么小,而付出的成本又是多么的微不足道。

如果把它比作一栋房子,我们就会看到地下室和一楼是由几个工匠在短时间内建成的。今天,整群工匠都在施工现场工作,但进度似乎已经放缓,甚至小小装修的成本也比造最初几层楼要高。

有人认为,近代科学界在过去几年产出的知识超越了历史上积累起来的所有知识。如果以发表的论文数量作为衡量的标准,那么这话可能是对的。但是,如果我们使用可以被称为"创造性"的标准,也就是看看我们做到了哪些先人无法做到的事情,那情况又将如何呢? 这是一个非常平常而务实的观点,但可能是一个非常有用的观点。

我们说过,如果按上述标准来衡量,那么物理学家特别是工程师的表现优于生命科学家,并且工程技术似乎是以指数式速度发展,而神经科学和医学的发展则可能只是按线性发展。

我们一直有点吹毛求疵,我们说过,我们对许多研究计划的产出和称为"有趣发现"的结果不满意,而病人还在那儿等待治疗。也许我们这样说对于那些在这些领域努力工作的许多人来说太过分了,并且可能会发现我们的批评不公平和令人沮丧。特别是对于参与大型科学计划的人来说,情况就更是如此,他们没有多少自由选择的余地。但是,我们并非专指哪些特定的科学家,而只是描述看到的情况,当然我们也理解在唾手可得的果实已经被我们的先辈摘走之后,剩下的问题有多么的困难。凡及还觉得大型科学计划如果从其本质上来说是工程技术性的,如果已有扎实的科学理论基础并以解决问题明确的重要任务为目标,那么这样的计划还是重要和可行的。多学科开发新的研究工具和大规模采集脑的基本数据也许对脑研究来说是重要的,并且需要巨大的投资和团队香农和图灵建立起来的以逻辑和数学为核心的范式似乎是破解这个谜团的正确方法。然而,他对这个问题考虑得越深入,对这种方法的数字本质以及脑作为计算装置的想法越使他感到不安。

弗里曼是卡尔遇到的第一位既对脑有极为深入的了解,又分享了这种不安的人。他甚至更进一步把"信息处理"原则称为是对脑的根本误解。通过我们的讨论,甚至凡及也对弗里曼的框架的意义有了新的认识,尽管他之前曾将弗里曼的著作翻译成中文。

试图理解弗里曼的观点,特别是他认为我们的脑并不计算信息,而是在持续的相互作用过程和"循环因果性"回路中提取意义,这一观点对我们两个人都有着至关重要的作用。这个观点可能并不尽善尽美,也可能并不适用于所有脑区和功能,当然也不会是这场辩论的结论。

我们在其他一些问题上仍然有不同的看法,例如主观性的本质,意识是否会从物质中涌现出来以及如何涌现,机器是否也可能有意识,我们对查默斯和丹尼特等作者也有不同的评价。但是我们都同意弗里曼工作的价值这一点远比我们之间的其他分歧更为重要。

对我们而言,弗里曼早已在《意识、意向性和因果关系》(*Consciousness, Intentionality, and Causality*)(1999)中提出了他的观点,他的观点可以作为一种有希望的理论框架,这一框架本可以在 HBP 和 BRAIN 倡议等研究计划中得到应用和检验。

我们的脑不是一种数字信息处理机器,也不是按照逻辑建立起来的,也没有进行数值计算,而是一种模拟装置,可以从噪声中提取出意义。这种观点对急于在更快的计算机上求解偏微分方程的数学家和程序员来说可能不是很吸引人,但我们希望它能成为寻找魔法城堡不同入口的年轻无偏见探险家一个有启发性的起点。

只要我们能激励几名年轻学生或许还有一两名有经验的研究人员尝试这种方式,那么我们的工作就已经值得了。

除了对脑、心智和人工智能之谜的质疑之外,我们还讨论了很多有关科学方法论以及如何在知识方面取得进展的最佳方法。尽管我们广泛地批评了一

跋

神经科学家更好地理解脑和心智呢？用更大更快的计算机进行仿真并由此改善神经脑模型对神经科学家理解脑和心智是否会有好处呢？

对于像脑这样高度动态的功能系统，我们甚至在物理层面都还很不了解，更不要去说更高的功能层次。我们都认为进行计算机仿真没有多大意义，然而试图这样做可能有助于改善所需的工具。

我们也最终同意，试图在硅片上通过逆向工程建立一种生物脑并没有太大希望。卡尔早就提出，工程师不会从脑研究的结果中获益太多，走他们自己的路而不用理会生物模型会有更好的结果。一开始，这在凡及看起来是一个很奇怪的观点，但后来他觉得这一观点并非那么荒谬了。

我们仍然无法回答许多令我们困惑的问题，甚至无法确定我们有共识之处是否就一定对。当然，我们并没有声称我们已经获得了任何新的科学见解。

※　※　※

关键在于我们两人在更好地认识这个领域方面都取得了进步。我们的这一段经历对我们来说都很有收获，但愿也能启发他人以类似的方式去探索这个领域。读者可以决定哪种立场较适合他们，也有些人可能对我们两人的观点都不同意。

出版我们之间通信的目的并非是要宣传某些理论，而是提出一种获取知识和探索个人以前未知领域的方法。

作为对那些对这种主观探索感兴趣的人的一种启发，我们要再次强调我们认为对我们最有帮助和最有启发性的观点。

在讨论所有这些问题时，有一个人给了我们最大的帮助，他给了我们一个非常有意义的理论参考框架。他就是凡及的朋友弗里曼，他在我们的通信过程中不幸逝世。我们推崇他为该领域最杰出的思想家和研究人员之一。可惜的是，尽管他很有名，但在我们看来在主流神经科学界他并没有得到应有的认可。

对卡尔来说，凡及向他介绍并不那么容易搞懂的弗里曼的工作是他在这场对话中最具挑战性和有益的经历。几年前，当他第一次面对神经系统时，他将大脑视为一种信息处理机器。他曾经相信过心智上传的可能性，冯·诺伊曼、工程师认为,当人工智能在高速计算机上编制出足够多的算法以最终执行我们心智所有的各种智能功能时,意识可能就会以某种方式涌现出来。当凡及坚持认为这不是一个显而易见的问题时,卡尔最初怀疑他是否会退回到了一种形而上学的立场上,从而偏离了讨论的理性基础。凡及不太确定卡尔是否抓住了让他思考多年的问题的核心。另一方面,卡尔认为凡及可能低估了硬件仍在发展的速度,并可能相信摩尔定律即将结束的流行观点。

我们共同对许多问题进行思考,试图对科学和技术中许多相关领域的发展有一个概括性的了解。通过反复交换意见,我们终于设法较好地理解了问题的所在,以及对方的立场。对意识这个困难的问题,凡及并没有后退到形而上学,而是提出了支持他立场的有力论据,这给卡尔留下了深刻的印象。

我们最终同意如果能百分之百地复制身体和脑的话,那么这一拷贝也可能包含主观性的内心状态,甚至是神秘的主观体验特性。但是凡及成功说明了这样的拷贝实际上是几乎不可能的。另一方面,卡尔不得不认识到,要想建立起能发展出意识和主观性的强大的通用人工智能系统,可能比他希望的要更困难。凡及也不得不承认他低估了 AI 的发展速度。

所以如果在今天问凡及,他可能仍然怀疑是否能实现强人工智能(在人工智能有意识的意义下),而卡尔仍然会选择一个肯定的答案。尽管我们从来没有在我们的信件中明确表达过这一点,但我们都有一种感觉,我们已经设法在对方的脑海中种下了一颗怀疑的种子。

无论如何,我们都同意,心智上传几乎是不可能的,即使是建造出人工无魂人,也要比库兹韦尔的信众所希望的花费长得多的时间。更不要说所谓的近在眼前的奇点了。我们也同意,要在机器、身体和脑之间建立神经联系比许多人认为的困难得多。

我们一直在思考的另一个问题是:工程师为什么在有关计算机和人工智能软件的概念上会取得如此巨大的成功,尽管它们都是基于显然非常不准确的脑模型之上?

如果工程师使他们的机器更像脑(brain-like),是不是对实现他们的目标会有帮助呢? 借助 AI 概念和工程师、数学家和逻辑学家的更多帮助,是否能启发

跋

订购书籍。凡及很难找到像卡尔这样一位朋友,愿意费时费力地思考、回答并和他讨论使他感到困惑的问题。凡及知道他找到了一位很好的朋友和老师。

这种不断澄清我们想说的话并真正理解彼此观点的过程,还要考虑到我们主观上以及文化上的差异,不仅其本身就是一种享受,而且还与我们共同感兴趣的主题息息相关。

在我们的讨论中,我们一再谈到是否有可能将一段话的内容(指的是它的意思)从一种语言准确且完整地翻译为另一种语言的问题,以及如何才能做到这一点。我们一次又一次地回过头来讨论这些问题,这不仅是因为凡及是一位经验丰富的翻译人员,他经常为这个问题而苦苦挣扎,而且也因为这是一个非常合适的实证测试平台。翻译问题与将内容从一个脑转移到另一个脑或机器所遇到的困难,在原则上是类似的。我们也花了很多功夫来讨论意识中的主观性和私密性以及由此如何产生意义。我们在通信中也一再谈到这些内容是否可以完全还原为物理过程,并且可以通过客观的理性研究来阐明。当然这和当前流行的下列讨论是相关的:借助于现代人工智能,我们是否可以制造出具有类似人类意识的机器。这一基本问题是我们在信中处理的两大谜题之间的自然联系。一个是脑的自然智能,即便在好几代最聪明的研究人员花费了巨大努力试图揭开这一秘密之后,我们对它的了解仍然很少。另一个是运行在计算机上的人造算法,也就是人工智能,计算机是完全不同于脑的机器,并且其能力以惊人的速度在不断提高。

实际上,人工智能在许多方面已经优于人类,但是人们往往忽视这一点,只是因为讨论集中在那些机器还不能做的事情上。

我们讨论的关键问题是:自主学习机器在所有智能领域(包括创造力和其他许多被认为是真正的人类的能力,如情感和幽默)中能否和人类一样,甚至超越人类? 如果能的话,他们最终能够发展出意识和主观性吗?

这是我们仍然持有不同立场之处,虽然我们两人都在对方论点的影响之下有所改变。在我们开始讨论之前,卡尔对主观性问题、意识可能在什么条件下出现以及如何出现等方面几乎没有想过。对他来说,它似乎更像是一个可能会消失的伪命题,就像物理学中旧的"生命力"(élan vital)问题消失了一样。许多就几乎必须为此道歉。也可能这种方法在欧洲更容易接受一些，而在中国则有可能被视为粗鲁和不恰当。

且不管我们在读者中引起了何种感受，重要的是要记住，在自然科学的黄金时代所取得的许多惊人进步可能都要归功于这种严酷的方法。只要读读书刊就可以看到，19世纪和20世纪之交科学家之间的辩论并不是基于客客气气的共识和你好我好大家好，而往往是很激烈的争论。通常朋友们竭尽所能为了辨明真相和知识而激烈争论，但这只是为了驳倒对手的理论而不是诋毁他的人格。当然，这样的争吵也可能有伤害并留下伤痕，但他们这样费力是为了找到更好的解决方案并为科学事业而战。

无论如何，我们（特别是卡尔）想要说清楚，如果我们在批评某种理论或观点时用了重话或隐喻，那么这只能以这种传统的体育精神来理解，我们的话只是针对这种论点，而并不是要抹黑某个人。

※　※　※

在卡尔收到的凡及的第一封电子邮件里充满了他无法回答的问题，其中有些问题他也不懂。但他为这些话题所深深吸引，他也赞叹他所遇到的头脑的严肃性，这个头脑总想弄懂并完全搞清楚。他也对其同伴尖锐的论点和勇气留下了深刻的印象，甚至对大权威也敢于提出尖锐的问题。即使在今天，卡尔也还不清楚这种非同寻常的执着和坚持不让步，直到一切都尽可能清晰和透明，究竟是凡及的天性呢，还是他在把许多当代神经科学巨著翻译成中文过程中获得的技巧。

无论如何，对于卡尔来说，凡及在与许多一流作者的讨论中磨炼了自己的技能，他会问一些难以回答的问题。当凡及把我们的英文信件翻译成中文时，凡及迫使作者澄清他想说的话，卡尔对此深有体会。凡及一次又一次地写信来确认他是否已经清楚了一段话的真实含义，有时甚至只是一个词。这个过程并不总是令人愉快和轻松的，因为它可能表明问题并不出在译者的理解有误，而有可能是由于对思想的表达不清楚，或者更糟糕的是这一想法本身就有问题。

凡及则非常欣赏卡尔的博学、认真和有问必答，为了回答问题，他甚至专门

跋

术在近年来的迅速发展抱有浓厚的兴趣，但是他很难找到一位像卡尔那样同时熟悉脑科学的信息技术专家可以向之请教。专家很少喜欢与外界讨论他们的业务，特别是当他们的意见可能受到质疑时。他们中没有多少人会耐烦别人像儿童问出来的那种基本问题，并以小孩都可以理解的方式表达他们的想法。

我们两人来自不同的领域，从我们的专业和文化背景来说，再也找不到更不一样的人了。但是，我们都充满了好奇心，也都崇尚理性思维，并且都有一个特点，就是对什么事都要质疑一番，特别是当问题涉及公认的学术见解和有人声称在科学或技术上取得了突破时。

虽然下面的这番话我们早已在序里说过一次了，但是由于其重要性，我们想在这里再说一遍：

我们都喜欢从孩子的视角来看问题，他们会提出简单的问题，以了解真相。有时孩子可以看到皇帝的新衣并不像所说的那样华丽。但是我们也不想过于夸大，因为说我们就是著名童话故事《皇帝的新衣》中那个勇敢说出看不到别人"看到"的东西的孩子，就未免太自以为是了。

然而就 HBP 而言，卡尔坚持认为，从很早开始，当其他一些人还在赞扬它的时候，凡及就认识到这个令人印象深刻的计划存在缺陷。

我们在早期的信件中花费了大量的精力来说明并使自己确信在 HBP 的概念中有多处错误，我们不应该对此计划期待过高。由此开始了我们的通信，它成为探索脑和心智及其与人工智能和计算机技术的可能联系的许多基本方面的良好试验田。

今天，在这个项目的名声在公众面前已严重受损之后，这种批评很常见，而我们过去的批评在一些人看来似乎有点像在打"落水狗"。也许现在一般性的批评甚至过多了，因为在我们看来，HBP 概念中也确实有一些有趣的部分值得再作尝试。

针对内容的激烈辩论曾经是常用的科学争论风格，就像体育比赛中的情况一样。但是今天这种非常成功的方法似乎已经过时，所以如果有人这样做了，

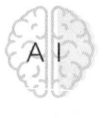

跋

2018 年 8 月

6 年前我们开始对话纯属偶然，我们探索性的讨论既没有计划也没有明确的目标。当时凡及在 HBP 这一著名而雄心勃勃的概念中偶然发现了一些自相矛盾之处，但他不能确定这种不安是否合理。起初我们只是像汽车修理工在类似情况下所做的那样，打开车盖好奇地进行检查，共同努力想了解究竟发生了些什么。

回顾以往，我们自己都对所发现的东西以及关于脑、心智、人工智能和大科学项目的讨论所走过的道路感到惊讶。

正如我们在序中指出过的那样，本书更像是一场随意漫游，从一个领域转悠到另一个领域，并在需要的地方深入下去。我们只是受到好奇心的驱使，并且当我们希望对事情理解得更为精确时或者当我们需要填补我们的知识空白时，我们花费了更多的精力。我们常常喜欢对知识追根溯源，包括追溯到我们两人极为不同的文化背景的历史中去。虽然我们这样的做法显得混沌无序，但我们感到，通过持续而且有时引起争议的辩论，我们获得了在更系统的方法下很难获得的见解。

回想起来，我们自己都对以前在某些事物、理论和概念上的看法以及在今天再看这些问题时有多么不同而感到惊讶。我们注意到，我们的观点逐渐发生了变化，随着事情发展而变得清晰起来。

例如卡尔认识到，看待脑还有更有希望的观点，而不是通常地把脑当作是一种信息处理设备。

在我们开始通信的时候，卡尔是（现在仍然是）一位神经科学的业余爱好者，他很高兴像凡及这样一位经验丰富的专家愿意与他讨论与最新研究进展有关的微妙问题。凡及则是（现在也仍然是）信息技术的门外汉，虽然他对这一技

当我看到所有这一切，并浏览了一遍英戈有关中国人工智能的材料，我可以很好地理解你的意思，因为显然中国的人工智能行业正在如火如荼地发展着，没有任何事情是在可能的阿尔法狗冲击之后匆忙发生的。

因此，我的印象是，我们两个以及世界其他地区的人，都会对年轻一代将会为我们带来什么惊喜而感到好奇。

晚安，我亲爱的朋友，祝好！

卡尔

我们讨论了你和我已经讨论过的一些问题,我真希望你当时也在场,因为你会听到他们对一些熟悉的名字和概念的意见,并且也会高兴地看到我们这些年来的讨论并未那么偏离正道。

所以我强烈建议到这样一种地方去旅行一次,我知道你不喜欢长途飞行,但对你来说有个好消息,因为你周围也有许多类似的 AI 巧克力工厂。英戈·霍夫曼(Ingo Hoffmann)是我以前指导过的一位年轻人和我的好朋友,他刚从中国旅行回来。他擅长人工智能,并主持着我所知道的有关此问题的最好的博客之一。[1]

英戈和中国有十分密切的联系,他到南京去参加"世界智能制造业大会"并作发言,他为此作了精心准备。

由他那儿我得到了许多有关 AI 产业发展的有趣信息,我建议你访问他的网站,在那里你可以找到一些文章的链接,例如李开复博士的文章。他的文章讨论的是我提到过的一些和人工智能有关的主题以及它们对中国经济发展的重要性,他比我更有资格来阐明这些问题。[2]

我发现在上海也有一个非常有趣的 AI 巧克力工厂。它的名字是西井实验室(Westwell Lab),他们开发了一种仿神经结构芯片,并声称已经克服了冯·诺伊曼架构的障碍。[3]

远处万里之外,我无法说他们有多好,但我的猜测是他们应该不错,无论如何也是雄心勃勃的,而且似乎资金充足。[4]

当然,还不止他们一家。实际上,AI 芯片看来似乎是中国的热点,寒武纪(Cambricon)是该行业著名的十亿美元独角兽。[5]

---

[1] https://ingo-hoffmann.com/

[2] https://www.eurasiagroup.net/files/upload/China_Embraces_AI.pdf 最近李开复发表了一本和我们讨论的许多问题有关的非常有意思的书:Kai-Fu Lee. AI Superpowers:China, Silicon Valley, and the New World Order[M]. Boston, Mass:Houghton Mifflin, 2018.中文版:李开复.AI·未来[M].杭州:浙江人民出版社,2018.

[3] http://www.westwell-lab.com/about_en.html

[4] https://www.chinamoneynetwork.com/2017/06/23/fosun-unit-leads-series-a-round-in-chinese-ai-chip-maker-westwell-lab

[5] https://www.chinamoneynetwork.com/2017/08/18/chinas-state-development-investment-corp-leads-100m-round-in-ai-chip-maker-cambricon

AI 的威胁。如果说 AI 带来的所有创新都没有问题，那就太天真了。从工程到法律体系的所有领域都将发生巨大变化。美国关于假新闻、滤泡[1]（filter bubbles）和控制社交网络的辩论都是我们所面临的问题的标志。

有些人可能会为此感到害怕，而另一些人则认为这是一个给这个世界带来新面貌的巨大机会。

在这个方面，我最近有一次非常令人鼓舞和别开生面的经历。我们前几天谈起过和 AI 有关的德国信息技术谷（Cyber-Valley）项目，以及令人惊叹的被称为 DeepArt 的 CNN（卷积神经网络）应用程序。碰巧我的女儿杰尔卡（Jelka）在一次会议上遇到了三位创始人之一的贝思格（Matthias Bethge），他也是 DeepArt 之父中的一位。

我女儿联系了他，这样我就有机会到图宾根附近对他和他的团队进行了一次访问，在那里我还遇到了另外两位信息技术谷的创始人——布莱克（Michael Black）和舍尔考夫（Bernhard Schölkopf）。好吧，你可以想象得出，这次访问对我来说就像是一个六岁小孩去参观巧克力工厂一样，在很多方面都很有意义也很有意思。

首先，我很高兴看到的这三位都非常谦虚、踏实和讨人喜欢，虽然他们在 AI 界是世界一流的。或者我应该说，因为他们是世界一流的，根据我的经验，真正优秀的人都是最和蔼可亲的，而平庸的人往往装得像神气总督一样。

另一件好事是我们很快就对 AI 下一步将要发生的事情以及哪些人值得注意进行了讨论。这个领域的领先人士在全球的圈子似乎很小，他们都相互认识。这似乎创造了一个友好而又充满竞争的氛围，这让我想起奥斯特瓦尔德和他的国际朋友共同奠定了现代物理化学的基础时描述的情况，后来整个工业部门就建立在这个基础之上。

他们清楚地认识到，作为开拓者，他们会开创新局面，并做出令人难以置信的发现和发明。这种意识将他们从世界各地联系到了一起。当你进入大楼时，你几乎可以像有心灵感应似地感觉到，有一种新鲜感和非常好的感觉。

---

[1] 指利用人工智能技术根据上网者的行为，定向向上网者发送他或她可能感兴趣的信息。——译注

机器在自行探索而不是遵循人类的指导时表现更好。

我敢肯定我们将会看到许多令人惊叹的事情,但请不要因为我的看法而去购买和 AI 相关的公司的股票,因为在经济中一切皆有可能,没有任何东西是理所当然的。☺这就像纳维耶-斯托克斯意义下的涡流一样无法预测。接下来的金融危机可能即将到来,也许像比特币这样的加密货币会崩溃,下一场赌博对许多人来说如此迷人和吸引人(因为他们不懂),这可能会带来不利之处。

但从长远来看,机器学习(ML)将起作用。我越想越想不出有哪些地方是机器学习不能应用的。实际上很难找到不受机器学习影响的领域。我从 ML 的应用中学到了这一课,这是我从未想到的。

一个例子是动态调制把热等离子体保持在托卡马克氢聚变反应堆中的磁场![1] 当你想到它时,你会发现在快速控制物理或化学过程方面都能从类似应用中获益。

另一个例子是最近由卡尔帕西以软件 2.0 的名义发布的工作。[2]

这个想法既威力巨大又令人惊讶:让机器读取你的代码并改进它! 这是与莫拉韦茨悖论中描述的现象相类似的另一种情形:任务越复杂,AI 做得越好!

现在关于人们失业的问题争论得很激烈,在媒体上你总会看到机器人将取代较低技能的人。我的猜测是,AI 的主要后果将表现在高端,正如卡尔帕西的软件示例中所表明的那样。但程序员不会把他们的岗位让给机器。

无论如何,我们在这个领域没有足够的合格人员,但是我们在大型计算机程序的安全性和鲁棒性方面存在着巨大的问题。从银行业到飞行控制系统,我们的世界所依赖的大多数程序都已陈旧而需要更新,正如某些老城的高速公路桥梁或下水道需要更新一样。

另一方面,在很长一段时间里,护士、厨师和草莓采摘者的工作还不会受到

---

[1] http://www.digitaljournal.com/tech-and-science/technology/google-optometrist-algorithm-brings-us-closer-to-fusion-power/article/498508

[2] https://medium.com/@karpathy/software-2-0-a64152b37c35

与生活平衡"的功能优化所取代。

我的朋友沃尔(Paul Wahl)是 SAP 的前任总经理,后来担任西贝尔(Siebel)总裁,在 20 世纪 90 年代后期来中国做生意,他也看到了类似的情况。他用一句话总结了他的观点,当我看到实际的发展时,我总是会想到他的那句话:

"如果你看看中国的年轻一代,并将其与我们的年轻一代进行比较,那么为我们的年轻人着想,你只能希望他们永远也不要在世界市场上较量。"

有些人认为中国将向西方学习,也会适应工作与生活平衡的模式。我不会为此打赌,为了欧洲的缘故,我宁愿遵循奈特的建议,希望至少在将 AI 用作促进全球繁荣的下一个助推器方面,向中国学习。

我们也可以向你们的投资者学习,不仅是企业巨头百度、阿里巴巴、腾讯或小米,他们在从孵化器和初创投资一直到战略收购都显示出非常聪明的策略。美的收购库卡(KUKA)[1]是一个很好,但也令人警觉的信号,因为没有一个德国玩家看到自动化领域中这块珍宝的战略价值。

你所问的所有关于 AI 在不久的将来的发展问题,我也都在问我自己。而根据我上个月看到的情况,我在做预测时变得非常谨慎。

我也不认为库兹韦尔的奇点已经接近,甚至不相信我们已经进入了通用人工智能时代。我们还只是在简单的人工智能方面开了个头,不过进展也是巨大的。在你的附言中所说的阿尔法狗元击败其早先型号的阿尔法狗就是很好的例子。但机器学习远不止于深度学习和神经网络。我也相信我们将看到老式人工智能(GOFAI)和各种混合系统的复兴,还有许多 AI 专用的硬件解决方案。

我认为我们还只是看到了一个开始,而真正重要的事情才刚刚来临,这些事情就相当于卡尔帕西和古德费洛在接受吴恩达采访时的看法以及哈萨比斯在《连线》访谈中所说的内容。无监督学习和生成对抗性网络(Generative Adversarial Networks,简称"GAN")的新变种实现了一种新型的机器学习,这些

---

[1] 全称为 Keller und Knappich Augsburg,是一家总部设在德国的工业机器人和工厂自动化制造商。——译注

Ⅲ-010 卡尔 147

器人公司一度开始成为头号无人机制造商,耗资一亿美元,最终却被中国的无人机巨头大疆公司(DJI)所击败。[1]

我不确定这是否是由硅谷的一位前员工在一篇文章中所说的"经典的硅谷傲慢"引起的,还是由于时机不巧,或者是由于一位缺乏经验的记者过于浪漫和天真而造成失败。安德森在书中把特斯拉树为这个新工业时代的榜样,在这里,硅谷的傲慢,如果不说是自大狂,很有可能起了作用。看看马斯克(Elon Musk)在大规模汽车生产中试图满足必要的质量标准时所面临的诸多问题,一些人已经开始在预言他的失败了。

我希望这不会发生,因为我总是更同情那些敢冒险并一往无前的人,而不是那些极力想维护现状而畏首畏尾的人。无论如何,我们德国人得特别感谢马斯克,因为他可能及时唤醒了我们的汽车产业,这仍然是我们国家繁荣的基础。

我希望当中国在开发人工智能方面付出如此多的精力,并将其工业提升到一个新的水平时,也会产生同样的唤醒效应。正如你以前所说,中国将不得不这样做,因为这可能是让13多亿人生存发展的唯一途径。当目标是每个人都应该过上美好的生活时,问题是如何提高生产力。这个问题是苏联一直没能解决的,最终导致苏联解体。我对中国的政治和经济状况很不了解,这就是为什么我觉得很难给出评估,而不沦为常见的陈词滥调。我告诉过你,我为在中国这个社会主义国家之所见所闻而迷惑不解,我见到了从未在加利福尼亚见过的最有创造性和最雄心勃勃的企业家。另一方面,我看到西方受过高等教育的精英们对资本主义制度和工业文化感到不舒服。然而,与此同时,这个制度使他们得以过上他们所喜欢的愉快而自由的生活。我很惊讶地看到,当我和这里的年轻人交谈时,即使是聪明人似乎也认为经济福利是理所当然的。他们关注的往往是如何以更公平的方式分享繁荣,而不是如何创造繁荣。

然而,当我与中国的年轻人交谈时,我的印象是,他们主要是想在生活中实现这个目标,并且他们明白,实现这一目标的最佳方式是努力工作和遵守纪律。在我们这里美德已经不合时宜,而被更高的政治正确性原则以及被称为"工作

---

[1] https://www.forbes.com/sites/ryanmac/2016/10/05/3d-robotics-solo-crash-chris-anderson/#65afe4073ff5

该文对此进行了清醒的分析,并对中国的计划是否现实作了评估:

"有一点很清楚:人工智能可能是在西方发明的,但你可以看到它的未来在世界的另一边成形。

中国的人工智能计划包括政府非同寻常的承诺,政府在最近宣布了对人工智能重要性的全面展望,该计划呼吁国产人工智能在三年内与西方开发的人工智能相当,要求中国的研究人员在 2025 年前取得'重大突破',并要求中国人工智能在 2030 年在世界上居领先地位。

有充分的理由相信该国能够实现这一愿景。在 21 世纪初,政府表示要建设高速铁路网,以刺激技术发展并改善该国的交通体系。这个铁路网络现在是世界上最先进的。"

奈特的话对他同胞的民族自豪感来说可能不太中听,但在我看来这是对情况的合乎实际的分析。人们称中国为模仿者的日子已经一去不复返了。通过公司收购获得的知识产权和授予中国公司的专利数量在不断增加,特别是在人工智能方面,这表明中国已经学会了如何玩这方面的游戏。

我也同意奈特对有关此举的工业意义所说的话。无独有偶,昨天吴恩达宣布在他的新公司 landing.ai 中决定将 AI 带入到生产行业,并率先从合作伙伴富士康开始。

吴提出了谷歌脑深度学习项目,后来又担任百度首席科学家,尽管他年纪不大,但他确实可以称作为 AI 的老手,而且肯定是一位不会胡言乱语的研究人员和经理,他知道自己在做什么。一段时间以来,我一直关注着他和他的学生的活动,我认为他现在正要做的就是我们在德国称作工业 4.0 的事业。

5 年前当安德森在他的《创客——新工业革命》一书中宣布新的工业时代时,他显得有点不够成熟。当时我们谈起过这件事,他的话在我看来是有道理的。然而他有一个假定错了,就是他认为革命将由 3D 打印机触发。但就自动化的总体趋势以及工业机器人的重要性而言,他是非常正确的。

然而,他想把失去的生产领域从中国带回美国却完全失败了。他的 3D 机

Ⅲ-010 卡尔

145

实。虽然这个规划在你的国家并未引起轰动,但它肯定在西方,特别是在美国造成了一些冲击波。

迄今已发表了许多研究和文章,在西方的反应从惊讶到印象深刻都有。有些人似乎是受到了侮辱,一个发展中国家居然也学会了玩这种工业游戏,现在正要和西方在先进技术方面平起平坐。

一个有趣的例子就是西格尔(Adam Segal)[1]于 2017 年 12 月 12 日向美国众议院"金融服务、货币政策和贸易小组委员会内务委员会(House Committee of Financial Services, Monetary Policy and Trade Subcommittee)"提出了最新报告"中国技术开发与收购战略和美国的反应(Chinese Technology Development and Acquisition Strategy and the U.S. Response)"。[2]

还有一些其他的评论,这些评论对待这一冲击的方式更符合美国人应对竞争的传统。奈特(Will Knight)[3]于 2017 年 10 月 10 日在《技术评论》(*Technology Review*)上发表了一篇文章,该文采用非同寻常的双语标题以英文和中文同时显示:

China's AI Awakening

中国人工智能的崛起[4]

副标题是:

"西方不应该害怕中国的人工智能革命,而应该复制它。"

---

[1] 美国有关中国技术政策的专家,是美国对外关系委员会中一个有关数字技术政策的项目负责人。——译注

[2] https://financialservices.house.gov/uploadedfiles/hhrg - 115 - ba19 - wstate-asegal‑20171214.pdf

[3] 奈特是 MIT 技术评论有关人工智能的资深编辑。——译注

[4] https://www.technologyreview.com/s/609038/chinas-ai-awakening/

封闭系统。问题在于真正的封闭系统只存在于书本或我们的幻想之中，在整个宇宙中并不存在完全封闭的系统。我不确定这是否是珀施师傅对热力学"定律"持怀疑态度的原因所在，也不知道这是否是他对人们所提出的任何被看作是亘古不变的教条的所谓真理一概持怀疑态度的原因。

我不得不承认，我也曾以一种天真的方式相信热力学第二定律，而给我上了一课从而医好了我这个毛病的正是哈肯本人。事情是这样的：布劳恩邀请阿迪和我参加他家中的聚会，在那里我们也见到了哈肯。我抱怨说，虽然我有绝世佳偶，但她不相信第二定律。事实上，她和我多年来都在争论这个问题，而我一直没能让阿迪成为热力学的信徒。我不得不承认，当涉及用思想实验来批判对量子力学的哥本哈根解释时，阿迪就像爱因斯坦那样富有创造性。所以我要求哈肯作为在世的最优秀的物理学家之一站在理性科学一边，并向这个固执的女人解释为什么怀疑第二定律是愚蠢的。让我非常吃惊的是，他倒是站边了，但不是站在我这一边，而是站到了阿迪那一边。他告诉我，我应该很高兴有这样一位聪明的佳偶，因为第二定律只适用于封闭系统，这意味着几乎永远不会成立。此事已过去20年了，我仍然对这一课怀有感恩之心。☺

当时我对生物学、自组织系统、相变、涌现行为或循环因果关系知之甚少，而我对我在物理学中所学到的一点知识却坚信不疑。这和我们一般在只有一些表面知识的事上的所作所为如出一辙。实际上，我们的大脑似乎是一个基于不完整、不一致、甚至是错误信息和见解就能采取行动（并且还感觉良好）的设备。只要有铁轨，我们的脑引擎就乐于在其上行驶起来，即使这只是空中楼阁。

有时需要很长时间才能传播新的见解。虽然普里高津和哈肯的类似工作对自组织系统的深入研究有很大的影响，但他们在主流神经科学领域里却默默无闻。和弗里曼一样，你在坎德尔的"圣经"中找不到他们的名字。

感谢你对中国《新一代人工智能发展规划》背景的见解。有意思的是，你的分析表明，该规划并非如纽约时报记者所说的那样是由阿尔法狗击败柯杰造成的"冲击"产生的，而是在一段时期以来就已存在的总体规划的一部分。

该计划可谓是雄心勃勃的，特别是考虑到时间表排得很紧，但对我来说，这比提出另一个像HBP或美国BRAIN倡议这样的脑研究计划要更合理，也更现

Ⅲ-010 卡尔

(事实胜于雄辩;中国人工智能的崛起;人工智能
的未来)

2017－12－15

亲爱的凡及:

谢谢你亲切和富有启发性的信!

我非常喜欢你引用"事实胜于雄辩"的中国谚语。

它非常确切地描述了我对科学和技术的务实态度,这让我想起了波普尔在自传中讲到的一个故事。在成为学者之前,他在维也纳跟着珀施(Adalbert Pösch)师傅学木匠活。当他们一起在车间里造家具时,这位多才多艺、有趣和自学成才的木匠师傅向这位年轻人讲述了他相当复杂但又务实的世界观。这一定是一段非常有意思的经历,因为波普尔声称他从这位木匠师傅那儿学到的知识要比从许多著名的哲学家那里学到的还要多。珀施质疑一切,甚至包括物理学中最基本的东西,比如"热力学定律"。波普尔在其自传的一开头引用了他有关永动机的一段话。用他的话来说:"人们说这造不出来,但要是一旦有人真的造了一台出来,人们的话就会不一样了!"[1]

不过,珀施师傅没有活着看到普里高津因其耗散结构研究而荣获 1977 年诺贝尔化学奖的那一天。他说明,在开放的化学系统中,能量的输入和耗散可能远离平衡,在此类系统中熵并不最大化,热力学第二定律也不再适用。

许多将热力学第二定律作为最可靠的物理定律的人,都忽略了它仅适用于

---

[1] Popper K. Unended Quest: An Intellectual Autobiography [M]. London and New York: Routledge, 1976, p.1.

附：我想你已经注意到，就在昨天，在《自然》杂志上宣布了阿尔法狗的一项新进展：阿尔法狗的新版本——阿尔法狗元，以 100∶0 的压倒性战绩击败了这个曾经战胜过李世石的阿尔法狗。另外，据说新版本不需要任何人类棋手的数据，它只是通过自我对弈从头开始学习。所需要的所有知识仅仅是下棋的基本规则和输赢的标准。只经过三天的训练，它就战胜了曾击败过李世石的阿尔法狗！结果是惊人的。据说阿尔法狗元可以用来解决其他领域的问题，如诊断医学图像。阿尔法狗元是人造通用智能，还是一组弱人工智能产品的集合？它在学习新技能时会忘记它所学到的所有旧技能吗？如何评估阿尔法狗元在 AI 发展中的意义？

图ⅢF10.1　自左至右：童勤业、弗里曼、哈肯、山口洋子、孟欣、顾凡及

（Acceleration Law）可能有一定的道理，尽管我仍然严重怀疑他的"奇点接近"的预言。我们有幸能目睹社会中这样翻天覆地的变化，我并不太害怕人工智能发展的后果。到目前为止，人工智能产品仍然是帮助人们解决问题的工具，它们没有自己的意志，甚至也没有自己的内心世界，我不认为它们在可以预见的未来会有。我只是在想：为什么人们要去创造一台有自己意志的机器呢？应该禁止这种研究，尽管这在现在还不紧急。我只是想知道未来 5 年内会发生些什么事情，在将深度学习方法广泛应用于各种领域之后，下一步会发生什么？在不久的将来，人工智能还会有什么其他突破吗？

凡及

该计划雄心勃勃。它提出了关键的任务和措施,其中 6 项重点任务是:
(1) 构建开放协同的人工智能科技创新体系;(2) 培育高端高效的智能经济;
(3) 建设安全便捷的智能社会;(4) 加强人工智能领域军民融合;(5) 构建泛
在安全高效的智能化基础设施体系;(6) 前瞻布局新一代人工智能重大科技
项目。

政府的财政支持和社会资本将广泛流入新一代人工智能的发展。试点项
目将进行探索,成功经验将得到推广。考虑到中国人口的规模,数据量巨大,我
们也有顶尖的超级计算机,这些都是中国发展人工智能的有利条件。

该计划还指出,中国人工智能整体发展水平与发达国家相比仍有差距。缺
乏重大原创成果可能是最大的问题,包括基础理论、核心算法、关键设备、高端
芯片等;急需优秀的科学家和技术人员;基础设施、政策法规、标准体系都有待
完善。

我对这个计划持乐观态度。中国有集中资源和力量开发一些关键事业的
传统。规划所涉及的领域看起来相当全面,预计到规划完成时中国将发生重大
变化。在我看来,该规划的重点是使用各种 AI 方法解决不同领域的实际问题。
这主要是一个实际的工程计划,但也强调基础理论研究,探索新技术和跨学科
研究。我认为这对作为一个发展中国家的中国来说是适当的。

你在上封信中所附的照片非常棒!正如你所说的那样,它"是循环因果关
系的发明者和主要倡导者的历史性见面"。2000 年,童勤业教授和我在中国杭
州组织了一次研讨会,我们邀请弗里曼和哈肯参加会议,也留下了张照片。很
明显,这两位巨匠彼此欣赏,他们在报告中都引用了对方的话。我有幸能结识
他们并多次获得他们的建议。

时间过得真快,从我们开始通信至今已有近 5 年时间。我们讨论了脑科学
和人工智能的许多悬而未决的问题,有些我们已经达成了共识,有些我们仍然
存在争议,还有一些我们根本就不知道该怎么认识。这两个领域发展得都很
快,虽然也许一个是按线性增长,而另一个则呈指数增长,这必然会影响到我们
社会和日常生活的每一个方面,这是我们的祖先所无法想象的。人工智能领域
的一些进展甚至超出了我的预期。我不得不承认,库兹韦尔的"加速律"

Ⅲ-010　凡及　　139

中也是如此。特别是当我们不试图复制生物时，就更是如此。"

我可能高估了 AI 遇到的困难，你上面的那段话已经对此做了批评，甚至在我们开始通信时你就提到过同样的想法。你的话和 AI 在这段时期里的迅速发展，使我改变了一些看法。正如一句中国老话所说"事实胜于雄辩"，AI 进步的速度和方式表明你上面的话是正确的。正如我们在开始时强调的那样，大自然和工程师们使用的方法完全不同。工程师可以从自然的方法中受到启发，但不应该复制生物的每个细节。

非常感谢你告诉我由我国国务院发布的《新一代人工智能发展规划》，我未能及时注意到这一则重要新闻☹。当然，人们对 AI 的快速发展印象深刻，特别是阿尔法狗战胜前世界冠军李世石。由于围棋在中国很流行，李世石也很有名，所以人们对这场比赛非常关注。就在比赛之前，包括棋圣聂卫平在内的几位顶级棋手都预测李世石会赢。因此，阿尔法狗的胜利使人们大为震惊。至于阿尔法狗击败中国冠军柯洁，他也是世界围棋第一人，倒反而没带来多少震撼，因为现在大多数人都已预见到了这个结果。因此，阿尔法狗击败柯洁，国务院几乎在同时宣布这一计划，可能纯属巧合。

事实上，中国政府早就关注人工智能。2016 年 3 月，在"十三五"规划中就提出要发展人工智能技术；2017 年 3 月，李克强总理首次在向全国人民代表大会所作的《政府工作报告》中强调了人工智能的重要性。接下来在 2017 年 7 月 8 日制订了《新一代人工智能发展规划》，并在同年 7 月 20 日公布了这一规划。[1]

该计划将发展人工智能提升到了国家战略层面，为中国发展新一代人工智能确定了战略目标。到 2020 年，人工智能的总体技术和应用将与世界先进水平同步，人工智能产业将成为新的重要经济增长点；到 2025 年，人工智能基础理论研究取得重大突破，部分技术和应用达到世界领先水平，人工智能成为带动中国产业升级和经济转型的主要动力；到 2030 年，人工智能理论、技术和应用总体达到世界领先水平，成为世界主要人工智能创新中心。

----

[1] http：//www.gov.cn/zhengce/content/2017－07/20/content_5211996.htm

Ⅲ-010　凡及

（阿尔法狗；新一代人工智能发展规划）

2017－10－19

亲爱的卡尔：

　　非常感谢你的美言。也非常感谢你对自由意志和决定论这一两难问题进行了深入分析，我曾试图解释这个问题，但是不如你那样清楚。我很高兴你的回应是正面的，这样我就知道我的想法可能并未误入歧途；我也很高兴我们都认为我们必须感谢我们的朋友哈肯、弗里曼和布劳恩，他们提出和/或发展了循环因果关系的概念，这似乎是解决这一难题的关键。可惜的是，我没能早点知道汉斯有关自由意志的工作，但很高兴我们的想法和他的很接近。在用 DeepL 将他的文章翻译成英文后，我应该读一读。你的论点非常有力，至少在我眼中看来是如此，我没有什么要补充的了。

　　DeepL 翻译的质量远远超过了我的预期！这真是太神奇了！我不得不承认，机器翻译水准的进步比我预期的要快得多。你的话看来是对的，现代技术按指数级数增长，而神经学和医学则只是线性的。我盼望 DeepL 有新服务将英语翻译成中文，以及将中文译成英文。正如你所说，我期望中国国内开发的一些中英互译软件可以做得更好。但是，到目前为止，我尝试过百度和科大讯飞的翻译，它们都和谷歌和必应翻译差不多，并不尽如人意。我不怀疑这些翻译有朝一日也能达到 DeepL 的水准，我只是希望他们能够快点改进。

　　你以下的话可能是给我们的长时间通信作了个总结：

　　"这让我想到，神经科学家是否高估了问题的复杂性。也许构建更强大的人工智能系统并没有他们想象的那样难，即使在一些被认为是特别困难的领域又：我和布劳恩通了一次很长的电话，结果发现他也是弗里曼的忠实粉丝。他告诉了我他第一次见到弗里曼的故事。这是 2004 年在土耳其库萨达斯（Kusadasi）举行的一次大会上，为了表达对他的尊敬，会议请弗里曼致开幕词。在接下来的问答环节中，布劳恩向他提出了一些关键问题，甚至在得到回答以后又问了更多问题并坚持己见。组织者似乎觉得攻击这样一位名人不合时宜。但是弗里曼似乎对此感到非常高兴，因为之后他走到布劳恩那儿，告诉他他的问题非常有意思，他希望更详细地讨论它们。因此，布劳恩在随后的晚宴上被邀请坐在弗里曼旁边。非常有趣，也非常典型地表现出两人的开阔心胸。

他给我发了一张这次见面的照片，你可以从照片中看到他们两个就坐在布劳恩的朋友哈肯和他的妻子旁边。布劳恩允许我与你分享这张照片，这是循环因果关系的发明者和主要倡导者的历史性见面。

图ⅢK9.1　自左至右：弗里曼、布劳恩、迈尔（Petra Mayer）、哈肯（承布劳恩允许发表）

力的高度敏感。

直到最近,西方许多人仍然相信他们的技术优势是从根本上得到保证的,其他人都不得不跟在后面学样。

我不得不承认,当我在1993年访问中国时,我也以为会看到一个发展中国家。我也有过那种由西方文化造成的典型偏见,这种偏见表现为自我感觉高高在上,并且深信现代技术只能来自西方,其他所有人都必须向我们学习。之前我曾看到过某种死气沉沉、低生产率和短缺经济的模式。我曾以为在中国多少也会看到同样的东西,但在你的国家逗留的十天时间已足以改变我的世界观。

以前给我印象最深的是加利福尼亚州,在那里我遇到了一种让欧洲相形见绌、看上去显得老旧疲惫的企业精神。在中国,使我感到非常吃惊的是,我遇到了充满决心和活力的企业家,他们的速度让加利福尼亚都显得缓慢。在一个社会主义国家里,我被这样的景象惊呆了。当时我对所看到的一切感到困惑不解,现在也还依然如此。尽管我曾试图理解,但我还远未真正懂得这个巨人的惊人崛起。

《纽约时报》的文章强调,当阿尔法狗击败中国的世界围棋冠军柯杰时,这对中国来说是一种"斯普特尼克震惊"(Sputnik-shock)[1],从而引发了这个AI规划。你也许可以告诉我这个想法是否只是出于记者的幻想,或者是否真的如此。无论如何,我很想知道你对这个大项目的重要性和可行性的看法,这个项目使得像HBP或美国的BRAIN倡议这样的其他计划显得非常小而且可以忽略不计。

我相信你也对这一出乎意外的举动感到兴奋,并对此有自己的看法。

祝好!

卡尔

---

[1] 当苏联成功发射人类第一颗人造卫星斯普特尼克时,这在美国引起了巨大的震动,由此导致美国制订阿波罗计划并与苏联争夺太空开发的霸主地位。这被称为"斯普特尼克震惊"。——译注

和动态闭环控制系统的影响。目前尚不清楚究竟要用什么方法,即使是创建这些新翻译系统的专家也不清楚。他们甚至感到惊讶,取得这样的进展竟是这么容易。在 YouTube 中,吴恩达(Andrew Ng)与业内的顶尖人士在"深度学习的英雄"的标题下有一系列有趣的访谈,这些访谈给出的信息是:"我们只是刚刚开始,我们还不知道接下来会发生什么,但会有很多事就要发生。"[1]

我们看到的在机器学习方面所取得的进展对我来说也是一个惊喜。我不会说库兹韦尔是对的,也不认为我们很快就会看到他所预测的奇点,但我不得不承认,进展比我想象的要快,我对此印象深刻。在我这里有几位年轻人,他们被 AI 领域所吸引,他们都报告说,大公司提供的一些框架很好用,并且他们通过使用有用的解决方案取得了比预期更快的进展。

这让我想到,神经科学家是否高估了问题的复杂性。也许构建更强大的人工智能系统并没有他们想象的那样难,即使在一些被认为是特别困难的领域中也是如此。特别是当我们不试图复制生物时,就更是如此。

这至少是我的青年才俊朋友们的感受,我不是告诉他们说他们正在努力实现不可能的事情,而是鼓励他们,让他们尝试实现自己的梦想。这也许会让我感到惊讶。事实上,至少上个月发生的事情就使我感到惊讶和印象深刻,而且留下深刻印象的似乎还不止我一个。

至少你的政府似乎也对此印象深刻,因为它已经采取了令人惊讶的行动,你当然已经注意到了这件事。中华人民共和国国务院于 2017 年 7 月 8 日宣布了一项巨大的工业界人工智能计划,按照该计划中国应该在 2030 年之前处于人工智能的领先地位。

这确实非常有趣,这一消息在全世界产生了巨大反响。纽约时报在 7 月 20 日发表了一篇评论。[2]

在我看来,文章的语气有点过度了,因为这种 AI 规划并非是一场工业战争的宣言。但是,这篇文章很典型地代表了美国和欧洲对中国不断增长的技术能

---

[1] https://www.youtube.com/watch? v=-eyhCTvrEtE&dist=PLfsVAYSMwsksjfpy8P2t_I52mugGeA5gR

[2] https://www.nytimes.com/2017/07/20/business/china-artificial-intelligence.html

不差,特别是当你考虑到文本被翻译了两次。对于"简单的"英德翻译,DeepL只有一处不太重要的缺陷。

尽管我不是一位翻译家,但我不能确定我是否可以做到这个地步。当然把英语翻译成德语再翻译回英语可能要比你从英语到汉语再回到英语的试验容易一些。

DeepL的创始人弗拉林(Gereon Frahling)曾经与谷歌合作过一段时间。他似乎受到巴赫达瑙、丘和本希奥2015年发表的"*Neural Machine Translation By Jointly Learning to Align and Translate*"一文的启发,使用了神经网络(此处应用了卷积网络,其设计初衷是用于处理视觉对象的),我在以前的邮件中提起过这篇文章。

但它能达到这个水平并不只是由于采用了CNN。我想其主要原因在于DeepL这几年来其在线字典Linguee从数百万文件中收集了大量数据,而且他们在创建自己的超级计算机方面也很有创意。据他们自己介绍,这台计算机在世界上排名第23位,并安置在冰岛。可能是因为那里的能源价格便宜,你总得设法在冷却机器方面少花点钱。

我不知道这些勇敢的敢捋虎须的科隆年轻人能独占鳌头多久,我也不知道那些庞然大物是否会很快就作出反击。[1]

无论如何,DeepL的这一成就与DeepArt非常相似,后者让你得以把你的照片创作成艺术作品,这些成就都清楚地表明了神经网络的有用性,以及AI工程师不仅只是在下棋方面取得了进步。

从机器人和机床控制到自动驾驶,从语音识别到在线营销或高速股票交易的预测性定位,在这些非常不同的产业部门中我都看到了类似的情况。

当维纳提出通过应用闭环控制来改进动态技术系统的想法而开创信息时代时,他早就清楚其应用领域几乎是无限的。虽然花费的时间比预想的要长一些,但在经过了70年之后,现代世界中几乎没有什么过程不受自动化机器学习

───────────────

[1] 2018年12月,DeepL宣布美国的风险投资公司Benchmark投资该公司。
https://slator.com/ma-and-funding/benchmark-capital-takes-13-6-stake-in-deepl-as-usage-explodes/

somewhat imaginative prose in my dissertation in Oxford (philosophy, not neuroanatomy), English became the international scientific language, it is up to us English native speakers to write works that can be read by 'a patient Chinese (sic) with a good dictionary'. The results of this self-imposed discipline speak for themselves: whether you are a Chinese, German, Brazilian — or even French — scholar, you insist on publishing your most important work in English, naked English, translatable with minimal difficulty, relying as little as possible on cultural allusions, nuances, word games and even metaphor. The level of mutual understanding achieved by this international system is invaluable, but a price must be paid: some of the approaches that seem to need to be done require informal metaphorisation and imagination, which attacks the barricades of closed minds with every trick, and if some of them are not easy to translate, then I only have to hope for virtuoso translators on the one hand and the growing eloquence of the world's scientists on the other."[1]

要想翻译丹尼特的这段话并不容易,尽管翻译并不完美,但在我看来结果

---

[1] 由于这段是机器翻译的结果,译者很难原汁原味地把它翻译成中文,而不留译者的痕迹。因此这一段就不翻译了,而把这段话的英文原作附在下面,通过对比,读者就可以判断这一翻译程序的质量了,两者确实惊人地接近:"First, there is a culture of scientific writing in research journals that favors-indeed insists on-an impersonal, stripped-down presentation of the issues with a minimum of flourish, rhetoric, and an allusion. There is a good reason for the relentless drabness in the pages of most serious scientific journals. As one of my doctoral examiners, the neuroanatomist J. Z. Young, wrote to me in 1965, in objecting to the somewhat fanciful prose in my dissertation at Oxford (philosophy, not neuroanatomy), English was becoming the international language of science, it behooves us native English-speakers to write works that can be read by 'a patient Chinese (sic) with a good dictionary'. The results of this self-imposed discipline speak for themselves: whether you are Chinese, German, Brazilian — or even a French — scientist you insist on publishing your most important work in English, bare-bones English, translatable with minimal difficulty, relying as little as possible on cultural allusions, nuances, word-play, and even metaphor. The level of mutual understanding achieved by this international system is invaluable, but there is a price to be paid: some of the thinking that has to be done apparently requires informal metaphor-mongering and imagining-tweaking, assaulting the barricades of closed minds with every trick in the book, and if some of this cannot be easily translated, then I will just have to hope for virtuoso translators on the one hand, and the growing fluency in English of the world's scientists on the other." ——译注

当你将许多大脑联合起来工作,驱散黑暗的效果最佳,这就是被称为科学的方法。☺

就在我思考你关于机器翻译未来的论点时,我为一项机器翻译服务深为惊讶,这项服务似乎能够提供超高质量的机器翻译。该项服务的名称是 Deepl.com,更令人惊讶的是它来自一家非常小的公司,而使我难以相信的是它来自德国科隆。

就在本周,这项服务对科技界造成了巨大的冲击,国际媒体对此进行了大量报道。许多记者对它进行了测试,并且迄今为止我所看到的所有情况似乎都证实了 DeepL 的大胆声称,即他们已经达到了机器翻译质量的新水平。从网页中你可以找到 TechCrunch 对此所做的更全面的分析[1]。

我还用德译英对它进行了测试,结果发现它确实非常有用,并且比我迄今为止使用过的任何其他机器翻译程序更有用。

该系统不仅是一台全自动翻译机,而且还是专业翻译人员的辅助工具。如果你不喜欢翻译结果中的某个词时,你可以点击这个词,系统就会提供你一系列同义词[2]。真的很好,唯一可惜的是,对普通话的翻译要到明年才能推出。☹[3]

我做了一个测试,类似于你翻译丹尼特有关翻译的引语,也就是从英语翻译成德语并再从德语译文翻译回到英语。

正如你从下面的结果中所看到的那样,比你从谷歌翻译得到的结果要好得多。

"Firstly, there is a culture of scientific writing in research journals that favours an impersonal, reduced representation of topics with a minimum of blossom, rhetoric and allusion. There is a good reason for the relentless sadness on the pages of the most reputable scientific journals. As one of my doctoral supervisors, the neuroanatomist J. Z. Young, wrote to me in 1965, when he turned against the

---

[1] https://techcrunch.com/2017/08/29/deepl-schools-other-online-translators-with-clever-machine-learning/

[2] 百度翻译也有这一功能。——译注

[3] 到2018年底还未推出此项服务。

Ⅲ-009　卡尔

的 DNA,从而在某个方面表现更好。这个结果造成了我们的祖先和其他哺乳动物之间的区别。

你和弗里曼说得很对:我们在一个互为因果的过程中建立并不断完善目标,在这一过程中逻辑子程序或策略导致变化并按自己的规则发展。这就是人类行为难以长期预测的原因。我们根本不知道下一个将改变游戏的想法究竟是什么,就像过去的轮子、铁、增值税、议会、电视、互联网或智能手机。也许在下一轮中,起这种作用的想法就是人工智能,它就像生物体中的超级酶那样加速整个自组织过程。

与耗时数十亿年缓慢的自组织进化过程不同的是,现在是意向性通过循环因果链在发挥作用。这非常像是另一种智能设计——不过这次并不需要有上帝!

当然,随机性仍然起着一定的作用,因为我们根本不了解我们在其中采取行动的整个系统,也不知道其元素的各种状态,因此也无法对此进行计算。

这种不可计算性与量子力学理论中假定的测不准原理无关。自从庞加莱在 130 年前处理了 n 体问题,因此奠定了混沌理论的基础以来,数学家和物理学家就知道在纯牛顿物理学中也存在确定性但不可计算的系统。因此,不需要靠量子理论引入更多的不确定性来解释人类行为的不可预测性或自由意志现象。

但许多人,特别是哲学家和哲学化的物理学家,似乎都爱上了这种基本的、神秘的、不确定性的想法。虽然几乎没有人真正懂得量子数学究竟是什么,但实际上几乎所有涉及这一问题的工作都要提到海森堡及其测不准原理。这非常像某种宗教仪式或向神殿进行跪拜,以此希望不会受到科学教派的神圣成员的雷霆一击。

但是像米德这样头脑清醒的工程师可以无需这一切就能工作,因为他们知道可以在如此不稳固的基础上建立起非常可靠和可预测的机器,以及非常成功地运行有用的算法。

如果从一个工程师的观点来看的话,我们的脑在理性推理方面非常笨拙和不可靠,因为进化主要是为了做其他的事情而塑造脑的。但是当用理性之光去冲破神秘的黑暗时,脑是非常有用的。你对自由意志问题的论证就是一个很好的例子! ☺

道上一样起到稳定器的作用。

当然,有人可能会想方设法提出一种观点,认为我们头脑中的想法以及我们的理性推理都是基于生理过程和我们脑的分子生物学。而这些可能都是由早先的状态"造成"的。但是,对运行算法的计算机也可以这样说。对这种情况可能没有人会认为算法的结果是由硅芯片中分子的物质状态引起的。当然,这种结果是由内置逻辑决定的。许多不熟悉软件技术最新发展(特别是机器学习)的哲学家不懂得这种内置的逻辑可以在循环学习过程中自行修改。计算机只能做给它编程了的事的旧范式已经过时了。神经网络在用足够多的数据训练之后,甚至可以以创建该系统和发明算法的程序员都不能理解也没能预见到的方式执行任务。这很像孩子们发明出他们的父母既不理解,也从未梦想到的东西。

关键是我们能够抓住想法和思想,并使它们在逻辑机器上运行,而不管这种机器是内部机器(我们的脑)还是外部机器(计算机)。就记忆和计算决策而言,我们的内部机器可能不如外部机器可靠,但无论怎么说,我们的内部机器是主观而有意向性的。

虽然我们许多的内部理性推理和决策都会受到情绪和各种生理微观状态的干扰,即使是这种主观理性的最轻微影响也会打破在物质上确定的简单线性因果链,但与此同时却也会使整个系统在宏观层面上非常稳定和可预测。

在我看来,把像举起一根手指这样原始的"信号-响应"因果链当作复杂的内部理性过程的模型似乎是很荒谬的,这些理性过程在我们的整个头脑中刮起风暴,经常或大或小地改变我们的思考。

许多神经科学家都只限于从心理学层面考虑脑的连接组,他们倾向于忽视社会层面的内容。但是这种观点在系统层面上是不完整的,因此他们有仅从片段解释整个系统的危险。这就像仅仅通过计算机的 BIOS 或操作系统来解释计算机的功能而完全忽略正在运行的应用程序和流经它的数据一样,这完全是徒劳无益的。但重要的部分是最高层次的内容,这决定了意义。

而这正是取得进展的关键。我们不断建立和完善我们的库,并将其传给他人和我们的孩子。我们不需要等待数百万年,直到进化的随机游走改变了我们像弗里曼对嗅觉系统和"再传入"现象的研究一样。

当他们以同样熟悉的方式谈论脑和神经元时,就好像他们谈的是他们客厅里的家具一样,你会感觉到这一点。他们都看到过数以千计的脑,并用从脑电到膜片钳的各种方法,对神经元进行过无数次信号记录。从他们两人那里,你都可以看到只有在深入研究时才能获得的那种能力。

因此,世界上没有多少人像汉斯那样深谙此事,而他所要说的话也是讲给真正懂行的鉴赏家听的。他的文章是货真价实的珍宝,但我怀疑他所提到的信口谈论这个问题的人中究竟有多少人能够懂得他的意思。但是,这场辩论早已由不得那些熟悉此事的人来决定,而是由那些仅对此进行推测,并只是从几个不能说明很多问题的实验中就得出奇妙和广为引申的哲学结论的人所决定。这场讨论似乎在很大程度上受到意识形态的驱动,我的印象是,谈论这个问题的许多人都有我前几天讲过的综合征的特点:他们似乎对他们最不了解的事情有最坚定的看法。☺

但是,对于科学辩论来说这种情况并不少见,即使是一些在实证上更容易进行检验的理论也是如此。人为造成的气候变化就是一个大家都知道的例子。

虽然对自由意志很难从生理细节的层面上来进行讨论,但你的论点更为简单,同时也更有力,并且对我很有说服力。

从长期理性思维的更高层面来看,信号通路和原始的运动控制回路(我们对之仍然不明白)并不是我们在世界上行事的合适模型。就像你说的那样,规划前程或婚姻和按照信号举起手指或按下按钮完全是两回事。

坚持认为线性因果关系不足以解释持续的监测过程,据此我们得以在很长一段时间里不断调整行为以追随变化着的目标,你的这种看法是对的。尽管像运动系统这样较低层次的许多过程可能是由硬件连线决定的,但在作为推理工具或应用程序的较高层次上的软件库则并非如此。这些工具可能是通过教育,或仅仅通过阅读自行开发,或从他人那儿得知,又或者是略加修改而来的。

我们的物质世界可能也是由微观的难以捉摸的原子建立起来的,然而它们在宏观层面上是非常稳定和可预测的。我们的社会行为也是如此。它受到规则和规章(格伦意义下)的稳定作用,这些规则和规章就像铁轨把火车保持在轨

Ⅲ-009 卡尔

（自由意志；线性因果关系；循环因果关系；机器翻译；中国《新一代人工智能发展规划》）

2017-09-02

亲爱的凡及：

谢谢，谢谢，谢谢!!

在谈到如何看待有关"自由意志"这个神秘的、争论不休的话题时，你不会相信我有多喜欢你的来信和你对循环因果关系概念在这个问题上有多大帮助的精彩说明。

我从来都不喜欢在这个问题上所进行的讨论，并与我们的朋友布劳恩一起对此进行过多次辩驳。他和我都不喜欢围绕着利贝特实验所宣扬的形而上学迷雾，并且认为有些人由此得出的过度引申是夸大其词和不可接受的。我认为我们三个人都认为，这个简单的实验以及海恩斯（Haynes）实验组在后来用fMRT做的实验可能意味着很多事情，但与人类是否拥有自由意志的问题无关。汉斯[1]写过一篇非常好的文章，讨论生理学家与哲学家就决定论、随机性和自由意志所进行的辩论中的种种混淆不清之处。它的标题是 "*Determinismus und Zufall in der Neurophysiologie — Die Frage des freien Willens*"（神经生理学中的决定论和巧合——自由意志问题）[2]。

汉斯利用了他在神经元信号处理的生理基础（直到离子通道水平）方面的专长。他一直致力于研究神经元通讯中的噪声和混沌问题，其深入程度可能就

---

[1] 布劳恩的名字。——译注

[2] https://www.hss.de/download/publications/AMZ-87_Homo_Neurobiologicus_02.pdf

些定律,因此确定性定律并非对所有现象都适用。在这里,我并非像一些科学家所说的那样,自由意志可以用量子论来解释。我的观点是,由于宏观物体通常是由微观粒子组成的,拉普拉斯确定性规律对于前者是成立的,而概率规律则对于后者成立,这意味着在等级系统中,不同的规律可能适用于不同的层次。考虑到心智处于等级中的最高层次,为什么自由意志就必须遵守拉普拉斯确定性规律呢?该定律适用于较低水平,但高于微观水平。另外,上面提到的因果关系链是基于线性因果关系之上的。事实上,对于像心智这样的复杂现象,我们必须考虑循环因果关系而不是线性因果关系,就像弗里曼强调的那样。按照循环因果关系,结果也可以影响原因,不仅等级系统中的较低层次对较高层次有贡献,而较高层次也会影响较低层次。所以,你不能沿着想象中的因果链一直追溯到起源。脑中的结构和相互作用非常复杂,很难想象你可以用线性因果关系来解释它的高级功能。正如弗里曼所说的那样:

"人类行为的决定因素不仅包括遗传和环境因素,还包括脑中的自组织动力学……"[1]

这种复杂性和循环因果性为自由意志留下了余地。

总之,如果自由意志被认为是选择的能力,那么有意识对自由意志并非必要,无意识的自由意志也可能存在,但仅限于短时间尺度和简单的动作。如果把自由意志看作为具有循环因果关系的等级系统在最高层次的涌现特性,那么确定性法则和自由意志也可以共存。但同时我也要强调,自由意志的自由程度是有限的,自由意志可以挑选的选择受到外部和内部约束。这就是我对这个问题的看法,你对此有何看法?

由于这封信已经很长,我不得不就此搁笔。

来自上海的祝福。

凡及

---

[1] Freeman W J. Consciousness, Intentionality and Causality[J]. Journal of Consciousness Studies, 1999, 6(11 - 12): 143 - 172.

方案中,进行选择并行动的能力。"[1]

接下来的问题是:这种选择究竟是有意识的还是无意识的? 自由意志是否只意味着有意识的选择? 人们倾向于认为答案是"是"。然而,利贝特(Libet)的经典实验提示,情况可能并非如此。利贝特要求受试者自行确定在某个时刻转动一下手腕,同时注视某个围绕圆形屏幕转动的光点。此外,利贝特也同时记录了受试者的脑电图。实验表明,甚至在受试者意识到自己的决定之前半秒钟,就可以记录到某种特殊的"准备电位"。因此,并不是大脑中有意识的命令或内心独白在发出开始运动的命令。如果坚持认为自由意志必须是有意识的,那么利贝特的实验似乎暗示自由意志只是一种幻觉。但是,如果我们将自由意志看作为选择的能力,那么他的实验仅仅表明,这样的选择似乎并不一定要是有意识的。此外,这个实验并不意味着意识在决策中没有任何作用,因为所有类似的实验只涉及简单的动作,你不能将结果推广到长期计划,比如当你年轻时对前程或婚姻所作的决定。总而言之,对于简单的动作,脑中会发生一些事件,它会从几种可能选择中做出决定以开始动作,而主体则要在稍后一秒钟左右才知道这个决定;对于长期计划,你找不到这样的准备电位,你首先意识到计划,可能是几个月或几年前,然后执行计划,但我们仍然不知道这种决策的细节。

在日常生活中,我们觉察不到这种微小的时间差异,我们容易以为我们觉知到所选择的动作是在行动之前,并将这种觉知误认为是行动的原因。利贝特的实验只是表明后一陈述是不正确的,但是,原因仍然是在主体脑中发生了某种事件,该事件选择了在给定时刻开始给定动作。至于意识的功能,它仍然是一个还没有解决的问题,它可能是长期计划的原因,但我们还不能肯定。

即使如此,对自由意志有一种反对意见是,根据决定论,每一个后果都是由前一个原因造成的,因此你可以找到一个因果链一直回溯到"大爆炸",一切都在开始时就决定好了,即使有些事看上去好像是从许多不同的可能性中挑选出来的。这个论点看起来很强,因为它基于一些物理规律。但是,我有两点反对意见。首先,我们知道确定性规律并不适用于微观层面,就像量子力学中的那

---

[1] https://en.wikipedia.org/wiki/Free_will

身的某些背景,如维基百科中的所有内容甚至更多。

库越丰富,翻译就越好。正如你在你的精彩信件中强调的那样,无论主体是人还是机器,都需要一个丰富的库来进行翻译。然而,即便如此,我仍然怀疑光有丰富的库是否就足以保证所有的翻译都是正确的。理解本身是相当微妙的,我不认为机器在可预见的未来能够理解。如果要想使翻译准确无误,机器作出的翻译可能仍然需要有真懂的人进行审校和修订。

尽管我在上封电子邮件中提到的谷歌翻译的例子表现不佳,必应和百度的翻译也差不多(信中没有显示),但这并不意味着机器翻译无法在可预见的未来取得明显的进展。你是对的,即使像我这样的人类翻译也经常犯错误,我也经常抱怨这样的翻译。机器翻译有可能达到普通翻译人员的标准,尽管为了使翻译完美,也许依然需要让人类专家进行审阅和修订。但是,那些差劲的人类翻译者其实也需要这样的把关☹。即便如此,机器翻译还是对我们有很大的帮助,至少对于漫游世界各地的游客来说就很有用。也许我不应该对机器翻译过于挑剔。

我完全同意你以下的话:"所以诀窍似乎是让机器做它最擅长的事情,而不是试图让它按照人类的方式来完成工作。"

现在让我把讨论转向有关自由意志的话题。

在采访了21位意识研究的顶尖科学家后,布莱克莫尔说道:"坦率地说,在我开始采访之前,我已经预料到,几乎每个人都会在理智上拒绝自由意志的想法,然而要是不多少相信这一点,那么连日常生活都很难过……这正如约翰逊(Samuel Johnson)那句令人难忘的话:'所有的理论都反对自由意志;而所有的经验都支持这一点。'"这似乎是一个难以解决的难题。我想告诉你我的想法,并渴望知道你的意见。

当然,自由意志的问题很难,我不可能像做综述一样触及这个问题的每一个方面。在这封信中,我只想讨论布莱克莫尔提出的两难问题。

不同的人对自由意志有不同的看法。首先,让我争辩说,自由意志不等于自由行动。没有人可以做他或她希望做的一切。由于受到外部或内部的限制,总有一些事是人无法做到的。因此,对我而言,"自由意志指的是在各种不同的

Ⅲ-009　凡及

（机器翻译；自由意志）

2017 - 07 - 15

亲爱的卡尔：

　　非常感谢你出色的来信，你在信中对翻译的本质和机器翻译的前景进行了深入分析。你的论据非常有力，我必须重新考虑我以前的论点。你是对的，虽然纯数学教科书中牵涉的文化底蕴很少，而在拉马钱德兰的书中则有大量和文化相关的内容，但这两种文本有一个共同点：要理解这些文本取决于读者头脑中相应的库，虽然这两种情况下需要的库很不相同。尽管纯数学教科书中的语法结构很简单，但没有适当数学背景的读者仍然无法理解，正如我在阅读莎士比亚的《哈姆雷特》之前不能理解拉马钱德兰的话一样，虽然他所用的语法结构并不复杂，用的词汇也很普通。因此，看来正确翻译并不等同于正确理解。翻译员即使不明白原作的意思，仍然有可能作出正确的翻译。

　　事实上，当我第一次翻译拉马钱德兰的话时，我不明白他的意思。我搜索了这段话在《哈姆雷特》原作中的出处，然后找出由著名的翻译家翻译的中文版中相应的译文，所以我的翻译是正确的。即便如此，我也不明白皇后为什么要这么说，为什么拉马钱德兰要用她的话作为反应形成（reaction formation）的例子。我想机器也可以很好地做到这一点。因此，我不得不承认，我的例子是不正确的，相反，这些例子表明，即使翻译人员不懂，他们仍然可以正确翻译。问题的关键是译者要有一个和原作者同样丰富的库，就像你所说的那样。也许最起码的库就只是一本简明的字典和一本语法手册；进一步地，可以是一本更好的字典，其中有大量的惯用法和搭配，这样可以根据短语的上下文对原文进行翻译；再进一步，是要考虑到整个句子、整个段落、整章、整本书的背景，甚至超出书本包阿汉硅神经元的应用只是众多应用之一,我无法从远处告诉你他的成就现在所处的地位。据我所知,其新颖性并非体现在对马霍瓦尔德的硅神经元从根本上有了什么新的设计,而是反映在更高级的架构之上。

但正如你正确地指出的那样,所有这些新硬件概念的问题,是有多少人愿意投入时间和精力去转向新的编程生态系统。只要在旧环境中仍然可以取得很大进展,现在情况似乎就是如此,那么这种可能性并不高。这就是为什么我相信我们还要经过一段时间后才会看到范式转换。

你关于欧洲最近的混乱以及你对经济前景和 AI 的作用所说的话都很有意思。但现在,这封电子邮件早已太长了,我宁愿将我的评论推迟到下一封信。

无论如何,我非常期待知道你有关自由意志的想法!

祝好!

卡尔

(a)

(b)

(c)

图Ⅲ K8.1　用 https：//deepart.
io/对凡及拍摄的上海街景（a）以
凡·高风格（c）艺术化后得出的
图片（b）。

的过程可以监督其他方法,并执行合理性检查,这有助于排除"愚蠢的错误"。

这篇文章中还有一张展示在此领域中诸多应用方法的图,表明还有许多和热门的深度学习不同的方法。

现在我们看到了一种热衷于深度学习的炒作,工程师总是倾向于使用最有效的工具并尝试用它来解决所有的问题。然而,我的猜测是,过一段时间,我们会看到更多的混合系统,它们非常适用于特定的问题。我们还将看到更强大的机器翻译系统,因为你的示例显示当今的系统尚未建立在最先进的技术基础之上。谷歌在建立好的围棋弈棋程序方面非常成功,但是它的翻译器与来自deepArt的绘画程序并不在同一水平上,后者可以把各种照片都变成凡·高风格的漂亮的油画。

一种基于卷积神经网络的翻译器可能不仅能够翻译出好的作品,还可以让它们具有诸如钱锺书、歌德、巴尔扎克或海明威等名家的作品风格。你可能不喜欢这个想法,但我确信它并不像听起来那么困难。

巴赫达瑙(Bahdanau)、丘(Cho)和本希奥(Bengio)在 ICLR 2015 上发表了一篇有趣的会议论文,题为"*Neural Machine Translation By Jointly Learning to Align and Translate*"[1],这篇文章正指向这个方向。

这是一篇技术性很强的文章,你不必仔细阅读它的细节。我从中得出的最有意思的信息是,最先进的翻译器还有哪些不足之处。我们得到的消息是,在处理文本方面还赶不上我们在图像处理方面已经看到的那种革命。所以我们可能会看到机器翻译方面的真正进步,或者正如我在这个领域的一位青年才俊朋友所说的那样:"到目前为止,我们看到的都还只是最初步的。"

像现在到处出现的各种仿神经结构芯片这样的专用硬件,可以在很大程度上加速这个过程,一如你正确地指出过的那样,当问题牵涉需要减少当今系统消耗的大量能量的时候,这种芯片的作用就更大了。对于诸如物联网(IoT)应用或无人机等依靠电池独立供电的情况就更是如此。

---

[1] https://arxiv.org/pdf/1409.0473.pdf

当今应用人工神经网络的最大问题是需要训练大量的数据。但是，如果它们得到适当的数据，也可以做出惊人之举，甚至在一些不久之前还被认为是机器解决不了的任务中也超越了人类。我提到过在区分两千种不同品种的狗的问题上，机器现在就要比人强多了。人脸识别是这种深度学习技术惊人力量的另一个例子。但是，在这两种情况下，我们都不明白神经网络是如何做到的。

这里有一篇有趣的文章，普林斯顿神经科学研究所的伯恩斯泰因（Aaron M. Bornstein）讨论了各种机器学习技术的预测准确性和可解释性之间相互关系的问题。[1]

而对这个"'什么'对阵'为什么'"问题的回答是：

"现代学习算法表现出人类可解释性与其准确性之间的折中，深度学习既是最准确的，也是最难解释的。"

所以诀窍似乎是让机器做它最擅长的事情，而不是试图让它按照人类的方式来完成工作。当工程师制造车辆、飞机或望远镜时，这种方法也非常成功，而这些工具的表现很快就超过其生物原型好几个数量级。

正如伯恩斯泰因指出的那样，这当然也有不好的一面，即工程师也不明白他们自己所创建的系统中究竟发生了什么，这是一种不愉快的感觉。当事情涉及有重大风险的应用时，这肯定是一个重要的问题，因为失误可能会导致人员伤亡或造成其他重大损失。

从系统安全角度来看，把各种方法结合在一起可能是有用的。可理解方法

---

[1] http://nautil.us/issue/40/learning/is-artificial-intelligence-permanently-inscrutable 关于人工智能中所用的不同方法，感谢 Dietmar Harhoff 告诉我们下列文献给出了最新的非常全面的概述：Dominique Cardon, Jean-Philippe Cointet et Antoine Mazières. La revanche des neurones. L'invention des machines inductives et la controverse de l'intelligence artificielle [M]. Réseaux, 2018/5（n° 211），p. 173 – 220. DOI 10.3917/res.211.0173

Ⅲ-008　卡尔　　　　　　　　119

效率高。

许多年前,我注意到过谷歌翻译和必应翻译都将英语短语"这辆汽车的汽油里程低"(The car has a low gas mileage)翻译成德语"Das Auto hat einen geringen Benzinverbrauch"(这辆汽车的燃料消耗低)。所以译文的意思与英文原文的意思完全相反。一分钟前我再次检查了一遍,只是想看看是否有人已经改正了这样的低级错误。这本来是很容易纠正的,但经过许多年后却仍然如故。好吧,也许是谷歌和微软的经理们并不太在乎公司汽车的耗油量吧。☺

为了对翻译机器公平起见,我不得不指出人类翻译人员也仍然常犯类似的小错误。这就是所谓的"假朋友"。在英德翻译中,一种常见的此类错误是把十亿和万亿混淆了起来。在德文中,一百万的一千倍称为"Milliarde",而一千个"Milliarden"则称为"Billion"。但不知道是出于何种历史原因,美文中的十亿(billion)是德文中的"Milliarde",而德文中的"Billion"却是美文中的一万亿。因此,德国人在阅读英文文本中见到 billion(十亿)这个词时很容易将其翻译成德文中的"Billion"(万亿)。这并非什么难事,每个人都应该知道这个容易犯的错误。但是你很难相信竟有多少德国记者犯这种错误,并报道了美国的"万亿美元"事件,他们无意中将这些事件夸大了 1 000 倍。

你的演示和我自己糟糕的机器翻译经验,似乎都支持你从维基百科有关"机器翻译"的文章中引用来的怀疑分析。不过,我不确定那种论点还能有多长时间保持有效。文章中的许多论点都涉及基于句法分析、语法规则、词典、本体和语义的传统老式 AI(good old fashioned AI,简称"GOFAI")。这些方法确实试图"理解"文本中的内容,而这往往是无望的。

但人工神经网络(artificial neural networks,简称"ANNs")在深度学习的视觉应用中非常成功,如面部识别或区别猫与狗,深度学习使用了完全不同的方法。它们并不想"理解",而只是要学会对象中的哪些元素相关从而使它们彼此相似。关于这些算法如何通过神经网络的许多"隐层"逐层改进辨识的问题,这即使不是无法理解的话,也是很困难的。甚至发明者也不明白他们所创造的工具是如何做到这一点的。在这个新的世界里,计算机只能做告知它做的旧范式已不再成立。代表知识的是这些网络所接受的数据,而不是算法!

118   人工智能的第三个春天   一位德国工程师与一位中国科学家之间的对话

中安装好丰富的库,以便他们能够以合理的方式处理问题。这意味着教育和实践,而这是需要时间和纪律的。这要耗费精力,但就生存和成功而言,这是最终起决定性作用的因素。一个领域越是封闭和有限,就越容易建立知识和规则,使人们能够在其中找到方向并成功地进行操作。世界各地的年轻人都喜爱运动和计算机游戏,这绝非偶然。两者都代表了高度结构化,明确定义和有限的微观宇宙,其熵极小。它们都非常适合应用规则和实践理性策略。科学(特别是数学)和工程技术也是如此,它们都高度标准化并有可靠的规则和规范。这里的内容非常专门和具体,而且极少有文化色彩。另一方面,我们拥有情感世界,在这方面我们拥有无限可能的解读能力。

当翻译家翻译一本书时,其难易在很大程度上取决于作品的内容。如果其内容牵涉的只是世界的一个高度结构化和闭合的低熵子集,比如不牵涉许多文化内容的科学和技术,而同时你又可以期待读者已经花了力气将基础知识加载到了他的库里去的时候,那么这种翻译是比较容易的。另一方面,我们也有充满文化色彩的作品,甚至在更极端的情况下还有押韵的诗歌,这种情况就要求翻译家能够重做原创性的艺术工作了。

因此,我相信 C.科赫建议翻译“应该看不到有译者介身其间”,只适用于比较简单的翻译,这种翻译只适用于有完善的标准化库的科学和非常专门的领域。

你用谷歌翻译器将丹尼特的文本首先翻译成中文,然后再把中译文反过来翻译成英文,以检验其翻译质量的测试令人印象深刻,结果很有趣。

我也有过类似的令人挠头的经历。但是有时候我也为其结果很有帮助而留下了深刻的印象,例如华硕的 PC 主板只有中文版的修理手册。而在其他一些情况下,如你所举的例子,我发现结果令人失望、无用甚至完全错误和误导。

举例来说,在美式英语中,用“汽油里程”(gas mileage)来定义汽车发动机的效率。它表示一辆汽车用一加仑汽油可以行驶多少英里,这意味着高汽油里程是一件好事,因为它表明热效率高。在德文中,这种效率的衡量标准是一辆汽车行驶 100 千米所消耗汽油的公升数。这种 L/100 千米的度量被称为“Benzinverbrauch”(燃料消耗),这意味着低燃料消耗值是好的,因为它表示热都独一无二,他们的心智是由非常不同的、主观的库构成的,那么为什么在许多情况下他们的行事又如此一致和高度可预测?

似乎有一种可能深深植根于我们脑的古老部分中的机制,这种机制导致某种类似于社会"锁相"(phase locking)的现象。这种现象类似于惠更斯(Christian Huygens)在观察彼此靠近的摆钟时所发现的它们很快就变得同步的现象。

我不确定以同样的节奏摆动是否真的是产生交流和语言的第一步,弗里曼似乎是相信这一点的。看看世界各地的足球场上欢呼的场景令人印象深刻,此时观众在座位席上创造出一阵波动传播开来,人们站了起来又再次坐下。

不需要任何语言,也不需要有人发命令或指挥,只要有小群人克服了初始的障碍并使他们旁边的人也这样做,即在稍微晚一点以后也站了起来,这个过程就自动发生了。波动就这样开始传播开来,这很像动作电位沿着神经元的轴突传播,并进一步通过神经网络传播。当人们通过电视看到过一次体育场上的情况,下次再开始这样的行动就要容易得多。你不必明白这一切究竟是怎么发生的,也不必考虑你应该怎样做。你只是和你旁边的人一起做就是了。这是一种自动的反射过程,虽然也有些人可以岿然不动,但这并不妨碍产生这样的波。这非常像某种音乐让人们随着节奏踏步或摇头。

总有些传言宣扬有关群体智能(swarm intelligence)的奇迹,以及在某些情况下它可能比理性思维更优越的说法。我早就说过我对这种想法是不大相信的,因为当我想到这样的集体现象和我们这种盲目的"做你旁边的人所做的事情"的行为产生过怎样的灾难性后果时,我宁愿把它称为群体愚蠢。对于鱼类或鸟类来说,参与群体或是一起行动确实可能非常有帮助,这使得攻击者更难捕捉到群体的成员,但对于人类而言,不用头脑而依靠他们旧有的古脑功能在我看来绝不是一个好策略。人类也是一种逃生动物,当危险的捕食者出现时,他们由于生物资源不足就不得不逃走。没有多少时间来验证是否真有狮子或鳄鱼在迫近,或是争论一下应该怎么做。你要是不逃的话,就只有等着被吃掉。

当你旁边的人逃走时,或者所在集体中的一员发出警报时,赶快逃是很聪明的。但是对付威胁还有更聪明的方法。然而,问题在于你必须在人们的头脑词时是否有相同的含义。

你所举的《哈姆雷特》的例子也说明了同样的问题。在这个例子中,你不得不牵涉大量的内容和意义。在谈话过程中或在阅读文章时,几乎不可能在对话者或读者的头脑中安装如此丰富的内容。我早就提到过这个问题的一个例子。爱因斯坦对"爱因斯坦教授,你能否用简单的语言向我解释清楚你的相对论"这个问题的回答是"不"。虽然提问者可能把这样的回答当作傲慢,但事实并非如此,因为这确实是不可能的。有时爱因斯坦不胜其扰而对类似问题作出诙谐的回答,后来人们(还常常自以为是地)引用这些话。那些对爱因斯坦所说的相对论究竟是什么意思一无所知的人,还将他们自己的主观解释强加到爱因斯坦头上,并且为自己掌握了爱因斯坦相对论的幻觉而雀跃。

不仅在复杂的科学问题上有这个问题,而且这个问题也与智力或高级知识无关。在日常语言中,甚至是涉及生活中最基本的要素,你也会发现这类在解释术语或整个概念时的模糊性和不确定性。爱、幸福、正义或健康食物也属于这一类,对这些问题你可能有各种各样不同的解释,这些解释的文化韵味就像彗尾那样广泛和模糊不清。我们的脑已经学会了怎样应付这个问题,有时候还很乐意将一个名字与一个模糊的含义联系起来,即使我们的脑并不确定这种含义是否与对方头脑中的含义相同。

尽管长远以来恋人们都海誓山盟他们永恒的爱情,但是他们实际上从来也不确切地知道这些誓言对彼此究竟意味着什么。现代的恋人和心理学家同样也不知道,但显然今天与150年前总有些不同。

一个领域越是专业化和标准化,它所用的术语和工作语言也就越专门。但是,正如我所说的,当你与另一个也耗费了大量时间在学习阶段建立起了这样一个库的人进行交流时,你就可以利用这一点。有时,这些知识是随着实际使用像计算机或智能手机这样的技术设备而获得的,实际应用引导甚至迫使我们学习基础知识并遵守规则。当一位 Python 程序员与另一位 Python 程序员进行交流时,他们都可以很容易地就卷积神经网络的问题进行对话,不管他们究竟来自上海还是来自卡尔斯鲁厄。

然而,对我来说,还有另一个比莫拉韦茨悖论更令人困惑的悖论。每个人

Ⅲ-008　卡尔

115

近法求解的程序,并以"sqrt"命名,那么他们以后在别的程序中对任何数都可以很容易地调用这个函数。这些工具就像生物过程中的酶或内燃机中的涡轮增压器一样起作用。系统获取的工具越是丰富和越有经验,它的学习能力也就越强,解决问题也越快。"思维敏捷"并不是因为在生理机制上优越才表现敏捷,虽然这在某种程度上可能也有所帮助,但主要是因为它在学习阶段建立起了如何获取知识和应用这些知识的方法和程序库。这种内容只有当该个体主观监控时才可以访问到。该主体才是该特定脑所独有的代码的所有者。即使你用X射线看到脑的每个原子,你还是看不到这些内容,这正如你对视频DVD做同样的事情时,你也无法看到它究竟是《飘》还是《卡萨布兰卡》。

与其他人进行沟通通常始于将测试信号(pings)[1]发送给另一个人,以了解接收方库中的内容。这不仅仅是检测有没有相关知识,也是收集关于解释、意义和情感内涵的信息。虽然在日常语言中有大量共同的词汇和短语,但每个个体脑中的含义和感受的内涵并不相同,而是高度主观性的。然而,人很善于发现他人头脑中的主观成分。

只要稍过一段时间,我们就能知道我们可以"调用"接收方头脑里的哪些对象,即使我们不敢确定我们是否可以分毫不差地让他们也知道完全相同的意义,我们至少可以合作得相当成功。沟通的内容越详细,就越容易做到这一点。但是,即使双方都努力建立起必要的库,并且使它们完全一致(full coherence),然而,一致并不就意味着和谐(harmony),因为虽然我们理解别人的意见,但别人的意见并不一定就是我们的意见。

当谈到理解别人的时候,似乎有一个类似于莫拉韦茨在AI早期就发现了的悖论。在参与者努力建立其特定的库之后,一位中国人和一位德国人很容易就仿神经结构芯片进行交流。而就最基本的文化和社会方面的沟通则要困难得多,因为这里需要调用的库要广泛得多。这方面的范围太广,并且模糊,以至于虽然某个对象具有相同的名称,但我们还是无法确定我们两个在用到这个名

---

[1] 程序员通常用这种技术来检测是否对对象有反应。Pings这一术语最初来自用声呐探测潜艇。
https://en.wikipedia.org/wiki/Ping_(networking_utility)——译注

当你(在先前的信中)问到我们是否可以确切地知道其他人心中的感受时，你讲到的正是这同一个问题的核心所在：

"对于动物来说，'他人的心智'仍然是一个没有解决的问题，也许从严格的意义上来说，这个问题对我们人类来说也没有解决，那么，我们如何期望在可预见的将来可以对机器解决这个问题呢?"

我们确实无法断定其他人对我们想要与他们沟通的问题是否也有相同的感受或理解。任何人际交流都有这种内在的不确定性。这取决于两个对话者心中的对象库有多相似。当然，仅仅在我的心中也有一个和你心中的某个带有相同标签的对象是不够的。当数学家进行交流时，他们得益于在他们的头脑中有许多明确定义的对象。但沟通之所以非常容易，只是因为他们利用了先前学到的东西，所以在交流时"明白"小数、质数、平方根、余弦或积分是什么意思。

如果在接收者的头脑里根本就没有这些东西，那么要在对话过程中建立这样的库几乎是不可能的。虽然从原则上说起来，也可以用简单些的术语来进行解释，但如果想在对话过程中这样做，那么时间就不够了。当你与某人讨论三角学问题时，你会立刻明白他或她是否知道正弦和余弦之间的区别。这正如在国际象棋下了几步棋之后，你就会明白对手是否懂得下棋。一个系统越封闭，变量和规则的数目越少，就越容易发现这种能力，但前提是你自己要拥有这方面的知识储备。这意味着你已经花了时间和精力把这些对象安装到了你自己的库中，从而将意义与语法联系了起来。你通过学习获得的丰富的对象库与技能并不是一回事! 费恩曼的巴西学生将所有必需的对象都存放到了他们头脑中，但却没有使用它们的技能。

这就是为什么我不愿意接受智力应该以"纯粹"与技能无关的方式来加以衡量的论点。储存在我们头脑库中的对象不仅可以包含与事实记忆相关的内容，而且还可以包含有助于获取更多知识并更有效而快速地利用这些知识的工具或方法。程序员在计算机语言库中添加像求平方根这样的函数时用的也是类似的方法。如果他们成功地编制好了一个求平方根的程序，比如说用牛顿逼

Ⅲ-008　卡尔

（机器翻译；人际交流；深度学习）

2017－05－01

亲爱的凡及：

我很高兴地听到，虽然你不是丹尼特的粉丝，但是你也喜欢他关于当一名翻译人员不容易的话。我忍不住给你发来"有一本好词典的耐心中国人"的话。当然，他也可以选择任何一种其他语言，但是他选择了中文，因为他想把他的话和塞尔的知名思想实验"中文屋"联系起来。但这又是一个翻译问题的例子。他的许多读者可能理解这一点，因此可能会报以微笑。如果他们不知道这一点，也不是什么真正的问题，因为即使未能欣赏到他这种微妙的暗示，读者对丹尼特这句话的主要意思还是可以理解的。所以，如果有一天你不得不把这段话翻译成中文，那么也不必为此做一个脚注。一位从来没有听说过"中文屋"的读者可能会把注意力更集中到论证本身。如果译者试图教他注意这个短语中的微细内涵，他甚至可能会感到恼火。

你用自己读英语和俄语数学书籍的体验作为例子，非常具有启发性。你在拉马钱德兰的《脑中魅影》一书中所提及的莎士比亚的《哈姆雷特》的例子，也讲到了翻译人员所碰到的一个真正的问题，这也很有启发性。

实际上，在上面三个问题中，"中文屋"的内涵非常小，可以忽略不计，像数学这样高度结构化和形式化的系统中很少用到文化典故，也很少有内涵，而像《哈姆雷特》这样具有巨大文化影响力的作品则内涵极其丰富，不过可以把所有这三个问题都放在一起讨论。

它们都涉及一个问题，这个问题不仅限于从一种语言翻译到另一种语言时有，而且在任何形式的人际交流中都存在。

构芯片中使用了硅神经元。如果我的认识是正确的,他的"神经元"是由模拟电路而不是数字处理器组成的。据我所知,米德的博士生马霍瓦尔德开发了一种用 CMOS 电路模拟离子通道的硅神经元。我不确定包阿汉是否使用了马霍瓦尔德的硅神经元或其变体,还是他自己开发了一种新型神经元。使用这样的"神经元"可以减少能量消耗。如你所知,我对技术不熟悉,我很难阅读关于微电子学的原始研究论文,你可以告诉我是否正确。

当然,我也注意到了欧洲的难民潮以及可怕的恐怖袭击,这些可以在电视、报纸和网站上频繁地听到或读到。你所描述的这些事情是严重的社会问题,远远超出了科学和技术的范围。纯粹的科学技术手段不能解决这些问题。我们都已年老退休,对此无能为力。☹

我同意你对 AI 对社会的影响以及你们国家在这场竞赛中的光明前景的乐观看法。虽然很多名人担心人工智能,并宣称 AI 机器人可能会奴役人类。从长远来看,要想确认或驳倒这样的预测是非常困难的,但至少在可预见的将来,我对这样的可能性表示怀疑。在我看来,人工智能领域中几乎所有的成就都属于弱人工智能,没有一个产品具有其主观世界。没有一种人工智能产品有其自己的意图、愿望和意志。它们只是为人类服务的工具。更有甚者,正如我们已经讨论了这么长时间,我们甚至不明白人类自己如何会有内心世界,意识是如何涌现出来的,或意识涌现的充分和必要条件是什么。对于这些问题,我们还毫无所知。我们怎么能期望在不久的将来就能创造出一个拥有自己内心世界的机器? 因此,AI 至少在可预见的将来只会成为帮助人类的工具。

最近,我读了一些关于自由意志的东西,因为这仍然是一个悬而未决的问题,当然在读后还不可能得出任何结论。如你所知,我对哲学知之甚少,我不确定我的理解是否合理。在一开始,我想就这个话题发表一些看法,但是,这封信已经太长了,如果你愿意,我可以在下一封信中再来谈谈这个话题。

一如既往的良好祝愿。

凡及

专门用于需要翻译绝对准确的场合。

……

长期为联合国和世界卫生组织做翻译的皮龙（Claude Piron）说过，机器翻译至多也只能做到使翻译工作中较容易的部分自动化；更难和更耗时的部分通常需要进行广泛的研究以解决源语言文本中的歧义问题，同时解决目标语言中有关语法和词汇最好的表达方法：

为什么翻译者翻译 5 页就需要一整天的工作时间，而不是一两个小时？……平均说来约 90% 的文字是比较简单的。但不幸的是，还有另外的 10%。这部分需要额外工作 6 小时才能译出。有些含糊之处必须要搞清楚。例如，源文本的作者是一位澳大利亚医生，他引用了"二战"期间公布的在"日本战俘营"（Japanese prisoner of war camp）中的流行病的例子。他谈论的是一个美军看押日本俘虏的战俘营呢，还是日军看押美国战俘的战俘营？"日本战俘营"一词在英语中有两种意思。因此有必要进行研究，或许还要打电话到澳大利亚去了解真相。

理想的深层方法需要翻译软件自己进行调查研究以消除这种歧义；但是这需要比现有 AI 更高级的 AI。浅层的方法只是对皮龙提到的那种模棱两可的英语短语进行猜测（也许是根据某一特定语料库中哪种类型的战俘营提到得更多），但是这常常容易猜错。根据皮龙的估计，如果"向用户询问每个含糊不清之处"，浅层方法只能使专业翻译人员的约 25% 的工作自动化，而更难的 75% 仍然需要由人工来完成。

机器翻译的主要缺点之一是：它不能像翻译标准语言那样确切地翻译非标准语言。

因此，问题在于机器不懂，它不知道正在翻译的文本的意义，只是基于语法规则逐字做字面上的翻译，或者从数据库找出尽可能接近的段落。无论如何，在我看来，机器只是做一些符号处理，而不是理解和用目标语言表达相同的含义。机器没有内心世界！这就是问题。我对吗？

让我对包阿汉的工作感兴趣的是他来自米德的实验室，以及他在仿神经结基本上没问题,可以尽可能地翻译,尽量少用文化典故,细微差别,文字游戏,甚至隐喻。这个国际体系所达到的相互理解的水平是非常宝贵的,但是要付出一定的代价:一些必须要做的思想显然需要非正式的比喻和想象的调整,每个人都打 closed 着封闭思想的路障这本书里的诡计,如果其中一部分难以翻译的话,那么我就只能期望一方面是大师级的翻译人员,另一方面是世界科学家英语流利程度的增长。"[1]

我也试着用必应和百度的翻译器进行翻译,其结果也是类似的。[2]

除了我加下划线之处以外,上述翻译似乎没什么问题,但是那些加了下划线的地方,有的句子或短语本身就不通。有一些如果只从语法的角度来看像是一个完美的句子,但你无法理解句子表达的是什么意思,或者其意思是错的。不正确的翻译大约超过了总数的二分之一。由于不知道这些翻译器是如何工作的,我无法猜测这些错误是如何造成的。为了更好地理解,我参考了维基百科中的"机器翻译"条目,[3]其中说道:

这就是机器翻译所面临的挑战:如何对计算机编程,使之就像人那样能够"理解"文本,并且"创建"一种新的目标语言文本,使其"看起来"就像是由人写的一样。

当前的技术还做不到在一切场合都可以应用机器翻译。虽然机器翻译的工作速度更快,但如果没有人参与,还没有哪一种自动翻译程序可以接近人类翻译的质量。然而,它能做的是提供原始文本的、一般性的、但不够完美的近似,从而获得它的"主旨"[一个称为"取得主旨"(gisting)的过程]。这对许多目的来说已经足够了,用它可以充分利用翻译人员的有限和宝贵的时间,使之

---

[1] 谷歌翻译器网站地址:https://translate.google.cn/

[2] 必应和百度翻译器的相应网址:https://www.bing.com/translator/? mkt=zh-CN
http://fanyi.baidu.com/

[3] https://en.wikipedia.org/wiki/Machine_translation

Ⅲ-008 凡及

C.科赫在他为《意识探秘》一书中文版序言中有关翻译所说过的话,这些话我在以前的信中提到过。现在我想再次引用它:

> "翻译任何文字都是一件极耗心力的工作,它需要译者首先理解纸面上文字背后的含义,然后才能将其组织润色成另一种语言。在一份成功的译著里,你应该感觉不到有译者介身其中——原作者与读者就像在直接进行交流一样。"

因此,按照 C.科赫的意见,翻译好坏的关键是译者能否充分理解原作,而不只是基于语法规则或统计分析,后者可能正是当今机器翻译所使用的方法。

在今天这样的人工智能热潮中,有时我会听到类似下面这样的话:"机器翻译非常好,翻译人员很快就会失去工作。"事情果真如此吗? 我不知道,但我现在至少有些怀疑。理解是内心世界的事,而不仅仅是符号处理。在我看来,现在的机器翻译只做符号处理,而不涉及理解。因此如果碰到意思复杂一点的文字,机器翻译就会出错。为了测试,我使用号称世界上最好的翻译之一的谷歌翻译器将你引用的丹尼特的引文原文翻译成中文,以下就是翻译的结果[1]:

> "首先,在研究期刊中有一种科学写作的文化,它确实有利于 — 以一种最少的繁荣,修辞和暗指的方式对这些问题进行非个人化的,简洁的表述。对于无情的作为我的博士研究生之一,神经解剖学家 J. Z. Young 在 1965 年写信给我,反对我在牛津大学的博士论文(哲学而不是神经解剖学)中有些奇特的散文,国际科学语言,我们应该用母语为英语的人写一些可以通过'一个有好词典的汉语(病态)'来阅读的作品,这个自律的结果是不言自明的:不管你是中国人,德国人,巴西人甚至是法国人,你都坚持用英语发表你最重要的作品,英语

---

[1] 在我们的英文原稿中,由于卡尔并不懂中文,所以译者把该段引文先用谷歌翻译器从英文翻译成中文,然后再把中文译文反过来翻译成英文,并与原文相比。但是在中文版中就没有这样的必要了,因此译者在此处作了点改动。——译注

Ⅲ-008 凡及

（机器翻译；AI 在可预见的未来都只是一种工具）

2017－03－20

亲爱的卡尔：

我对你在我们讨论的所有事情上都能有深刻见解的能力很是"嫉妒"，这次是关于反讽和幽默。你发现"反讽也是人类在童年时代学会的最后一件事"。你所举的打破另一个玻璃杯的孩子的例子很有趣，但从中我们确实也能学到很深刻的东西。我忍不住笑了起来。

虽然诚如所言我不是丹尼特的粉丝，但我不得不赞赏他关于科学写作和翻译的评论。他的话正是我作为"有一本好词典的耐心的中国人"的体验。当我还是一名大学数学专业的学生时，阅读英语甚至俄语的基础数学课本并不难（当时我们在大学里外语只学俄语，幸运的是，我在高中学的仍然是英文而非俄文）。当然，这里我只是指纯数学教科书的语言很容易，而不是指内容本身。数学课本中最难理解的部分是公式，但它根本不需要翻译，而需要读者事先有相应的背景知识。它的语法很简单，总词汇量小，数学术语的英语（或俄语）和汉语词汇之间几乎一一对应，也几乎没有歧义，极少用文化典故、细微差别、文字游戏和隐喻。我想不出还有比数学课文更容易翻译的材料了。这可能是最适合机器翻译的文本。

毕业后，我进入了一个新的领域 ——生物控制论，情况完全不同。词汇非常丰富，文本的语法很复杂。所使用的词经常有许多不同的含义，其确切含义取决于上下文，这里所说的上下文不仅限于一个句子，也可能是一个段落、一章、整本书，甚至作者的文化背景。我无法单靠语法分析和简明的英汉词典来读懂书！经过这么多年的实践，现在我可以翻译得比以前好多了，我完全同意有这样先进的技术。

那里的人们似乎还与创造一个更加严肃的项目信息技术谷（Cyber Valley）有关。[1]

当然，这还不足以构建大型企业，它还不是谷歌、脸书、京东或腾讯，但至少是一个令人鼓舞的开始。

所以我还是有希望的，我们还会有足够的养老金，全球邮轮业也会依然红火。☺

我想知道如何从中国的角度来看待欧洲和德国的发展及其前景。

祝好，晚安上海。

卡尔

---

[1] http://www.cyber-valley.de/en

是软件公司 SAP SE[1]。在亚洲,我们首先在日本看到了很多软件公司,接着是韩国,而现在轮到了中国以惊人的速度发展。我在美国的一些朋友曾一度非常有信心地认为,美国依然可以在很长的时期里保持世界技术超级大国的地位,但是现在不得不怀疑起积极帮助中国成为世界工厂是否是个好主意。

现在人工智能、自动化和机器人技术正处于这一切的中心,每个人都在问最终技术的下一个热点会在哪里。很多人相信这将是中国,美国正在失去优势,欧洲有成为这场竞争中的输家的真实危险。但像在经济中总是会发生的那样,这很难预测。我记得在 20 世纪 80 年代也有过类似的讨论。那时看起来好像日本要超越美国和欧洲。但这并没有发生,主要是因为日本经济由于其内部问题而几乎停滞不前。

但有一件事是肯定的:人工智能、机器学习和机器人将再一次使许多行业发生天翻地覆的变化,其影响甚至可能比 40 年前的微电子技术还要大。虽然很多人又一次担心那些在夕阳企业中工作的人会失业,但在这个问题上我较为乐观,因为类似的情况我以前就看到过了。

在所有这些我们看到过的创新阶段,有更多的新产品、服务和就业机会被创造了出来,甚至多于消失的岗位。想想个人计算机、互联网和智能手机的例子吧。然而,问题是谁将领先,谁能够负担得起将他们的退休人员送上豪华游轮去世界各地?

要是在一年前,我会说欧洲确实有落伍的风险,并且还告诉一些雄心勃勃的企业家朋友,如果他们真的想在技术上领先,那么应该想一想是否要到美国去发展的问题。

但有一些指标显示德国可能有机会参加人工智能竞争。至少我能看到有新一代的研究人员和企业家真正想要为之奋斗。当我在这里看到 https://deepart.io/时,我还以为它是谷歌的产品。

但其实这并不来自谷歌,而是来自蒂宾根大学(University of Tübingen)的一个很小的衍生公司,离我的出生地不远。我以前不知道德国在深度学习方面

---

[1] https://en.wikipedia.org/wiki/SAP_SE

Ⅲ-007　卡尔　　　　　　　　105

声——没有人喜欢听到听起来不舒服的真话。☺

实际上,年轻一代的主导思想似乎是,西方优越的经济形势无论如何都由其基于现代科学和技术之上的优势产业来保证。但是时代到了还需要考虑其他参数的时候,例如对生态的影响、公平贸易,以及通过合理使用人力和物力资源迈向更美好世界所需的其他内容。

现在,欧洲的上空布满了乌云,正面临着一场有关非常基本的生活问题的混乱讨论,讨论的主题包括如何在欧洲的家门口阻止战争和恐怖主义,以及如何应对大量难民。议程的范围从可能的军事干预和打击中东恐怖主义到建筑隔离墙以阻止难民潮,到在欧洲接收更多难民并将他们融入现代工业世界。

因此,就欧洲应该如何置身于一个更加困难的现代世界中,这里正在进行着一场激烈的辩论。当然,年轻一代最关心的是走哪条路。

这也与工业 4.0 的启动以及人工智能和机器人在这种向新兴产业过渡的过程中发挥什么样的作用有很大关系。这种产业虽然可能不像库兹韦尔的奇点那样激进,但无论如何都还是非常令人兴奋的。

在最近的这段时间里,我和一些年轻的研究人员和企业家朋友越来越多地讨论这个问题。尤其是年轻的企业家朋友对经济繁荣的源泉非常了解,他们担心欧洲尤其是德国可能会因这一转型而受到损害,因为我国大工业的骨干依赖于汽车和其相关的供应商这样一个生态系统。他们目前仍然做得很好,但随着向电动汽车转换,以及已经出现了像特斯拉这样的竞争对手,有很多理由值得担心。

问题在于,正如一位美国朋友(略带讥讽)所说,德国"正在制造世界上最好的 19 世纪产品"。他想说的是我们不应该错过创新之舟。而我们的大工业界的人士确实并非因为最具创新性而闻名。

这种情况在中型企业中要好得多,在这些企业里我们拥有众多全球技术的领先者。德国缺乏的是对整个工业的新支柱,就像德国在 19 世纪所创建的那样,今天这样的支柱企业正在美国和亚洲尤其是在中国崛起。

第二次世界大战之后,德国在技术领域只创造了一项重要的新工业,那就久前还受到了极为热情的欢迎,现在却变成了战争和毁灭的噩梦。数百万难民陆续并正在前往欧洲以逃避威胁和悲惨的命运,并以德国作为他们的首选目的地。

生活在和平与繁荣中的这代欧洲人,在他们一生中第一次遇到了这样重大的恐怖和威胁,这些他们以前只有从书本、电影或者大学研讨会中才得知。一夜之间,人们意识到我们幸福的生活条件的基础是何等脆弱,而且被他们认为理所当然的一切又多么容易濒于险境。人们还认识到,帮助其他需要帮助的人是一回事,但要为如此众多的有着完全不同文化的人提供食宿和融入社会是另一回事。许多人也开始清楚认识到,我们以前一直在关注的重点其实往往只不过是一些微不足道的官僚主义问题,或者有些人所称的"奢侈问题(luxury problems)"。

特别是在英国选民选择退出欧盟,美国选民现在又选出了一位在其政纲上宣传"美国优先"的总统之后,许多人都在考虑欧洲在今后的经济和世界政治中所扮演的角色。与此同时,这又对欧盟及其行政机构造成压力,引发许多成员国的巨大离心力,导致政治上的不稳定,而这些在不久以前都还被认为是不可能发生的。

当今德国的许多人都把日益增长的经济标准,慷慨提供优质的食品、商品和各种服务视作当然。同时,每个人都可以得到高质量的医疗和牙科护理,几乎可以免费获得高质量的教育。而且工业生产力高到可以让普通人过上被其他许多地方的人视作是天堂般的生活。

世界各地的休闲和旅游业由于欧洲的繁荣而兴旺起来,尤其是德国给其退休的公务员和教师提供了非常好的待遇,游轮公司几乎无法提供足够多的船只将其运送到全球最具异国风情和舒适愉快的地方。这些地方大多数其他人只能从好莱坞电影中看到。

20世纪初的德国是以研究、科学和技术为基础的现代工业的发源地,在那里勤奋的人民遵循众所周知的普鲁士纪律,努力改善自己的生活水平。

按照一位广受欢迎的哲学家的话来说,德国现在最显著的成就就是让奢侈民主化。不过,在他说了此话之后,他就不再那么受欢迎了,并且招来一片责骂与你所说类似的事情,但他的分析没有你那么广泛。他在12月见到了IBM的莫德哈,并且对真北芯片和莫德哈个人印象深刻。我周围的其他人对真北芯片或其他专用AI芯片是否能在市场上取得成功则远没有那么肯定(类似于你引用的杨立昆的意见)。你下面的这段话又一次讲到了问题的关键点:

"至于这样的系统是否能发展成为新一代计算机,这将取决于有多少人愿意投入时间和精力来学习使用这种新的生态系统。"

说得好! 在过去的50年中,这一直是IT发展的关键因素,这一因素比技术本身及其可能带来的好处还要重要得多。

至少在目前,年轻的技术精英和他们的AI初创公司并不太关心这些新奇的设备,并且依然按照谷歌铺就的TensorFlow道路奋勇前进,他们用一大堆GPU来计算它们的神经网络。谷歌作为公司来说在技术上可能并非最强,但它在营销方面极为擅长。

但在这方面,没有什么能一劳永逸,因此光靠营销可能很快就不够了。

不管怎么说,你的调查都鼓励我做更多的研究,并在我能够给你有说服力的回答之前,我还要向我年轻有为的朋友们寻求建议。特别是我必须仔细看看由米德的学生包阿汉设计的这款芯片,这是我以前没有注意到的。

谢谢你告诉了我这一点!

我是米德的忠实粉丝(尤其是因为他对哥本哈根正统量子理论的有力反对),所有来自他那一派的东西都值得仔细观察。

我也很喜欢你关于《神经元》杂志的特刊上有关大型脑计划的报道,至少到目前为止还没有让我们两人落到傻瓜的地步。至少当下还没有。☺

当我想到所付出的巨大努力和流入这些计划的巨额资金,并将其与可能的成果进行比较时,我就乐不起来。

但正如上面所提到的,我们现在面临更严重的情况,而不只是可能把10亿欧元浪费在一个前途未卜的HBP身上。

你一定听说过,在"阿拉伯之春"以后,欧洲陷入混乱。"阿拉伯之春"在不对我在我的牛津大学博士论文(哲学而不是神经解剖学)中所用的多少带有点想象色彩的散文体。英语已经成了科学的国际语言,我们这些以英语为母语的人在写东西时必须要让一位'有耐心的中国人(原文如此),靠一本好字典'也能读得懂。这种自律的结果不言自明:不管你是一位中国的、德国的、巴西的、甚至是法国的[1]科学家,你都坚持用英语发表你最重要的作品,所用的英语都是最必要、最容易翻译的,尽量少用文化典故、微妙的差别、文字游戏、甚至隐喻。这个国际体系所达到的相互理解水平是非常宝贵的,但是也要付出一些代价:有些想法如果想要说明清楚,就需要用一些非常规的隐喻,巧妙地应用隐喻和想象以引起读者头脑里的共鸣,而非直白的字面描述和理性解释。如果其中有一些不那么容易翻译,那么我就不得不一方面把希望寄托在技艺超群的翻译家身上,另一方面则寄希望于全世界科学家的英语流利程度越来越高。"[2]

顺便说一句,"耐心的中国人"后面的"(原文如此)"是原来就有的。☺

丹尼特说明了科学翻译工作的难度,他支持了你之前说过的立场。因此,我希望你会赞赏丹尼特所说的话,尽管我知道在其他方面你并不是他的粉丝。

非常感谢你给我发来了有关仿神经结构芯片领域发展的出色总结。收集这些信息本应是我的职责,但我必须说,在过去的几个月里,我一直忙于其他事情,所以忽略了这个领域中许多我本该注意的事。和往常一样,在这些不断变动着的 IT 工程领域中,一直会有很多事情发生,如果你有 2 个月不去阅读最新消息,你可能就会赶不上事态的发展。

所以我非常高兴和感谢你的所作所为,这给我留下了很深刻的印象。你总是坚持说你不熟悉技术,但你是一位优秀的研究人员!你发现的正是专家们现在正在讨论的内容。我在芯片制造业研究部门工作的一位朋友也向我报告了

---

[1] 法国人以对自己的语言为傲而闻名,认为法语是所有语言中最美丽也最严谨的语言,流传的一种说法是法国人不太愿意用其他语言。因此这里要强调"甚至法国的科学家"也用英语写作他们最重要的作品。——译注

[2] Dennett D C. Intuition Pumps and Other Tools for Thinking [M]. New York: W. W. Norton & Company, 2013, p.11.

的元事情'。"[1]

而幽默和讽刺就属于这种"深入元层次"。它是我们的脑在进化过程中最后添加到组织架构中的层次之一。反讽也是人类在童年时代学会的最后一件事。脑忙于掌握基本的层次,因此没有剩余的能量(或结构)去处理心智中的额外(元)层次。要经过很长的时间之后才能学会处理一个以上的层次,并掌握额外的幽默和讽刺层次。甚至有很多成年人始终都没有学会这一点,有时有些达不到这种元层次的人却在基本层次上显示出卓越的智能。

为了更好地认识我小外孙脑中心智的发育状况,我找了些有关儿童和发育心理学的书籍来读。这些书籍描述了这种情况如何分阶段发生。我所学到的是,即使是最聪明的3岁或4岁儿童,他们也不懂得讽刺,我们对此不必感到惊讶。例如,当他们不经意把桌上两个杯子中的一个扔掉时,你告诉他们:"你为什么不把另一个杯子也打碎了呢?"不要指望他们会像成年人一样笑起来。他们很可能真的把你的话(语含讽刺)当真了,并把另一个杯子也给打碎了。

"直觉泵"很有趣,虽然他有关模因的论述令我有点失望。丹尼特对模因的阐述,比道金斯和后面布莱克莫尔在讲及此概念时所说的要更有帮助和更清晰。但是他们中没有一个人认识到建立"规章"的重要性,而格伦在多年前早就看到了这一点。但是,我想在另一封信中再来讨论这个问题,要是你愿意的话。

在这里,我想向你说说我在丹尼特的这本书中找到的另一处发现,这会引起你作为专业翻译人员的兴趣。

"首先,在研究期刊中有一种科学写作文化,这种文化喜爱(实际上是坚持)用第三人称的角度简明扼要地进行表述,而尽可能少修饰、修辞和暗示。在最严肃的科学期刊中这种严肃的干巴巴的叙述风格是有充分理由的。神经解剖学家扬(J. Z. Young)是我的博士论文的评审人之一,他在1965年写信给我,反

---

[1] Dennett D C. Intuition Pumps and Other Tools for Thinking [M]. New York:W. W. Norton & Company, 2013, p.9.

Ⅲ-007 卡尔

（深入元层次；科学写作文化；仿神经结构芯片；
新工业革命）

2017－02－05

亲爱的凡及：

感谢你亲切的节日问候。欢庆鸡年的到来！

我祝愿你和你的家人来年万事如意。你已经向我描述过春节有多么热闹而忙乱，我希望你们开局大利，并且一切顺利。也许到处都很忙乱，但并不是所有的忙乱都像庆祝新年那样令人愉快。

我稍后就要来谈另一种混乱，但让我先告诉你，对于你对我下面问题的回答我有多么赞赏。我在上封信中问你是否真心称我的上封邮件为"深信"，还是语带讽刺。

其实我也希望你会矢口否认语带讽刺，就像你信中所说的那样，这不仅仅是因为我喜欢被认真对待（有谁不喜欢这样呢），而且还因为这让我有机会告诉你我在丹尼特所著《直觉泵和其他思维工具》（*Intuition Pumps and other tools for thinking*）一书中的一个有趣发现。这是我在寻找他对"模因"（Memes）这个概念有没有什么新贡献时偶然发现的。

举例来说，"深入元层次"（going meta）原则是我们的心智在进化过程中发展出的工具之一：

"关于思考的思考，谈论谈话，对推理进行推理。元语言是我们用来谈论另一种语言的语言，元伦理学是对伦理理论的鸟瞰检验，正如我曾经对道格（Douglas Hofstadter）说过的那样，'无论你做什么，我都可以做它际上是一个模拟视网膜头三层解剖结构（感光细胞、水平细胞和双极细胞）和功能原理的芯片。这一人工视网膜能够产生和生物视网膜类似的输出信号。该实验室后续开发出来的人工视网膜和人工耳蜗，成为仿神经结构工程在医学上得到实际应用的第一批成果之一。他们的另一项成果是用 CMOS 电路直接模拟离子通道，并在此基础上构建出硬件的"硅神经元"。他们的这些工作成为仿神经结构工程的发轫之作。

和精力来学习使用这种新的生态系统。虽然仿神经结构芯片的前景看起来很光明,但我不确定它是否能成功地成为新一代计算机。我不禁想起日本的第5代计算机计划。从理论上讲,这个想法并没有错,他们也开发出了一些生态系统,但是没有多少人愿意使用这种新型计算机,最终项目失败了。仿神经结构芯片最终是否能发展成为新一代的计算机,还是只是又一个日本的第5代计算机?我不知道。然而,如你所知,我对技术不熟悉,你可以告诉我,我的印象是否合理,并告诉我你的意见。

好吧,我的信已经够长了,在我的脑海里,你那里的节日音乐已经响起,我想我应该停下来了。

圣诞快乐,新年快乐!

凡及

背景专栏Ⅲ F7.1

米德和仿神经结构工程

美国计算机科学家米德是大规模集成电路的先驱,也对脑如何进行计算的问题深感兴趣。他说道:"我对于动物视觉系统的机制越来越佩服。我老是对自己说:'我永远也想不到这一点,但是这确实是一个好主意。'"这样他就在20世纪80年代末提出了仿神经结构工程(neuromorphic engineering)的概念。最初这一概念指的是采用由模拟电路构成的超大规模集成电路(VLSI)模仿神经结构来实现相应的神经功能。现在这一概念已不限于模拟电路,而可以是用模拟电路、数字电路或混合电路构成的 VLSI 甚至软件系统来建立神经系统的模型。

20世纪末,正是通过学习视网膜的神经机制,米德指导的一位博士生马霍瓦尔德[Michelle(Misha)Mahowald]研制出了"硅视网膜"。这种硅视网膜实

"它试图在今天的无机硅技术范围内尽可能地接近脑。"

这似乎表明如何在"脑样"(brain-like)和"脑启发"(brain-inspired)两者之间进行权衡仍然是一个问题。对真北系统也有其他方面的怀疑意见。2014年，深度学习的三位主要先驱之一，脸书的人工智能研究组负责人杨立昆声称，该芯片难以使用卷积神经网络的深度学习模型运行图像识别等任务。然而，在2016年的后续文件中，IBM表示它能够用其仿神经结构芯片快速而准确地运行卷积网络。无论如何，我认为真北系统很值得注意。

斯坦福大学的包阿汉（Kwabena Boahen）是仿神经结构工程之父米德（Carver Andress Mead）的学生，他构建了一种称为"神经格"（Neurogrid）的类似芯片。也许他用的神经元模型是米德实验室研发的"硅神经元"的改进版，这种神经元模型在物理上使用 CMOS 电路模拟了一些离子通道。尽管他的神经元比莫德哈用的要复杂，但仍然比马克拉姆用的简单。我对包阿汉的工作也很感兴趣。米德实验室有使用神经芯片解决实际问题的传统，如开发人造耳蜗和人造视网膜，但不是人造全脑。让我们来看看包阿汉下一步会做什么吧！

无论如何，在上述所有项目中都有一个共同点，它们都使用脉冲发放神经网络(spiking neural networks)来节省能量，这受到了脑机制的启发。所有这些系统都是由硬件直接实现的，而不是在传统计算机上进行仿真。至于它们使用的神经元元件，它可以简单到像点神经元那样，或者复杂到像具有类似模拟离子通道的硅神经元那样。哪种更好？我不知道。尽管对于应用目的来说，要把模拟神经元的准确程度提高到与马克拉姆所做的那样，似乎并不切合实际。

莫德哈在他 2016 年发表的一些论文中报告说，他们开发了一个硬件和软件生态系统，包括模拟器、编程语言、集成编程环境、算法和应用程序库、固件（firmware）、深度学习工具、教学课程和对真北芯片的云支持。由于芯片架构是全新的，他们必须从头开始编译整个软件系统。他们甚至建立了一个虚拟 SyNAPSE 大学，教导他人使用他们的系统。在我看来，这样的系统可能对实时的模式识别、动作控制以及在节能成为关键问题的情况下具有应用前景。至于这样的系统是否能发展成为新一代计算机，这将取决于有多少人愿意投入时间除了你在上封信中提到的贵国海德堡大学以及其他单位的研究人员开发的 BrainScaleS 系统之外，HBP 下的另一个"脉冲发放神经网络架构"（Spiking Neural Network Architecture, SpiNNaker）系统也已在曼彻斯特大学开发。另外，美国也有几个实验室致力于开发类似的芯片。也许最著名的是莫德哈开发的 IBM 真北（TrueNorth）芯片，这种芯片有超过百万的"神经元"和 2.56 亿个"突触"。莫德哈就是马克拉姆想要把他"倒吊起来"的那个人。这还只是美国国防高级研究计划局（DARPA）支持的"仿神经结构自适应可塑性可扩缩电子系统"（Systems of Neuromorphic Adaptive Plastic Scalable Electronics, 简称"SyNAPSE"）项目的一部分。

我想你已经注意到最近莫德哈和他的同事发表了好几篇有关真北芯片的论文。真北芯片运行所需的功率仅为 70 mW，比完成同样任务的冯·诺伊曼计算机低四个数量级，速度为每秒每瓦 460 亿突触操作。这里的关键点在于处理单元和存储器不像冯·诺伊曼计算机那样分开，因此节省了在它们之间连续传输所需的能量。此外，其元件输出的是脉冲，从而节省了冯·诺伊曼计算机在高低电平频繁切换期间消耗的巨大能量。有一个演示表明，这样的系统可以实时识别不同的对象，例如行人、骑车者、公共汽车、汽车等。[1] 相反，马克拉姆的蓝脑不能用于执行任何实际的智能任务，还耗能巨大。当然，莫德哈的神经元是一种整合—发放神经元，比马克拉姆所用的神经元模型要简单得多，这使马克拉姆非常愤怒，并把莫德哈的话称为"耍花招"或"欺骗"。所以莫德哈后来要强调：

"让我们把事情说清楚：我们还没有建立起人脑或任何脑，我们只是制造了一台受脑启发而来的计算机。这台计算机的输入和输出都是脉冲。从功能上来说，它只是将输入脉冲的时空流转换为输出脉冲的时空流。"[2]

然而，在另一场合，他又说道：

––––––––––––––––––––

［1］ Service R F. The Brain Chip[J]. Science, 2014, 345(6197)：614 – 616.

［2］ http://www.research.ibm.com/articles/brain-chip.shtml

非常感谢你推荐我看电影《机械姬》，这我以前还没听说过，我要尽力找来一看。你对我非常了解，我一直对主观性问题感兴趣，我认为这个问题无论对神经科学家来说，还是对人工智能工程师来说都是最困难的。我相信神经科学家早晚可以阐明行为的神经机制，并且机器也总有一天可能会通得过图灵测试。即便如此，我们可能仍然不知道主观性是如何产生的，也许仍然不得不把它当作一种不可还原的涌现性质。一台机器可以有表情和智能行为，但我们仍然不知道它是否有感受、意图、欲望、意识等，我们仍然不知道它是否有一个内心世界。也许它有，只是我们不知道；也许它只是一个无魂人。对于动物来说，"他人的心智"仍然是一个没有解决的问题，也许从严格的意义上来说，这个问题对我们人类来说也没有解决，那么，我们如何期望在可预见的将来可以对机器解决这个问题呢？

在你的信中，你还提到了大科学、仿神经结构芯片和其他有趣的话题。我也想谈谈这些问题。

在上个月 2 日出版的一期《神经元》(*Neuron*)杂志的特刊中，各国脑计划的组织者介绍了包括欧盟 HBP、美国 BRAIN 计划、中国脑计划、日本 Brain ／ MINDS 以及澳大利亚、加拿大、韩国等国家在内的脑计划。在该刊上也发表了一些相关主题，包括由 C.科赫和琼斯(Allan Jones)署名的题为《神经科学中的大科学、团队科学和开放科学》(*Big science，team science，and open science for neuroscience*)的论文。这些论文的内容似乎支持了我在以前的电子邮件中讨论过的想法。我想我们无需由于这些新进展而从根本上改变观点。

你上次介绍的"莫拉韦茨悖论"非常有趣，我不知道这个名字，虽然我从霍普菲尔德(John Hopfield)在 20 世纪 90 年代的一篇论文中也读到过类似的想法(对不起，我忘了论文的标题)。我记得直到几年前，人们总是使用模式识别和动作控制作为例子来说明计算机的缺陷。人们曾经说，即使是 3 岁的孩子也可以很轻松地完成这些任务，但对于计算机来说却很困难。但是，由于机器学习的快速发展，情况似乎正在发生变化。尽管如此，为了完成这些任务，计算机需要消耗大量的能源，而且这样做也很耗时。这可能就是许多实验室争相开发仿神经结构芯片的背景。

中写道：

"人们曾经给出有关智能的许多不同形式的定义，其中包括逻辑、理解、自我觉知、学习、情感知识（emotional knowledge）、计划、创造力和解决问题等的能力。也可以更普遍地把智能描述为感知或推断信息的能力或倾向，并将其作为知识应用于在某种环境中采取适应性的行为。"[1]

它还列出了其他几个定义，其中没有哪一个是精确并能得到公认的，因为没有人能下这样的定义。上面的话列出了许多要素，也包括学习和解决问题的能力。也许"一些自动化行为或以前学得的技能"仍然应该被视为智能行为。在主流人工智能中，人们也只是让机器去做某些这样的技巧。事实上，在王培对智能所作的分类中，最低一级的智能也是没有学习能力的智能。但是，我在上一封电子邮件中提出的问题仍然存在，就是哪种技能才可以被认为是智能的。显然，并非所有的技能都可以称为智能，否则，泥蜂也有智能。如果以前只有人脑才能做到的技能是智能的，那么其他动物就都没有智能。所以问题仍然存在。当然，从应用的角度来看，只要机器能够解决我们要它解决的问题，这台机器是否可以被称为智能并不重要。正如你在上封信中所说："正如在学究气时代的学者们试图去做的那样，通过'正确'的定义来研究任何自然现象都是徒劳的。只有当用经验手段来检验理论时，定义才会多少有点用。"也许目前我们应该对智能有一个粗略的理解感到满意，就像我们对意识概念所做的那样。克里克关于意识曾表示，即使没有公认的定义，我们仍然可以研究意识。至于定义问题，还是让哲学家去研究吧。

我同意你的观点，即学习能力是智能的一个高级要素，不管它有多重要，它不可能是智能的全部。尽管如此，根据学习的能力，王培将智能划分为四个不同的类别的想法对我来说仍然很有趣。这里面似乎确实有些东西，但我无法解释清楚。

---

[1] https://en.wikipedia.org/wiki/Intelligence

Ⅲ-007　凡及

（智能的定义；他人的心智；仿神经结构芯片）

2016－12－23

亲爱的卡尔：

我非常高兴和兴奋地阅读了你极好的来信。

我和你有同样的感觉。虽然我们知道"由于崇拜一个人而对他所说的一切都照单全收是不对的，但否认他所说的一切也不对"。并且"我将尽量避免这个错误，但要完全避免也很难"。正如中国的一句老话"知易行难"，我也是如此。是的，我们必须尽最大努力公正地看待其他人的想法，即使他们以前有许多观点是我们所不同意的。我们不应该对任何人抱有偏见，尽管有时很难做到这一点。

非常感谢你告诉我帕金森定律。尽管我不属于"年轻一代"，但我很遗憾地告诉你，对我来说，与"帕金森"相关的唯一内涵也是以此命名的疾病。我从来没有听说过帕金森定律，但我们常常可以发现由这一定律所描述的现象。

我很抱歉为了想表现一点幽默感而把你的上一封信称为"深信"，结果却让自己出丑了。你知道我非常欣赏"深蓝""深度学习"和"深心"，我只是想让我的信件能够生动一点，并且表达我对你信件的赞赏，丝毫没有讽刺的意思！事实上，你在上封信中提到的很多事情让我深思，特别是有关那些我从未听说过的哲学家的事。与你相比，我很遗憾如此无知。☹

非常感谢你指出我下面一段话中的缺陷："我特别欣赏王培将智能与技能区分开来的想法。无论技巧多么复杂，如果不能改进，只是固定在同一水平上，那就不能称之为智能。"尽管我仍然坚信第一句话，但第二句话可能太绝对了，这可能不是王培的原意。你说得对，智能有许多"元素"或方面，不应该把智能与其中的一个元素等同起来，比如说学习能力。我刚刚查了一下维基百科，其主题并不新鲜,自"星际迷航"以来经常会拍一些这样的片子,但我相信你会喜欢电影中处理主观性这一你特别感兴趣的问题的方式。所以如果你有机会的话,我愿意向你推荐这部影片;如果你看不到的话,维基百科会告诉你是怎么回事。☺

　　祝好,晚安上海。

卡尔

以及如何在其中生存的长达 10 亿年的经验。我认为,我们称之为推理的精心考虑过程是人类思想最浅薄的表层,它只是因为有年头更久远、功能更强大,但通常是无意识的感觉运动知识的支持才得以有效。我们在知觉和运动方面都是优秀的奥运选手,在这方面我们是如此优秀,以至于使困难的任务看起来似乎很容易。抽象思维则是一种新的伎俩,可能还不到 10 万年,我们还没能很好地掌握它,但就其本质而言其实并不那么困难,只是当我们这样做时似乎很困难而已。"[1]

我长期以来一直也有类似的想法,并很高兴找到莫拉韦茨的这一引文。我认为这可以与格伦关于像工具和观念(这可以总结为一些规定,这些规定起着导引系统或支持系统的作用)之类的人类发明所起重要作用的想法联系起来。

与海德格尔不一样,格伦在英语世界中被忽视了,但其思想却在"扩展心智"(Extended Mind)[2]的标题下传播了开来。扩展心智涵盖了格伦的一些论点,但不是全部。总体而言,与格伦早在 50 多年前就已经得出的人类学观点相比,扩展心智的说法反而是倒退了一步。

我还没有来得及做完功课,把所有一切都汇集在一起,如果你有兴趣,我可以在下一封信中一试。

关于这个问题,今天我就到此为止,最后提到一个你可能也有兴趣的智力上的游戏。

正当我考虑王培有关哥德尔、冯·诺伊曼和图灵意义下的逻辑推理对实际 AI 系统的重要性的论点时,正巧看了场电影《机械姬》(*Ex Machina*)。这是一部 2015 年的科幻惊悚片,讲的正好就是有意识的人形机器人,以及如何确定某个对象究竟是人还是机器的问题。故事情节是,一位程序员被聘来对一个人形机器人做终极图灵测试。我不想在这里透露太多,因为结果出乎意料。这个

---

[1] Moravec, Hans. Mind Children[M]. Cambridge, Massachusetts: Harvard University Press, 1988, p.15 – 16.

[2] "The Extended Mind" 是 Andy Clark 和 David Chalmers 在 1998 年发表的一篇文章,他们在该文中提出外界环境中的物体从功能上来说也是心智的一部分。他们认为由于外界物体对认知过程起重要作用,因此心智和环境就构成某种"耦合系统"。这样,心智就扩展到了外部世界。——译注

随时间在不断提高。要知道一个经验丰富的脑比一个裸脑更聪明,这一点是很重要的,这不仅因为有经验的脑收集了更多的事实性知识,而且因为它提高了可以更快地掌握事物的能力;这就像一个经过训练的人工神经网络要比一个经验较少的同类神经网络更聪明一样。

另外还有一个与这个智能问题有关的难题,这个问题与自然和人工智能之间的差异也有关系,对此我想在这里提一下。人工智能研究人员和工程师在传统上主要是对智能的高级功能感兴趣,这在人工智能的早期阶段就已经有了一个实际问题,这就是所谓的"莫拉韦茨悖论"。

当像莫拉韦茨(Hans Moravec)[1]、布鲁克斯(Rodney Brooks)[2]和明斯基(Marvin Minsky)这样的工程师在 20 世纪 80 年代开始建造智能系统和机器人时,他们曾预计人类的高级智力功能要比较基本的功能更难以仿效。但令他们惊讶的是,他们很快就了解到情况并非如此。他们认为困难的部分比较容易模仿,而他们认为容易的部分却很难解决。因此就有了莫拉韦茨悖论这个术语。

莫拉韦茨写道:

"让计算机在智能测试或玩棋类游戏时显示出成人水平的表现是相对容易的,但在感知和移动性方面却难于或根本不可能让计算机学会 1 岁婴儿的技能。"

你可以在维基百科页面上看到很好的解释[3]。

莫拉韦茨找到了很好的理由来解释这种奇怪的情况。他指出:

"在人脑巨大、高度发达的感觉和运动部分中编码了的是,对于世界的本质

---

[1] Hans Peter Moravec(1948—),奥地利裔加拿大人工智能和机器人学家,也是一位未来学家。——译注
[2] Rodney Allen Brooks(1954—),澳大利亚机器人学家。——译注
[3] https://en.wikipedia.org/wiki/Moravec%27s_paradox

在这方面要把重点放在智能这一更高级的元素上,而不只是自动行为或以前学过的技能。我之前曾说过,正如在学究气时代的学者们试图去做的那样,通过"正确"的定义来研究任何自然现象都是徒劳的。只有当用经验手段来检验理论时,定义才会多少有点用。因此,我不想对王培的智能定义是否正确进行争论,这一定义是他的理论所需要的(我想这一理论有其本身的价值)。我的观点是,在每一个定义中,你总是会引入某些理论假设。

说到智能要高于纯技能,其目的是说某种智能更高级。在历史上,这通常被用来宣称人类是一种独特的物种,拥有如此卓越的智能,以至于和任何其他物种迥然不同。在宗教历史上也相应地提出过类似的问题,即除了人类之外其他生物是否也有灵魂的问题。

许多意识讨论也是以类似的方式开展的。对许多人来说,动物也有灵魂或机器人可能有意识的想法似乎是难以忍受的。我记得和一位朋友讨论过这个问题,他是一位科学家,也是一位坚定的宗教信徒。我们进行了辩论,他坚持认为人类拥有灵魂,但低等物种则没有。当我不接受他的论据时,他停了下来,看着我说道:"你的话就好像在说,我们只不过是另一种猫或狗一样。"对他来说,这是不可想象的,他将我的立场降低到了荒谬的地步,以此证明我错了。

当然,我并没有将你的论点和王培的观点与这位宗教信徒相提并论。正如我所说的,在智能定义中强调学习的非技能部分在他的理论框架中是合理的。

为了确保我没有误读王培,我再次仔细阅读了你 7 月 14 日的来信,我偶然发现了你引用的王培的一句话,它在我第一次阅读时被忽略了。你写道:

> "'智能'不是'解决具体问题的能力',而是'获得解决具体问题的能力的能力'。"

这句话的重要之处是"获得能力的能力"!

我非常喜欢这一"获得能力的能力"的想法!它远非只是一个定义,因为它隐含着一种理论观念。它接近我自己的观点,即我们总是通过经验不断提高我们的认知能力。这是一个动态的过程,在此过程中我们的学习和理解能力总是基雄(Ephraim Kishon)[1]多年前就提出过一个解决这个问题的办法,同样既巧妙、简单而又便宜。

他建议允许把"博士"和"教授"用作出生登记册中的名字![2] 如果真的能听取这一主意,那真是妙极了。在我的国家中有多少天赋上稍差一点的学者可幸免于不快乐的科学生活,有多少非常有才华的年轻人能够作为快乐的厨师、木匠、机械师和园丁丰富我们的社会? ☺

好吧,让我们看看 HBP 如何在新的和谐规则下发展吧。

感谢你把我的上一封信称为"深信",尽管我不能确定在这夸奖中有多大的讽刺成分。☺

阿迪说,我的(很多)缺点之一是,总是在一些问题上好为人师,即使他们在这个问题上早已知道了,甚至比我自己了解得更好。☺你是一个非常有礼貌的人,永远不会抗议,所以我对任何试图向你解释那些你知道得比我还多的事情表示歉意。

对于我的道歉,我应该补充一点,向别人解释事物的这种方法是增进我自己对事情理解的最好方法。与其他人讨论也是如此,尤其是在有争论的时候。

因此,我非常喜欢你反对我对智能定义所表达的怀疑,这种定义试图将学得的技能与纯粹的学习能力分开。首先我必须承认,当你为王培辩护时,你可能是对的。从你所引用的内容可以看出,他似乎也知道我想要提的问题。我所说的与王培的原文(你可能是从他的中文文章中得到的)没有关系,而是与你在2016 年 2 月 24 日的来信中所说的内容有关:

"我特别欣赏王培将智能与技能区分开来的想法。无论技巧多么复杂,如果不能改进,只是固定在一水平上,那就不能称之为智能。"

我可以理解,为什么王培和霍金斯(在这个问题上你也引用过他的话)等人

---

[1] Ephraim Kishon(1924—2005),以色列作家、戏剧家、电影剧本作家、奥斯卡奖提名影片的导演。——译注

[2] 在西方,名字只能从有限的名字集中选取,并且名在前,姓在后。学术头衔也同样在姓的前面。如果一位姓史密斯(Smith)的人的名字恰好叫 Doctor,那么他的全名就成了 Doctor Smith,正好和史密斯博士完全一样。——译注

（1）"官员想要培养的是下属,而不是对手"。

（2）"官员们彼此为对方创造工作"。

如果帕金森定律中有那些真正想去开创新领域的人要遵从的,那就是其中的第一条。

排除对手或竞争者是最致命的危险,因为它也排除了争论和讨论。这就像在深度学习网络中阻止了纠错(或反向传播)功能一样。

这也就是为什么我不赞同一些人"对从马克拉姆的独裁失败中应该学到什么教训的看法",他们认为要为每个孩子在游乐场中都分配指定一个只属于他一个人的、安全的地方,以避免受到竞争的影响。如果这样做的话,这将会是一个更大的失误,因为其后果几乎可以保证是很少或根本不增加知识。

我看到在这方面有一种倾向,就是像 C.科赫这样聪明而能干的大科学组织者,在统治他默默获得的新帝国时所采取的办法。比好斗成性和过分雄心勃勃的马克拉姆聪明得多,他给每个孩子一份公平的份额,并邀请每个人都和谐地参加聚会庆祝。如果在其职业生涯、在著名杂志上刊登论文以及去好地方参加学术会议方面,有谁想要获得超过可预见的一份,那么这种人还是不要在那些由有毒巧克力制成的大科学城堡中工作吧。当然不会有很多人抵制得了这样的诱惑。120 年前只有少数一些人是这样,今天也还是这样。

好吧,关于这个问题我们已经讨论过很多次了,因为在此问题上我可能过于浪漫主义,除了希望年轻学生做出疯狂之举,不遵循委员会和教授的计划之外,我想不出还有其他更好的解决方案。☺

当然,我不得不承认,总得有人做数据收集、数叶子和其他不那么激动人心的工作。一所大学里如果充满了爱因斯坦和哥德尔,那也将难以生存。今天在德国,每一年中有 50% 以上的当龄青年从高中毕业考大学(其中多数人都能如愿以偿),而在 1900 年这一比例只有 1% 到 2%,所以这也并没有什么不对。

所有的父母都希望他们的孩子能够在社会上出人头地,并且以为取得学术地位就有了保证,但是对这个问题也有其他的解决办法。广受欢迎的讽刺作家控制。它们就像波一样,以某种我们不知道的方式开始和结束,但你不知道其原因和发生的时刻。唯一可以预测的是,正如"帕金森定律"中所描述的那样,这些资金总是随着官僚作风的膨胀而增加,同时却伴随着生产力的下降。帕金森定律是自然科学中与热力学第二定律一样可靠的另一条定律。这条定律简单说起来就是:"扩大要做的工作,用以填满完成工作所需要的时间。"

当年轻一代听到"帕金森"一词时,其唯一含义就是众所周知的无法治愈的可怕的神经疾病。50年前,情况与此不同,每个人都将其与专门研究组织和官僚机构的帕金森(Cyril Northcote Parkinson)[1]联系起来。他观察了官僚主义是如何膨胀起来的,以及官僚主义者如何确保他们的预算和为他们工作的人数稳步增长,他把这些观察结果总结在一系列以"帕金森定律"为名的幽默文章中。[2] 这些文章在当时非常受欢迎。不仅在西方,而且在苏联,那里的人们怀着很大的兴趣去阅读对他们非常熟悉的一种情形的描写。假如一个庞大的官僚机构一直试图控制经济中的所有事情,那么唯一的后果就是他们越是努力,效果就越差。[3]

对我来说,帕金森作品中所显示出来的智慧,要比过去几十年里我们看到的一些在科学上只是略有改进的工作更应获得诺贝尔经济学奖。但是不知道出于什么原因,帕金森定律几乎已经被人遗忘了。我曾经见到过一些拥有经济学和管理学博士学位的现代研究人员,他们从来没有听说过帕金森定律。很难理解为什么这样一个人类洞察力和知识的瑰宝会遭人遗忘。也许是因为它太清晰易懂?也许是因为这对一些研究人员来说就像是一面不受欢迎的镜子,他们不愿意承认他们的研究已经成为一个庞大的官僚主义产物?也许官僚主义会赢,并且成了多数,如果每个人都是吸毒者,谁会抗议吸毒?

正如我所说,对此你无能为力。帕金森发现官僚作风有两大成长因素:

---

[1] Cyril Northcote Parkinson(1909—1993),英国历史学家和作家。其成名作是畅销书《帕金森定律》(1957)。——译注

[2] Parkinson C N. Parkinson's Law: the Pursuit of Progress[M]. London: John Murray,中译本:帕金森.官场病:帕金森定律[M].陈休征.译.北京:生活·读书·新知三联书店,1982.

[3] https://en.wikipedia.org/wiki/Parkinson%27s_law

Ⅲ-006 卡尔 85

曾注意过它,但不够仔细,主要是因为我对 HBP 的印象并不太好,其原因我们在前面已经讨论过了,所以对它的态度也就比较负面。但这种态度如果不说是傲慢的话,至少也是不够恰当的。☹

有时候,我们容易怀疑来自某个特定来源的所有内容,这可能会让我们对一些有趣的事情也视而不见。就像你前几天说过的那样,由于崇拜一个人而对他所说的一切都照单全收是不对的,但否定他所说的一切也不对。我将尽量避免这个错误,但要完全避免也很难。谢谢你的提醒,我对这个问题又仔细想了一下,并启发我得出了一些想法。这些想法可能会和构建某种新概念的模拟-数字混合型机器的项目有关,这个问题我现在正在和我的两位朋友讨论。你还让我意识到有这方面知识的人就在我附近。谢谢你!! 而提醒这一点的竟然不得不通过上海,这听上去不是很滑稽吗?! ☺

当谈到你总结出来的三大科学事件的结果时,我同意你清醒的分析和总结。我也看不到有任何真正的进展,特别是 HBP,当你将其结果与承诺进行比较时,尤其如此。当然我们必须公平对待那些勤勉工作的人,但莱瑟姆的问题"你可愿意花 10 亿欧元来做这些事?"正是我们应该追问的问题。

但正如你在调研的第三部分中所讲的那样,脑研究已成为许多政府和亿万富翁的宠儿,而且有资金不断涌入这个领域。有时候这会形成一股热潮,就像足球场里大家都熟悉的一阵阵欢呼声一样,每个人都必须参与到原子物理、太空计划、癌症研究、纳米技术或任何当时炒作的事情中去。现在轮到了"神经"时代,看起来没有任何一个有自尊心的工业化国家的政府可以承受赶不上潮流的风险。

感觉上好像迈入了一个新的淘金时代(bonanza-time),而且在这种时候,总是有太多的金钱会流入假定的金矿,并且吸引更多的人。因此,有关科学进步,我们总得问自己一个问题:给一名重病患者召来 1 000 名医生是否管用,抑或召来 1 万名医生会更好些吗?

如果你要问我从这些运动中可以得到哪些经验教训或者有什么替代方案,我答不上来。至少没有人提得出什么实际的解决方案。这种运动的产生就像雪崩或海啸那样的自然事件。你可以(在事后)描述和认识它们,但无法预测或

Ⅲ-006　卡尔

（大科学；帕金森定律；智能和技能；莫拉韦茨
悖论）

2016-11-18

亲爱的凡及：

感谢你的来信，特别是你关于脑研究领域大科学项目最新进展的详尽报告。其中有些我是知道的，但是也有些我还没有听说过。在过去几周里，我更关注 AI 的进展情况，目前在 AI 中正在发生很多重要的事件，因此你的总结正合时宜，对我也很有帮助。我发现了这两个领域之间的某些联系，这对我来说非常重要，因此我要特别表示感谢。它与现在世界各地到处出现的"仿神经结构芯片"的新概念有关。实际上，它们并不像其名称那样"神经"，但其中有些也确实含有相当睿智的想法。在这种情况下，我惊讶地发现"仿神经结构混合系统脑启发多尺度计算"（BrainScaleS）[1]计算机硬件中有一些模拟部件，它们专门用于运行 HBP 的脑模拟软件。

在你有关 HBP 的报告的第（3）点中提到了"仿神经结构计算系统"，这使我更仔细地考察了这种机器的架构。我很高兴地发现就在我家附近，海德堡大学（Heidelberg University）和德累斯顿技术大学（Technical University in Dresden）的人员也参与了制造构成这种机器核心的 VLSI-CMOS 晶圆。前一段时间我

---

[1] 仿神经结构混合系统脑启发多尺度计算（Brain-inspired multiscale computation in neuromorphic hybrid systems, BrainScaleS）项目是 HBP 仿神经结构计算平台中的另一个项目。由德国海德堡大学（Ruprecht-Karls-Universität Heidelberg）的迈尔（Karlheinz Meier）领衔，欧盟 10 国 18 个实验室共同参与。它的硬件是在一块晶片上生成的模拟式大规模集成电路。每块直径 20 cm 的晶片上有 384 个芯片，每个芯片包括 512 个脉冲发放神经元和 12.8 万个突触。这样，整个晶片上就共有约 20 万个神经元和 4 900 万个突触。——译注

想法吗?

关于大科学还有许多问题需要讨论,但是这封信已经太长了,我不得不在这里停下来。

最好的祝福。

凡及

HBP、美国 BRAIN 倡议和艾伦脑科学研究所之外，其他计划都正在启动或准备推出，现在对这些大型计划下结论还为时尚早。如果只考虑正在进行的项目，那么这些计划似乎都集中于数据收集工作（如绘制各种图谱）和新技术开发。这些工作可能为脑科学研究的突破提供重要基础，但还不是突破本身。另外，制作各种图谱的工作量很大，例如，HCP 花了 6 年时间，只是为健康的年轻人建立了宏观连接组的数据库，对于其他年龄段以及各种患者的宏观连接组数据库还有待进一步的工作。但是这也许还是类似的数据收集任务中最简单的呢！

在我看来，脑研究应该如何进行的问题仍然没有解决。组织大项目可能只是其中的一种方式，还要考虑这样做在脑研究过程中应该发挥什么样的作用的问题。即使对于这些大项目而言，如何组织它们也仍然是一个问题。HBP 给了人们一个严肃的教训，改革后的计划能否进展顺利，让我们拭目以待吧！

一个根本问题是如何评价脑研究的现状，脑研究是否已经进入了工业时代，或者仍然主要是以某种手工业的方式在进行，就像美国物理学家鲁克斯（Michael Roukes）所说的那样："我们仍然处于神经科学的手工业时代，每个人有他们自己的秘诀。"[1]

至于数据共享，HCP 和艾伦脑科学研究所可能是很好的例子，但是他们的数据都是图谱类的，并且都得到了政府或亿万富翁捐赠的大力支持，那么那些"手工业式"的神经科学家也能提供他们的研究数据吗？20 世纪末，美国科学家科斯洛夫（S. H. Koslow）提出了一项美国人类脑计划（US Human Brain Project），要求在全球范围内共享数据，而术语"神经信息学"一时在科学界也广为流行。这在当时曾经红火了一阵，还建立了国际组织以促进和协调全世界相应的活动。然而，它很快就销声匿迹了，今天当人们再谈及神经信息学时几乎都不再提及它了。我不知道这个计划究竟发生了些什么事，为什么它这么快就销声匿迹了。可能数据共享对于大科学来说更容易，但对于"手工业式"的科学来说则不然。现在，当人们再次讨论神经信息学时，我们是否应该总结一下科斯洛夫的人类脑计划的经验和教训，以避免类似的失败？我不知道，你有什么

---

[1] Reardon S. Global Brain Project Sparks Concern[J]. Nature, 2016, 537: 597.

Ⅲ-006　凡及

81

到他在 2009 年和 2012 年两次提出的承诺：他可以在 3 年内仿真整个鼠脑，而不仅仅是一个微回路！此外，正如一些科学家所指出的那样，他们的微回路模型没有任何有意义的功能！使用更简单的模型也几乎可以重复他们仿真的现象。

并非所有的科学家都赞扬这篇论文。你的同胞神经科学家黑尔姆斯泰特（Moritz Helmstaedter）批评说："并没有真正的发现。把大堆数据堆砌在一起并不能创造出新科学。"英国神经科学家莱瑟姆（Peter Latham）说："工作量惊人地大，但是在有关脑如何工作的问题上并没有告诉我们什么新东西。"他同意他们的工作可能是有助于对一些假设进行检验，但是"你可愿意花 10 亿欧元来做这些事？这才是问题之所在"。[1]

当然，对 HBP 的前景作出任何结论可能还为时过早，特别是在它经过了重大改革之后。现在 HBP 似乎将成为具有 IC 技术平台的永久性国际基础设施，这些平台被专门用来研究脑、认知神经科学和受脑启发的计算，这远离了马克拉姆当初提出的目标——在可预见的将来创建人造鼠全脑，甚至人造人全脑。

第三个事件是世界各地的许多国家已经或正准备启动他们自己的脑研究计划，国际合作与协调也在考虑之中。今年 4 月，14 个国家的科学院发布联合声明，呼吁国际脑科学研究合作[2]。

上个月，洛克菲勒大学、哥伦比亚大学、美国国家科学基金会（NSF）和卡夫利基金会召开了一次会议，讨论如何协调世界各国的脑计划。来自欧盟 HBP、美国 BRAIN 倡议、艾伦脑科学研究所、中国脑计划、日本综合神经技术脑疾患成像研究（Brain Mapping by Integrated Neurotechnologies for Disease Studies, Brain/ MINDS）计划和以色列脑技术（IBT）计划的代表参加了会议。他们介绍了各自的目标、成就、衡量 10~15 年内的影响、面临的挑战，以及为解决可重复性、数据可用性和资源共享问题所需要作出的努力。这次会议让科学家们有机会对大型脑计划进行回顾，探索国际协调与合作的可能性。然而，除了欧盟

---

[1] http：//www.sciencemag.org/news/2015/10/rat-brain-or-smidgeon-it-modeled-computer

[2] http：//sites.nationalacademies.org/cs/groups/internationalsite/documents/webpage/international_172183.pdf

31 日期间支持 HBP 8 900 万欧元。这意味着 HBP 已被允许至少在未来两年内进入其第二阶段——"正式实施阶段"（operational phase）。根据 HBP 官方网站[1]，HBP 在第一阶段——"起飞"阶段（"ramp-up" phase）取得了以下成就：

（1）已完成年轻小鼠新皮层微回路的数字化重建。

（2）建立了一个 HBP 网络合作实验室（HBP Web Collaboratory），注册用户可以访问其软件和数据库。

（3）已经在海德堡大学（Heidelberg University）和曼彻斯特大学（University of Manchester）的成员实验室中开发出仿神经结构计算系统（Neuromorphic computing systems）。

如果与原始提案中规定的本阶段结束时的标志性成就相比：建立信息交流技术平台，利用它们处理一些重要的实验数据并开始服务于神经科学界，已取得的成就似乎实现了其基本目标。但是，如果仔细察看一下，那么它似乎刚刚及格。尽管 HBP 呼吁全世界的神经科学家使用他们的平台，包括其脑仿真工具、可视化软件和可远程访问的超级计算机，但并不清楚神经科学界的反应会怎样。一个平台是否成功最终取决于它是否能带来一些科学或应用方面的突破。HBP 在第一阶段取得的标志性成就是去年在《细胞》杂志上发表的一篇论文[2]，该论文报道了对幼年小鼠 1/3 立方毫米皮层组织所做的仿真，其中有 3 万个神经元和 4 000 万个突触。参与该文所述研究的科学家共有 82 位，总计合作了近 20 年。他们使用了 207 种不同的神经元模型，考虑了 13 种不同的离子通道。由于缺乏实验数据，对突触联系作了假设，结果仿真了这个微回路内的活动模式，并与实验数据相似。这项工作得到了 C.科赫的赞扬："蓝脑计划对一块大鼠体感皮层的数字化重建，提供了迄今为止对一块可兴奋脑物质所进行的最完整的仿真。"[3]

马克拉姆宣称他终于实现了自己的承诺。然而，他显然忘记了或不愿意提

---

[1] http://www.humanbrainproject.eu/——译注

[2] Markram H, et al. Reconstruction and Simulation of Neocortical Microcircuitry[J]. Cell, 2015, 163：456-492.

[3] Koch C, Buice M A. A Biological Imitation Game[J]. Cell, 2015, 163：277-280.

最低级的;然后是只有在训练阶段才具有学习能力的智能;再后是随时都能学习的智能,但是其技能的提高是线性的;最高级的是始终都能学习,且其技能的提高呈指数式增长,尽管我们到目前为止还举不出最后一种智能的任何例子。就技能而言,尽管如果主体具有更多已习得的技能,它将更容易适应新的情况,但正如王培所主张的那样,很难定义应该具备什么样的技能才算是人工智能。同样,如果我们将智能定义为只有人脑才有的技能,那么这意味着除了人类之外,任何动物都不具有智能;而如果我们将智能定义为任何脑的技能,那么按照这种定义,连泥蜂也有智能了。☺我不确定你是否会同意我的观点。

在你的信中,你也谈到了大科学的问题,上个月有几个与此主题相关的重大事件,所以我想和你分享这些消息并且知道你的想法。

第一件事就是人类连接组计划(HCP)宣布已于今年6月完成了第一阶段工作。正如我们在前面提到的那样,HCP仅研究正常年轻成年人在解剖学和功能上,不同的皮层区之间以及和皮层下结构之间的宏观连接。据他们的报告,他们取得了以下成就:

(1)改进了四种不同的磁共振成像(MRI)技术(结构磁共振成像、任务激活功能磁共振成像、扩散磁共振成像和静止状态功能磁共振成像),以观察不同皮层区之间以及和皮层下结构之间的解剖和功能连接。其空间分辨率、时间分辨率和信噪比都得到了改善。

(2)采用上述四项技术,共收集了1 100位受试者的各一小时数据,其中包括来自300个家庭的双胞胎。

(3)将每个大脑半球分为180个不同的区域,这比传统的布罗德曼(Brodmannian)分区要精确得多。这对于宏观连接组学的研究可能具有基础性的意义。

(4)他们的数据和分析工具都已经向神经科学界开放。[1]

第二个事件是上个月12日欧盟与HBP协调员之间签署了"特定资助协议"(Specific Grant Agreement)。欧盟承诺在2016年4月1日至2018年3月

---

[1] http://humanconnectome.org/ccf

拷贝。如果这是真的,那么为什么将一个人的连接组拷贝到一台计算机里就会是心智上传了呢?

你绝对是对的,心智只能在身体和社会环境中或如你所说的"规章"的背景下存在,并且存在于和它们之间的相互作用中。因此,我对"盒子里的大脑"(brain in a box)理论持怀疑态度。如果一个孤立的脑只是靠一些营养物质维持,我很怀疑这样的脑在时间久了以后是否还会有任何心智。想想那些长时间被关在黑暗单人牢房里的囚犯的故事吧,他们中的大部分人几乎都失去了理智。

你讲过传授技能和知识的过程是一个比通过 DNA 遗传更快的适应过程,这让我想起了拉马钱德兰博士的一句话。他强调说,由于前一个过程,人们学会了杀死北极熊并立即取下它们的毛皮温暖自己,而不是等着进化出毛皮。要是这样等待的话,那么人早在长出厚毛之前就已经因寒冷而灭绝了。人类也因此而得以快速地适应恶劣的环境。

到目前为止,我几乎同意你在上一封信中的所有评论。然而,我对你下面的这段话有些不同意见:

"我对下面这样的智能定义持怀疑态度的原因:这种定义忽略了学习技能并试图分离出纯粹的学习能力。……我不喜欢这种纯粹的/与经验无关的智能的想法。"

对不起,我可能在上一封的电子邮件里对王培的论点有点误导,其中我写了一些关于技能和学习能力之间的区别,并强调了学习能力对智能的重要性。事实上,正如我在前一封的电子邮件中所说,王培将智能定义为"在知识和资源不足的情况下的适应能力",其中资源意味着时间、速度、内存容量等,这和你在上一封电子邮件中强调的是一样的。因此我认为在这一点上,你和王培之间并没有分歧。问题在于,在其他地方,王培强调了学习能力对智能的重要性,而非技能。也许这不是他对智能的正式定义,因为这两种表达的含义似乎并不完全一样。也许他只是想表达智能有某种等级结构。那些没有学习能力的智能是

Ⅲ-006　凡及

Ⅲ-006　凡及

（智能与学习能力;人类连接组计划完成第一阶
段工作;HBP进入正式实施阶段;各国脑计划）

2016－10－15

亲爱的卡尔:

非常感谢你的"深信",我读了三遍以掌握你经过深思以后得出的精髓。其
中一些我从来没有像你那样想过,有些我也有某种类似的感觉,但无法像你那
样清楚地表达出来。总之,你的信非常有启发性。

在你信的开头,你指出:"但是,当你感受到这种共鸣时,总会有崇拜某个科
学家的危险并相信他所说的一切。无论如何,我们容易相信,当一个人在某个
领域表现高超时,他或她所做或所说的一切都是对的。"

尽管很少有人会否认批判性思维和理性思维的重要性,但人们往往容易犯
你所讲的那种错误,我有时也这样。有时,人们正好相反,他们会拒绝他们不喜
欢的人说的每一个字。你早期对海德格尔工作的态度可能是这方面的例子,然
而,你并不属于我刚才描述的那种人。你在他的工作中找到了某种合理的东
西,你甚至在格伦的工作中还发现了对他的合理思想的一种更好的表达。我很
抱歉对这些哲学家一无无知,甚至都没有听说过他们的名字。☹

格伦很早就指出了人与人之间技能和知识传授的重要性,这点非常有意
思。同样有意思的是,你将口头语言和书面语言与"心智上传"进行了比较。是
的,当我看到一些关于心智上传的文章时,不禁会想到阅读生动的小说,特别是
一些已逝者的自传。我在感觉上似乎作者还活着,并在向我讲述他的经历。这
是一种心智上传的形式吗? 我知道不会是所有人都同意这种说法。这种生动
的体验只是在我心中,而不在书中。当然,自传也是作者在写作时思想的一个

（leaf counting）的倾向，但不想在这里透露太多。首先是因为我认为你会喜欢读他所写的内容，其次是因为这封邮件已经太长了。因此，我要把我对机器学习和 AI 的最新发展的评论放到以后再说。

但有一件事我必须在你开始生气之前按照承诺完成，也就是我是怎么认得艾伦的故事。当时我在数据库软件企业工作，艾伦已经从微软退休了，并开创了一家名为 Asymetrix 的新公司。他们有一个非常好用的名为"工具书"（Toolbook）的软件工具，比起通常的编程来说，它允许以更简单的方式设计包括视频和音频在内的交互式多媒体用户界面。这在今天非常普遍，但在 20 世纪 90 年代初却很轰动，当时还没有人听说过互联网和浏览器。我们的首席技术官韦格曼（August Wegmann）是一位真正的计算机科学天才，他设法将他的杰作，我们的数据库产品 Adimens 与工具书结合起来。当时在同类中它是世界上最有效益的开发工具之一，特别适用于自助服务终端和自动柜员机（ATM）。因此，我与 Asymetrix 以及艾伦进行了联系。不幸的是，微软在当时开始使用一种新的编程语言"Visual Basic"，这种语言更像是传统的编程语言，它把工具书看作是一个令人不快的竞争对手。我不知道艾伦在此事上试图说服比尔·盖茨方面有多艰难，但无论如何，微软忽视了这一发展，并且在稍后也同样地忽视了互联网以及浏览器的重要性。具有讽刺意味的是，IBM 却对解决方案感兴趣。当时微软成了 IBM 的严重威胁，虽然这两家公司仍然是 OS/2（也就是 PC 版操作系统的 IBM 版本）的正式合作伙伴。最终，IBM 收购了我们的银行 ATM 多媒体解决方案，然而将其装到 OS/2。在互联网出现后，微软重新找到了我们，我们合作开发了一个完整的银行业务解决方案，ADI 称之为银行 NT（Banking NT）。我和比尔·盖茨于 1997 年共同向法兰克福银行集团提交了一份报告。但这是另一回事了。[1]

早安!

卡尔

---

[1] Schäfer, Waldemar. Gates Präsentiert Deutsche Bankensoftware-Partner [R]. Handelsblatt, Nr. 200, 17 / 18.10.1997, p.22.

第一篇讲的是创造力如何工作的问题,第二篇则是关于游戏在发展智能方面所起的重要作用。关于后者,我现在有了机会可以相当密切地进行观察,我看着我的外孙亚历克斯在他开始走路、说话和用他够得到的一切东西来玩之后,其脑中的连接组天天都在飞速发展。

我听说过科什莱尔的天才,并且知道他像他的朋友冯·诺伊曼一样也是出生于奥匈帝国的多才多艺的人之一。但我从来没有读过他的任何一本原著,现在趁机购买了《创造的艺术》一书。不过我还没有读完,因为它有800多页。而且我也不能确定他的双重联想(bisociation)理论(就是把两种联想联系起来)对解释幽默(人类独有的一种感觉)和"尤里卡"(eureka)效应[1]到底能起到多大作用。但我能说的是,这确实是西方文学中的珍宝和杰作,我强烈建议你在长假中阅读,即使只是作为一种智力享受也可以。

在此期间,我还部分地阅读了弗里曼和科兹马的新作:《大脑皮层中的认知相变——通过建模神经场来加强神经元学说》(*Cognitive Phase Transitions in the Cerebral Cortex — Enhancing the Neuron Doctrine by Modelling Neural Fields*)。

这篇新作包括了对弗里曼的研究的最新总结以及他和科兹马合作完成的工作。我不知道你是否已经读过这本书。它的大部分内容你应该都很熟悉,但它也包含了神经科学界的许多相关人员的"评论",向弗里曼致敬并讨论他对该领域的贡献,其中也包括了你的朋友凌瀚思。这本书非常符合德国大学里著名的"纪念文集"(Festschrift)的古老传统,在那里,学生和同事在一位教授退休或庆祝生日时向他表示敬意。

当我读完之后,我会告诉你我更多的读后感。现在我只是建议你读一下韦尔伯斯(Paul Werbos)[2]的评论(第19章),因为在那里你可以找到关于像BRAIN这样的大科学项目或艾伦脑观察站的工作究竟是非常有意义呢,还是没有意义的问题的答案。韦尔伯斯似乎对这种巨额资助非常了解和经验丰富,并对此有着非常明确的看法。我主要赞同他所说的关于科学有一种"数叶子"

---

[1] 指因找到某物,尤指问题的答案而兴奋地高呼:"我发现了(我找到了)"的这样一种效应。——译注
[2] Paul J. Werbos(1947—),美国人工神经网络专家,他以提出反传算法而闻名于世。——译注

记得确切的措辞,但它的意思大概是这样的:"在生物学中,智能就是动物能认识到在什么时候应该放弃以前一直认可的策略。"所以可怜的泥蜂就要被排除在外。☺

另一个定义是,对智能来说,重要的是"不行-补足-行了"(incompetence-compensation-competence)。这是我自己对德国哲学家马夸特(Odo Marquardt)所创造的一个术语的私人变体。本来他的意思是描述当今哲学所处的令人遗憾的状态。以前哲学对万物都适用,而今天哲学主要关心的是补足其不适用之处。这不仅仅是一个有趣的术语,因为在社会交往中(以及在科学中),我们经常不得需要处理许多未知的东西,通常非常聪明的人习惯于精确地计算一切,甚至精确到小数点后 10 位,但当他们没有必需的信息来做他们通常的计算时,他们就给卡住了。在同样的情况下,不那么聪明的人可能会表现得更好,因为他们习惯于不作太多的计算和推理就采取行动。

第三个定义是由来自特拉维夫的贝特-哈拉希米(Aharon Beth-Halachmi)提出来的。他是我的好朋友,也是我见过的最聪明、最有见识的人之一。他强调了智能和聪明之间的微妙差别,他说:"一个聪明人(clever person)永远不会把自己置于只有有智能的人(intelligent person)才能找到出路的境地。"☺

最后一种智能定义是我在读了分属敌对阵营的科学家之间激烈的辩论后提出来的。我并不要求版权,因为我确信一定有别人早就有过这种想法:

"没有人确切地知道智能究竟是什么,但每个人都认为自己有足够多的智能,而其他人则没有。"☺

智能的一个关键因素是创造力,但是这一点常常在智能定义、智商测试和人工智能中被忽略掉。

我很高兴地看到,在弗里曼和科兹马(Robert Kozma)的新作中,他们提到了科什莱尔(Arthur Koestler)[1]的《创造的艺术》(*Art of Creation*)和赫伊津哈(Huizinga)的《游戏者》(*Homo Ludens*)。

---

[1] Arthur Koestler(1905—1983),匈牙利裔英国作家和新闻记者。——译注

霍金斯在《智能论》中给出了一个非常好的解释,即人脑可能从类似的效应中受益。他提出模型说,只要感觉输入不是新的,它就快速地通过新皮质层中各层,并且只有在有新的或重要的东西时,人脑才会付出更大代价的注意和/或存储过程。

我明白为什么王培有兴趣要将已学得的技能与"纯粹"的学习智能(learning-intelligence)分开,因为这对检查他的通用人工智能的纳思(NARS,非公理推理系统)的性能会有所帮助。但是,尽管我很欣赏他在《人工智能中的三个基本误解》中很有见地的分析,但我不喜欢这种纯粹的/与经验无关的智能的想法。

无论如何,试图像过去一样通过"正确"的定义来解释现象是没有任何用处的。它最多从经验意义上对测试理论或多或少地有些用处。然而,有趣的是,看看在各种学科中存在有多少种与我们的脑表现相关的定义。

我所知道的关于"智能"问题的最好和最全面的一篇文章中,就有约 70 种不同的智能定义。深心公司的创始人之一莱格(Shane Legg),是亨特(Marcus Hutter)[1]和施米德胡贝(Jürgen Schmidhuber)[2]的学生,他在其博士论文《机器超级智能》(*Machine Super Intelligence*)中收集了这些定义。他的论文是一篇非常有趣的阅读材料,比他的联合创始人之一哈萨比斯的博士论文更令人印象深刻。你可以在维基百科网站上有关莱格的条目中找到其链接。[3]

在莱格列出的所有定义中,令我惊讶的是,许多研究人员并不考虑对我来说最为重要的一个方面:时间和速度。

所有的智商测试,无论他们测量什么,都要在有限的时间内完成,这很有道理。也可能有时间无关紧要的智能任务,但在大多数竞争激烈的情况下,速度很重要,因为如果你太慢,其他人会快过你,而你就会输掉。

令我惊讶的是,我最喜欢的定义并不在莱格列出的 70 种智能定义之中。

我最喜欢的一个定义是我在多年前读到过的,但不记得是谁写的了。我也不

---

[1] Marcus Hutter(1967—),德国计算机科学家。——译注

[2] Jürgen Schmidhuber(1963—),德国计算机科学家,专长为人工智能。——译注

[3] https://en.wikipedia.org/wiki/Shane_Legg

果你不把脑放到社会环境中去考虑，如果你不考虑装在脑里的和规则相关的内容，你也就不能理解脑和心智。我们把这种基本上是一套什么能做和什么不能做的规则的知识当成是主观的，但实际上它是从我们的祖先或我们的同伴那里获得的。这并不会让我的祖母感到意外，但令人惊讶的是，有多少脑研究人员的作为就好像不用考虑这些学习所得的内容，光研究孤立的脑就能解释脑功能似的。

这就是为什么我对下面这样的智能定义持怀疑态度的原因：这种定义忽略了学习技能并试图分离出纯粹的学习能力。

我们所获得的每一点新知识，总是根据我们在之前学到的东西进行了处理，并存储在我们的头脑中。它甚至通过连接组而在物理上也与之相关。

在我们的头脑里，并不像计算机那样可以有空白的磁盘，也不能像机器那样用千兆赫兹或 RAM 空间那样的指标来检查原始的空白脑的性能。即使是计算机，也不能做到这一点。要想检查机器的性能，你必须先以任何一种 BIOS 来启动它。只有当机器在启动过程中经历了一段快速运动的"童年"之后，它才会"知道"这是一台 Mac、一台 Windows PC、一台 Linux 服务器或别的什么机器。即使是相同的硬件，这些系统在其硬件的基本使用方面也会显示出非常不同的行为，这取决于程序员在编写驱动程序时付出了多少努力。随着你加载的每一层程序，和输入更多数据到其数据库中，这一进程也会继续下去。有些数据库有令人称奇的特征，即存储在其中的数据越多，工作越快。如果数据来自有限的对象范围，这是很容易理解的。一个"经验丰富的"地址数据库早就已经有了一个国家的大部分邮政编码，比起一个什么都不知道的空数据库来说，当更新索引表[在关键树（key-trees）[1]中插入新的节点]时所需要的工作量要少得多。系统越"富有经验"，那么某个邮政编码早就已经收集在数据集中的可能性就越大，因此比起经验不足的系统来说，其学习优势也越大。

――――――――

[1] 关键树又称索引树，是为了能从数据库中快速找到其中的某个元素的组织方法，例如按字母顺序排列的索引表。英文词典就是这样组织的，你为了找某个单词，例如 book，第一个字母是 b，所以你得到 b 起头的部分去找，而不必去看前面的 a 部。到了 b 部，第二个字母是 o，所以你可以略去所有第二个字母是从 a 到 n 的单词，如此等等，直到找到 book。如果把这样的搜索过程画出来就像一棵不断分枝的树，故名。其中的每个分枝点就称为节点。――译注

的祖先在大约一万年前发明的一种技术。我们把这种技术称为语言,并以口头和书面形式使用它们。这两个过程都非常缓慢和不精确,但从长远来看却非常成功,以至于我们要有那么长的童年和学习阶段也都还是值得的,这也是人类与其他所有哺乳动物的不同之处。这对人类学家来说当然不是什么新见解。但是,格伦对社会层面上的自组织的详细分析是新的,同时他也认识到学习的一般规律,如上文中他的引语所说。

格伦感兴趣的对象不是细胞或神经网络,而是"规章"(institutions)。这类似于人类发明的锤子或小刀那样的工具,不过是在社会层面上。这就像许多集合在一起的规则(capsuled procedures),不过比道金斯多年后提出的"模因"(Memes)更强也更重要。正式婚姻就属于此,法律制度、议会、政党、政策、军事、宗教、学校、税务系统、银行、信用卡支付、停车标志、交通灯和网络安全甚至有组织犯罪和腐败都属于这一范畴。总而言之,也就是在现实世界中一切和我们生活管理有关的集合在一起的规则组。当我们想要在我们出生的特定环境中取得成功时,必须要考虑到它们。我们可以在把它们代代相传时,适应它们并做出修改。这是一种比修改 DNA 更快速的适应过程,并且这个过程中包括通常进化无法提供的真正有目的的设计。我不想再深入介绍格伦的人类学和制度理论而让你感到厌烦,但我相信这可能有助于填补我称之为"第二绑定问题"的空白。

如果没有这些在个体脑之外定义、存储和传授的规则和规范,那么我们就无法理解人类的行为。即使我们能够绘制出精细到神经元和原子的脑图谱,我们对这些决定和约束我们行为的力量还会一无所知。如果没有这类知识,光是生理学和心理学知识还不足以解释人类的行为,而这些知识必须在每代大脑中都重新安装。当然,神经科学家至少在原则上也知道这一点,但在他们禁止让其进入他们的方程的边界条件后,往往就忘了这一点。[1]

但是,就像如果你不把脑放到身体里面考虑,你就不能理解脑和心智一样,如

---

[1] 如果要求偏微分方程的特解的话,需要给出初始条件和边界条件。这里是说神经科学家在解决他们的问题时不考虑那些规则(尽管他们原则上也知道这些规则是有影响的),就像数学家在解偏微分方程时在边界条件中漏了点重要的约束,因此所得的解就有了些问题一样,但是神经科学家却好像忘了这一点。

格伦将此过程描述为一种循环过程。循环过程求助于心理媒介,也就是种种知觉,从物理部分出发,通过自己的运动,进入事实层次(factual level)[1],然后再返回。其结果是,格伦并不把这种行为认为是二元论的:发生的过程不能被分为生理和心理部分。所有部分都彼此密不可分,并且在同一过程中不断地一起工作。他用下面一段话介绍了他的行动概念:'我要讲,行动本身就是一个复杂的循环运动……并依赖于反馈,行为发生了变化。'"

这些话还是在 1957 年说的!

我怀疑有多少格伦的作品(甚至是究竟有没有他的作品)已经被翻译成了英文[2],因此没有多少国外的科学家注意到他的思想。我确信你可以进行更好的翻译,我认为将格伦的核心思想翻译成中文是值得的,因为它非常具有启发性。年轻的研究人员,不论是在神经科学还是在机器学习方面工作的,只要他们关心从心理学到社会学和人类学等更高层次的绑定问题,那么他们都可能从他的观点中受益。

简而言之,格伦所提出的是薛定谔最早在《生命是什么》(What is Life)[3]一书所提出的思想的延伸:即有关物质和能量自组织原理之谜。物理学家薛定谔提出了细胞如何执行其主要任务的问题,用他的行话来说也就是如何"输出熵"。用更通俗的话来说,也就是细胞如何能在混沌中建立和维持秩序。所以他转向脑低端的微观层面。格伦则在脑的高层和更高层面提出同样的问题。他问道,是哪些社会工具和发明可以让人类组织得如此完善,以至于比其他动物更具优势,尽管这些动物更强壮,身体上自带适用的工具。

所以,格伦感兴趣的是存储在脑中的与行动相关的内容,也就是可以传授给他人的部分:技能和知识。虽然很多人都渴望心智上传,但我们往往忽略了我们

---

[1] 事实层次是指物质作用于外界真实世界的层次,而与内心的精神层次相对。——译注

[2] 后来我才发现至少有两本格伦的主要作品有英文版:
Arnold Gehlen. Man in the Age of Technology[M]. New York: Columbia University Press, 1980.
Arnold Gehlen. Man: His Place and Nature in the World[M]. New York: Columbia University Press, 1988.

[3] 有中译本:埃尔温·薛定谔.生命是什么[M].吉宗祥,译.广州:世界图书出版广东有限公司,2016. ——译注

懂。记得当我们在大学举行研讨会时,有批判理性主义(critical rationalism)[1]的顶级支持者参加,他们把"海德格尔主义"当作是吹牛和欺骗的原型。在公开场合,他们以较有礼貌的方式表达了他们的厌恶情绪,但是在会后附近酒店举行的传统聚会上,我记得有不少知名哲学家把海德格尔的论述称为"夸张做作的废话"或"假作智能的垃圾"。作为一名年轻的科学家,还刚想在这个领域为自己定位,我对这种强烈的言辞印象深刻,并且深深受到影响。后来我发现海德格尔作品中的内容并非都是垃圾,人类学家格伦(Arnold Gehlen)以更清晰和更有用的方式提出了和他类似的观点。

我认为弗里曼之所以知道海德格尔和梅洛-庞蒂,是因为德雷福斯(Hubert Dreyfus)[2]使他们在美国出了名,并且令人感到惊奇的是,这还使不少美国研究人员,甚至是顶级的计算机科学家,如威诺格拉德(Terry Winograd)[谷歌联合创始人佩奇(Larry Page)的博士生导师]也对相当空洞的欧洲哲学感起兴趣来了。

如果有人能够更为清楚地把格伦的观点解释给他们听的话,我想他们可能会印象更深刻一点。只要看一下德文《维基百科》中有关格伦的词条中以"行动周期(Handlungskreis)"开头的一节就行了。[3]

对不起,这只有德文版,但我知道你可以理解它。

在这里,我大致地翻译了这里面的一段话,希望能够清楚地表明格伦谈论的是循环因果关系,以及与弗里曼后来称之为"再传入"非常相似的过程:

"他以下面的例子来描述行动周期(action circle):当你试图用钥匙打开生涩的门锁时,你就得把钥匙来回活动。你会发现钥匙是向前进比较顺利还是后退时比较顺利。所以在这些尝试中你会体验到成功或失败,你会得到反馈。如果你对此反馈作出回应并改变你的行动,你将会取得预期的成功,锁打开了。

---

[1] 批判理性主义是一种知识论哲学(epistemological philosophy),它认为科学理论应该能经受理性的批评,要是这种理论含有经验内容的话,那么应该对它可以进行检验,并有可能证伪。——译注
[2] Hubert Lederer Dreyfus(1929—2017),美国哲学家。他的主要兴趣是现象学、存在主义、心理学和人工智能中的哲学问题等。——译注
[3] https://de.wikipedia.org/wiki/Arnold_Gehlen

诱惑，即使对事情一无所知，他们也会热衷于在公开场合对一切都横加评论。

当哈萨比斯在评论诸如像马斯克（Elon Musk）、沃兹尼亚克[1]（Steve Wozniak）、霍金（Stephen Hawking），甚至还有一些好莱坞演员等众多科学家和名人所发出的 AI 危险的警告时，非常有礼貌地指出了这一观点（见我上一封邮件中引用过的那篇《连线》文章）：

"这些人在这个方面实际上并没有做过，所以他们谈论的只是从哲学和科幻小说角度出发产生的忧虑，他们对于这些能力究竟能做什么的问题几乎一无所知。"

事情确实如此，这也是一个很好的例子，支持了我长期以来有关深深植根于所有人心智架构中的核心特征的一个假设：

"我们对某个问题的理解越少，我们对它的看法就越坚定"。

当然，弗里曼并没有超出他熟悉的领域。所以，为了表明即使对于我的朋友和偶像，我也试图抱着批评的态度，我不得不说，他作品中的某些部分对我来说很陌生。我不同意他对海德格尔[2]（Heidegger）和梅洛-庞蒂[3]（Merleau-Ponty）的赞赏态度。当然这并没有太大关系，因为我们都以不同的方法激发自己的创造性，并使自己的心智产生新的想法和见解，但问题是他没有发现一个更好的解决方案，其实这个方案早就有了，而且他可能会更喜欢。

对我而言，胡塞尔[4]（Edmund Husserl）传统中的现象学[5]（phenomenology）总像是哲学中最不具吸引力的部分之一，尤其是其海德格尔变体更是晦涩难

---

[1] Stephen Gary Wozniak（1950—），美国发明家和电子工程师，苹果计算机公司的共同创始人。——译注

[2] Martin Heidegger（1889—1976），德国哲学家。他被广泛地认同为 20 世纪最有创造性和重要的哲学家之一。以其对现象学和存在主义的贡献而闻名。——译注

[3] Maurice Merleau-Ponty（1908—1961），法国现象学哲学家。——译注

[4] Edmund Gustav Albrecht Husserl（1859—1938），德国哲学家，创立了现象学学派。——译注

[5] 现象学是 20 世纪最重要的哲学流派之一，由德国哲学家胡塞尔正式创立。现象学是对经验结构与意识结构的哲学性研究。——译注

Ⅲ-005　卡尔

（格伦的循环过程；社会规章和行为；智能的
定义）

2016－08－29

亲爱的凡及：

非常感谢你亲切的来信。信虽短却往往包含很多内容,你的这封邮件也是如此。它讲到了 3 个相关领域以及这些领域中的重要问题,坦率地说,解决这些问题都很困难。☺

一个是王培定义的智能问题,另一个是现代人工智能中的范式转换及其可能带来的巨大后果,第三个是你在艾伦脑研究所公布了他们的研究结果后提出的 BR（大研究）的组织问题。我不确定我是否能够给出所有人都能满意的答案,但希望我能够在拖延很久之后告诉你"我如何遇见艾伦"的故事。☺

但让我先告诉你,我有多喜欢你所说的"心有灵犀一点通"。事实上,当我读到弗里曼作品的精髓时,我感受到了这种共鸣。但是,当你感受到这种共鸣时,总会有崇拜某个科学家的危险并相信他所说的一切。无论如何,我们容易相信,当一个人在某个领域表现高超时,他或她所做或所说的一切都是对的。

爱因斯坦成为现代科学最重要的明星可能是最众所周知的例子。尽管他确实是一位真正的物理天才,但在许多其他领域中并不都是天才。无论爱因斯坦对一位记者说了些什么,第二天都会成为头条新闻,（尤其是）当他在远远超出了他的能力范围之外这样做的时候,有时甚至是说了些纯粹的废话时更是如此。他喜欢开这种玩笑,因为他很有幽默感,有时他只是为了自己开心而故意这样做。其他缺乏这种自嘲能力的名人常常经不起自己名气的是有智能的,尽管它们的行为没有任何灵活性。你的看法如何呢?

好吧,现在让我再来谈谈另一个话题。你可能已经注意到了昨天在许多科学媒体上的报道,其中发表了艾伦脑科学研究所在 4 年前发起的艾伦脑观察站(Allen Brain Observatory)宣布的成就。在今天的《自然》杂志上还发表了题为《发布大脑数据金矿》(*Brain-data gold mine released*)的报告。[1] 艾伦脑观察站公布了 360 次实验中 25 只小鼠视觉皮层的四个区域中 18 000 个神经元对各种视觉刺激反应的活动数据,这些刺激中也包括好莱坞电影《历劫佳人》(*Touch of Evil*)的片段。

这是历史上第一次在细胞水平上公布神经元群体活动的实时数据,这些数据显示神经元在执行认知任务时如何协调其活动。艾伦脑科学研究所是美国 BRAIN 倡议的一个成员单位,因此这个报告对于该计划来说肯定是个好消息。艾伦脑科学研究所采用一种类似于工业组织的科学研究新方式运作,他们组织来自不同领域的专家团队共同围绕一个目标工作,并向公众公开他们的数据和分析工具。有人称之为"大科学""团队科学"和"开放科学"。在我看来,如果工作量非常巨大,但程序相对固定,这种方法对数据收集来说是合适和有效的。艾伦脑基因表达图谱、艾伦小鼠脑连接图谱和艾伦脑科学研究所发布的艾伦细胞类型数据库就是这样的例子。人类连接组计划可能是另一个例子,但我不知道它的确切进展。然而,所有这些工作都得到了政府巨额资助或亿万富翁捐赠的大力支持。此外,我不认为这种方法可以扩展到大多数科学研究领域,因为这些研究的主要目标不仅仅是收集数据,而是强烈依赖于洞察力和创造力。我们之前的信中曾多次就此进行过讨论。

一如既往地致以最良好的祝愿。

凡及

---

[1] Shen H. Brain-Data Gold Mine Released[J]. Nature, 2016, 535: 209 - 210.

这样解释是否合理？

另外，王培在他的文章中提出了另一个我非常感兴趣的论点："智能"是学习的能力，而不是解决某些具体问题的能力，后者属于技能。"智能"不是"解决具体问题的能力"，而是"获得解决具体问题的能力的能力"[1]。他将"智能"系统分为四类：系统具有恒定的技能水平，但不能从其经验中学习，这种系统即使可能具有非常高的技能水平，而从他的观点来看也不具有智能，例如深蓝国际象棋弈棋系统就是如此；系统学习能力有限，可以在学习阶段提高技能，在经过训练阶段后达到恒定水平，阿尔法狗就是这种系统的例子；系统始终具有学习能力，可以学习新技能，并一直不断地提高技能，人类属于这一类；系统学习能力按指数规律增长，这正是库兹韦尔和其他人所想象的，但直到现在我们甚至举不出一个这样的系统的例子。[2]

从他的观点看起来，大多数现今的"智能"机器，如深蓝，都应该排除在 AI 以外，尽管它们构成了传统 AI 的主体。在主流的人工智能中，人工智能被定义为机器具有只有脑才能做的某些工作的能力。王培反驳说，根据这个定义，在一台机器完成这项工作之后，就不能再说这只有脑才能完成了，因而不应该再被认为是 AI，这个定义本身就有内在矛盾。虽然人们可能会反驳说，定义可以修改为系统可以做一些过去只有脑才能完成的事情，以避免这种矛盾，但是，根据这个新定义，即使是只能进行加法或减法的系统也应该被认为有"智能"，而大多数人并不认可这一点。机器学习，如深度学习属于他的第二类，而通用人工智能则属于他的第三类。他说如果有哪种系统属于他的第四类，那么这样的系统应该被称为"人工神"系统，而不是人工智能系统。尽管他将他的第一类排除出人工智能的论点可能违反了公众的普遍认识，但我认为他的想法以更清晰的方式描绘了人工智能的范围。尽管人们可能会争辩说，为什么应该把只会做加法的机器排除出 AI？你可以接受它，并承认这样的系统处于 AI 的最低级别。然而，这可能与智力的定义有关，如果是的话，那么法布尔的泥蜂也应该被认为

----

[1] 参见：王培.人工智能迷途：计算机的高技能等于高智能吗？[EB/OL].赛先生,2016-02-17.——译注
[2] 参见：王培.人机大战赛前思考：计算机会有超人的智能吗？[EB/OL].赛先生,2016-03-09.——译注

Ⅲ-005 凡及

（"智能"不是"解决具体问题的能力"，而是"获得解决具体问题的能力的能力"；人工智能的分类；艾伦脑观察站；大科学）

2016－07－14

亲爱的卡尔：

尽管你之前并不认识弗里曼，但你对他去世的深切悲伤令我深为感动。正如中国的一句老话所说："心有灵犀一点通"和"惺惺相惜"！在我看来，弗里曼就是你所描述的那样的人。要是他能读到你的电子邮件的话，他一定会很高兴又找到了一位真正理解他的同行。

你的来信回答了一些困扰我很长时间的关于人工智能的问题。这些问题是：会有第三个 AI 严冬来临吗？深蓝和阿尔法狗所使用的方法有本质区别吗？如果有的话，那么其区别是什么？如何评估 AI 的最新进展？我不得不承认我对计算机工程和人工智能知之甚少。更糟糕的是，我从来没有下过国际象棋和围棋，这也严重影响了我对人工智能未来发展的判断和理解。与我相比，你，我亲爱的朋友，曾经开发过 Adimens，并且甚至在深蓝与卡斯帕罗夫鏖战之前就已参与过开发国际象棋弈棋系统！你是这个领域的专家，所以我很幸运能够找到恰当的人来回答我想要知道的问题。

你说即使是深度学习神经网络的程序员也不知道他们自己设计的系统中的"魔术师是如何做到这一点的"，这非常有意思。为什么？我刚刚读了王培博士的几篇文章，他解释说，是网络中成千上万个参数决定了阿尔法狗应该怎么走，而这些参数是由整个训练过程的历史决定的。因此，即使你复查整个训练过程，也很难搞清楚落子的原因。我猜想，也许脑中的操作也是如此。你认为当然还有很多工作要做,但在这个领域确实有了真正的进步,而不仅仅是公关噱头。而最好的消息是,新一代最耀眼的年轻天才感受到了这个领域的吸引力。

因此,当我说我相信人工智能从长期来说会取得成功时,你说得对,我的意思就是指强人工智能。至于意识问题我不能确定,但最终这可能会是一个伪命题,你可能不喜欢听到这种说法。☺

但是,我们完全同意王培所说的那些话的重要性。与此同时,我读了他更多的作品,并发现他确实很有意思。他也是一位独立而深刻的思想家,似乎很喜欢长时期钻研同一个难题。我再次感谢你将我的注意力吸引到一位有意思的研究人员和他的想法上去!我不知道你是否看过他的文章《对人工智能的三个基本误解》(*Three Fundamental Misconceptions of Artificial Intelligence*)[1]。如果你还不曾读过的话,那么我强烈地推荐给你读一读,因为这是在这个问题上我读到过的最好评论。水晶般地清澈,真的很棒——你会爱上它,值得作一整轮的讨论。

但黎明又来临了,我不得不停笔。

早上好,向上海致以最好的祝愿。

卡尔

---

[1] Wang P. Three Fundamental Misconceptions of Artificial Intelligence[J]. Journal of Experimental & Theoretical Artificial Intelligence, 2007, 19(3): 249 - 268.

完美地下棋。

当然这在封闭的公理系统中特别适用。但它也可以应用于开放系统,在这些系统中,神经网络通常很快学会完成一项任务,如从 2 000 种不同种类的狗中识别特定犬种。这需要大量数据来训练系统和大量的 GPU(或像 Asic 芯片那样的专用硬件)。

在这篇文章中,哈萨比斯和他的朋友很好地描述了新旧 AI 之间的区别。当然,绝大多数这类应用程序都仅限于特定的问题。虽然无法知道隐藏在网络"隐层"中的知识,因此程序员实际上也不知道"魔术师是如何做到这一点的",在目前也不可能将一种应用中得到的知识用到另一种应用中去。

但这从原则上说起来并非不可能,像卡尔帕西(Andrej Karpathy)[1]和古德费洛(Ian Goodfellow)[2]那样的人工智能界中年轻的莫扎特可能会找到方法来提取这些知识,建立更通用的人工智能以及更好的方法来训练网络。[3]有时候,只是一点小想法就可以引起像古德费洛所取得的那种巨大的进步,他用两个网络相互竞争,因此使他们更快、更高效地磨炼他们的专业技术。该方法被称为 GAN(Generative Adversarial Networks,生成对抗网络)[4]。据说古德费洛与朋友一起去酒吧后提出了这个想法,并彻夜编程得到了解决方案。我不确定这是否属实,但我喜欢这个故事,这个年轻人确实在一夜之间成为 AI 的超级巨星。

---

[1] Andrej Karpathy,2015 年获斯坦福大学计算机科学博士,师从李飞飞。2017 年 6 月起成为特斯拉公司的 AI 主管。——译注

[2] Ian J. Goodfellow,美国人工智能专家,现在是谷歌脑的研究科学家。——译注

[3] 事实上,在人工智能系统之家传输知识是目前人工智能研究中最活跃、也是最有前途的领域之一。参阅:Tyukin I Y, Gorban A N, Sofeykov K I and Romanenko I (2017). Knowledge Transfer Between Artificial Intelligence Systems. https://arxiv.org/abs/1709.01547

[4] 生成对抗网络是无监督式学习的一种方法,通过让两个神经网络相互博弈的方式进行学习。该方法由古德费洛等人于 2014 年提出。生成对抗网络由一个生成网络与一个判别网络组成。生成网络从潜在空间(latent space)中随机采样作为输入,其输出结果需要尽量模仿训练集中的真实样本。判别网络的输入则为真实样本或生成网络的输出,其目的是将生成网络的输出从真实样本中尽可能分辨出来。而生成网络则要尽可能地欺骗判别网络。两个网络相互对抗、不断调整参数,最终目的是使判别网络无法判断生成网络的输出结果是否真实。生成对抗网络常用于生成以假乱真的图片。此外,该方法还被用于生成视频、三维物体模型等。——译注

Hassabis)是 2014 年被谷歌收购的深心公司的创始人之一,他使用 Breakout[1]这个计算机游戏的例子很好地描述了这种方法。该游戏与我们在 20 世纪 70 年代都玩过的老式电子游戏"乓"(Pong)类似。

在接受《连线》(Wired)杂志采访时,他讲解了神经网络如何学习游戏,指导它游戏的唯一目标就是最大化游戏得分:

"在经过 30 分钟和 100 场比赛之后,成绩还是非常糟糕,但它知道了应该用球棒去接球"……"1 小时后,虽然从定量上来说有了起色,但仍然不够出色。但 2 小时后,它或多或少地掌握了游戏方法,即使当球速很快时也是如此。在 4 个小时之后,它找到了一种最佳策略——在墙的边上挖一个隧道,并以超人的精确方式将球打回去。这是一个连这个系统的设计者都不知道的策略。"[2]

对于不熟悉现代神经网络的人来说,很难相信这种可能性,而且更难以理解程序是如何做到这一点的。比有可能做到这一点更令人惊奇的是,程序员也并不确切知道程序是如何做到这一点的。

神经网络一开始就像一块石头一样愚蠢,而且(不像深蓝)对世界,也对围棋一无所知。最初它的神经元以任意的权重开始,然后开始四处摸索,非常像弗里曼的蝾螈一样地探索世界。开始时几乎没一点成功,但在每个周期后程序都调整神经元的权重(增加或减少),从而知道哪些变化使得输出的分数变好或变差。这需要大量计算(偏导数),当有许多神经元和许多层时,计算量呈爆炸式增长。但只要有足够的时间或计算机能力,系统仅仅靠摸索就可以学会如何

---

[1]《Breakout》是一款由雅达利开发及发布的街机游戏,此游戏参考了 1972 年雅达利街机游戏《乓》,于 1976 年 4 月发布。游戏开始时,画面显示 8 排砖块,每隔两排,砖块的颜色就不同。游戏开始后,玩家必须控制一块平台左右移动以反弹一个球。当那个球碰到砖块时,砖块就会消失,而球就会反弹。如果玩家未能用平台反弹球而使球触到底线,那么玩家就输掉了那个回合。当玩家连续掉 3 次后,玩家就输掉整个游戏。玩家在游戏中的目的就是清除所有砖块。当球碰到画面顶部时,玩家所控制的平台长度就减半。另外,球的移动速度会在接触砖块 4 次等情况之后加速。——译注

[2] https://www.wired.co.uk/article/deepmind （June 22, 2015）

是想用最新一代的计算机来解释脑及其构筑。在 GOFAI 时代一直都是这样做的,直到深蓝战胜卡斯帕罗夫而达到高潮。对于 IBM 来说,这是一次巨大的公关成功,但在这次令人印象深刻的表演之后并没有发生太多事情。

仅仅是为了赢得国际象棋而建立一个专家系统,其代价是非常高昂的。而当你想赢别的棋,你就必须再次从头开始构建知识库,比如说围棋的知识库,更不要去说其他更重要的事了,例如从组织样本中检测不同种类的癌症、进行手写体识别或从视频中识别某个人。

如果要靠单个主控单元监督和控制整个过程,决定下一步该做什么,并告诉子单元该做什么,那么你就需要大量计算机。当事件空间呈指数级数增长时,计算机的速度很快就跟不上了。在国际象棋之后 AI 弈棋系统的下一个“果蝇”是围棋,它在复杂性上要高出好几个数量级,这是因为它的棋盘不是 8×8 格,而有 19×19 格。

把许多通用 CPU 并行组织起来的想法在很长一段时间内颇为流行,特别是在日本,但从未真正实现过。并行计算机组织 CPU 的代价太高,而且还需要一种新的编程方式,而它所带来的好处至多也只是线性的,并且下一代的传统计算机也能做到这样的好处,所以没有人愿意在这方面进行投资。

然而,并行计算的真正突破并非来自通用 CPU,而来自 GPU。GPU 是游戏机中的主要部件,用以处理图片和视频。GPU 大概可以算是最不智能的一类计算机了。它们只能完成很少的任务,比如移动一个像素,但是它可以并行进行并且极端快速地完成。而 GPU 也非常适合做一类非常古老的人工智能(神经网络)所需的计算。

然而现代人工智能变得如此强大,并非因为我们有了快得多的机器和更智能的算法,而是因为人们改变了机器必须在事先就装备好所有现有知识的想法。

阿尔法狗在战胜李世石比赛时使用的神经网络与其前辈深蓝完全不同。它们甚至不能算是亲戚,也许就像鱼和鸟那样的不同。不同之处在于,阿尔法狗的深度学习网络并不需要人类专家告知下围棋的所有规则、策略或技巧,也不需要从数据库中进行学习,它只是通过下棋来学习。哈萨比斯(Demis

区域转移到另一个区域会使应用程序变慢,并且在读取和写入过程中也容易出现错误,在运行过程中发生更改还可能会导致逻辑问题。在数据库中如何处理这些过程直到今天还依然是信息学中一门微妙的艺术。这些问题的根源就来自目前占主导地位的冯·诺伊曼机器的设计之中。这里两种独立的物理存储设备必须通过将数据和程序代码从慢速内存中传送到快速(和耗能)的短期内存中进行连续更新,反之亦然。

速度还不是这种方法的唯一问题,另一个问题是程序代码和数据存储在同一个空间中。这就像在同一个锅的不同部位煎鸡蛋和煮汤一样。如果汤沸腾了,就可能把鸡蛋搞糟,从而导致整个菜都给毁了。如果在为数据变量保留的内存区域中写入超量,这就可能会破坏存储在其旁边的程序代码,并导致程序做蠢事甚至崩溃。事实上,这是黑客用来劫持他人计算机的最常见方式。

还有其他的计算机架构,比如哈佛架构(Harvard-architecture),它甚至比冯·诺伊曼架构的历史更悠久,其中代码和数据分别保存在存储器中物理上彼此分离的部分中。它用于嵌入式系统和微控制器,这不仅是出于安全原因,而且也出于速度的原因,因为这样你就可以并行地读取、写入数据和代码。

在现代多 CPU 架构中,我们看到许多新的冯·诺伊曼架构和哈佛架构的混合形式,这种架构经过优化之后专门处理特定问题的数据流和程序代码流。如何管理数据传输的存储和总线结构,这一直是尝试为问题寻找正确架构的工程师所关注的重点所在。处理这个问题还有大量其他可能性,其可能变种数之多可以和生物系统相媲美。和生物系统的主要区别是工程师不必等待数十亿年,而是可以有意识地塑造和不断调整他们的系统;当他们这样做时,他们使用上一代的机器来塑造和改进下一代。

我在这里提到这些技术性问题,是因为这些问题与神经生物学家和 AI 工程师审视脑及其智能表现时所需要解决的共同问题有很大关系。脑也有长时记忆和短时记忆,神经生物学家总是想知道这些内容是如何从一个部分转移到另一个部分的。这两种类型的存储机制差异很大,这当然是一个重要问题,但实际上我们对于脑如何完成这项工作知之甚少。

在过去的 70 年中,当工程师和生物学家并行工作时,总是有一种倾向,就和其他制造商销售。在我的两名学生为新的雅达利 ST 计算机做了改进之后，它成了最先使用图形用户界面的数据库程序之一，并成为雅达利机器上的一种标配。因此，雅达利的德国主管和卡斯帕罗夫团队、特别是他的朋友弗里德尔（Frederic Friedel）签订了一份合同。弗里德尔想要建立一个新的国际象棋程序和包含棋谱在内的国际象棋数据库，并正在寻找存储棋局和落子的最佳工具。当时个人计算机相对较慢，在半兆字节的 RAM 空间范围内工作存储非常有限（即使是今天的智能手机的容量也要比这大 8 000 倍）。所以我们与弗里德尔讨论了有关国际象棋落子的一种非常特别的标记法，这种标记法要非常紧凑以便尽可能多地将数据存储在数据库中，并快速加载到计算机的内存中以进行快速计算。当时我想不出有什么更好的方法来克服计算机性能的局限性，所以弗里德尔提出的标记法对我似乎很有说服力。[1] 当时我并没有认识到，人工智能还会有好得多的方法，不过这还得等上 30 年。我眼光太差而看不到这一点，但像欣顿那样的其他人则看到了。不管怎么说吧，虽然我发现弗里德尔非常令人喜欢和超级聪明，但是我在编制国际象棋程序方面并不能看到有什么真正的商机，因此就没有再参与这方面的开发。他在一位技能超群的程序员的帮助下实现了他的想法，并设法将“ChessBase”打造为世界上最著名的国际象棋程序和数据库之一。他仍然沉醉于国际象棋，1997 年卡斯帕罗夫和深蓝对战时，他是卡斯帕罗夫团队中的一员。

但打那以后时代已经发生了很大的变化，为了弄清楚我们已经看到了什么样的范式转换，让我首先来解释当时所面临的一些挑战。

在弗里德尔发明他特有的国际象棋标记法（这一方法几经修改之后一直沿用至今）的时代，计算机最大的问题是从稳定但缓慢的长时记忆库中把大量数据加载到快速但易变的 CPU 的工作内存。直到今天这仍然是计算机在性能方面的一个瓶颈，但当时这是一个巨大的问题。将内容从物理存储区域中的一个

---

[1] 感谢 Gert Hauske 教授审读了我们的手稿，他提示这一想法是否源于香农（Claude Shannon）早在 1949 年有关国际象棋程序的重要文章中的想法。就卡尔记忆所及，这种标记法是一种专用格式（proprietary format），可以肯定国际象棋数据库（ChessBase）的开发者确实知道香农的著名工作。Shannon C. Programming A Computer to Play Chess[J]. Philosophical Magazine，Bd. 41, 1950, Nr. 314.

也下得更快的人类棋手。它试图通过下列优势来弥补缺乏直觉的缺陷：对以前棋局的超强记忆，对更多步数作快速而合理的计算并避免错误，尤其是那些有时甚至会发生在最好的人类棋手身上的明显错误。国际象棋棋谱中充满了甚至是大师也犯过的致命错误的例子，他们在时间不够或激动之中没能看出原本很容易发现的危险或机遇。即使是早期的国际象棋程序也决不会犯这样的错误。

当时的国际象棋对 AI 研究人员来说就像是某种模式生物，类似于生物学实验室中的果蝇和秀丽隐杆线虫。而原来的范式是让计算机充分理解问题的事件空间并将其教到专家水平。那是本体论（ontologies）、基于专家和知识的系统的时代，程序员的重点是语义和语言。人工智能的主要问题似乎就是如何以最好的方式告诉机器人类对该领域所知的一切。

二十年前，机器只是亦步亦趋地做程序员告诉它要做的事。在 GOFAI（美好的老式人工智能）时代，程序员试图收集有关领域中所有可能的知识，在上面所讲的例子中就是有关国际象棋的一切知识。其目的是建立一个庞大的知识库，理解规则，从专家那里吸取专业知识，并利用机器访问大容量的存储器。

尤其重要的是"开局数据库"，其中存储了所有已知的开局的走法，包括某种走法获胜可能性的相关信息。这其实也是人类棋手用来准备比赛的方法。一名优秀的棋手可以知道数千种棋局（单是有名称的开局就超过 1 300 种！），并且根据剩余的棋子数量，还可以预测 5 步，有时甚至更多步落子。机器当然可以做得更好，如果时间不受限制，那么从原则上说起来，整盘棋都完全可以算得出。这在 20 年前是一个大问题，但是现在，即使使用基于 GOFAI 的软件，计算机也可以击败任何人。

还在深蓝与卡斯帕罗夫之战的 10 年之前，我就恰好面临过这个问题。当时卡斯帕罗夫是雅达利（Atari）计算机公司的广告合作伙伴。我的公司 ADI 软件公司（ADI Software）开发了一款数据库管理程序 Adimens，该程序在当时非常流行。它由苹果、DEC、惠普（Hewlett Packard）、利多富（Nixdorf）[1]、西门子

---

[1] Nixdorf Computer AG（1968—1990）是德国的一家计算机公司，1990 年为西门子公司收购成立西门子-利多富信息系统公司（Siemens Nixdorf Informationssysteme, 1990—1999），后又改成西门子利多富资讯系统（Wincor Nixdorf），专营银行业软硬件。——译注

的一个显著特点。他认识到,在神经网络之上还有更多层次需要去发现和认识,这个网络使蝾螈在其嗅觉系统的帮助下找到方向。他对嗅觉系统非常熟悉,对这些层面的研究一直进行下去,直到我们能认识到整个脑是如何工作以及意识怎么会产生出来为止。

我相信他是工作到了最后一天,因为我看到了几个月前弗里曼在加利福尼亚大学伯克利分校演讲的一段录像。他在讲话中对听众总结了他大半生所研究的理论。他的讲话就像你一样生动睿智,看着他把自己的思想和理论雕琢得有多好真是令人高兴。他运用了再传入的优化原则,发现这对于改进他自己的"不断逼近周期理论"(theory in continuous cycles of approximation)也非常重要。这个人有那么多的激情、自律和坚持,真是令人难以置信。

一位巨人走了,这确实是一个令人悲伤的日子!

但是下一代研究人员已经准备好站在他的肩上,有可能到达并发现他所看到的大陆。

他比大多数人都更清楚,现代人工智能和机器学习中使用的神经网络与其生物学名称几乎没有什么共同之处,年轻一代的 AI 工程师可能会从他的工作中获得比他预期的还要多的收益。

这一领域正在发生范式转换,但是这一次,大概不会像我们过去所看到过的那样,人们对 AI 的炒作和兴高采烈又导致下一个 AI 严冬。

你引用许博士的书《深蓝揭秘》正好切中这一现象。我以前没听说过他的下列说法:1996 年是作为棋手的人赢得了胜利,而 1997 年则是作为工具制造者的人赢了。

这句话在 20 年前是绝对正确的,因为卡斯帕罗夫不得不与许多专家联合起来的智慧进行竞争。他们把自己所有的知识和专长都融入了软件和数据库之中。尤其重要的是包含"开局手册"的数据库,它允许机器在下棋开始阶段选择已知和已评估的最佳走法,而无需进行太多计算。

下一步是要预见尽可能多的步数,对棋局进行评估,根据一些早就确定了的原则经过合理的计算挑出最佳走法。所以工程师们试图在下棋开始之前,就尽可能多地把自己的知识装到机器里面去。机器就好像是一位知识更丰富、棋

Ⅲ-004 卡尔 55

Ⅲ-004　卡尔

（人工智能领域正在发生范式转换；从深蓝到深度学习；并行计算）

2016－05－27

亲爱的凡及：

今天早上我很高兴在收件箱里找到了你的电子邮件，但是你的朋友弗里曼过世的坏消息让我非常难过。失去一位朋友总是很糟糕，但是这次情况更糟，因为你也失去了一位智者，他再也不能回答你许多迄今还没有答案的问题了。

虽然我从未见过弗里曼，但我对他的思考方式非常熟悉和有认同感。我以前也听说过他的名字，但是只有在你把他的工作介绍给我之后，我才认识到他的工作有多么重要。他不仅在神经动力学领域，而且在系统生物学和许多其他领域都是一位重要的思想家。你经常阅读一些有思想的作家写的有趣书籍，但是很少能碰到像弗里曼这样能打动你的情况，因为他总有些话会以他的方式丰富你的知识和拓宽你的视野。我早就在谈及《意识、意向性和因果性》这篇虽然篇幅很长，但内容丰富而紧凑的文章时，提到过这个印象。它写得就像是一份遗产，仿佛想要把他40年来对同一谜题不断钻研的结果，记录、总结和整理出来说给下一代研究人员听。

给人留下深刻印象的不仅仅限于他对这个领域的洞察力和能力，也还在于他的高度谦虚。他从不假装知道自己不懂的东西，也从不把在一个层面上奏效的好主意夸大为整个系统的基本原则。在这点上，他正好和那些假装知晓一切、夸夸其谈的家伙及炼金术士们截然不同。他明确地指出，对于脑是如何工作的问题的许多方面，我们甚至还毫无所知。他强调要把注意力集中在"介观"层面，即分子微观层面和整个身体以及脑宏观层面之间的层面，这是他的工作信号元件联结的非符号处理方法。后者主要指的是神经网络研究。大体同一时期,神经网络研究在经受了自身"两起两落"之后,同样迎来了第三次崛起,"深度学习"(deep learning)成为再度崛起的关键。它是一种多层学习算法,在网络不同层次上自动抽提对象的不同特征,尽管这一思想早就有了,但层次一多,就要求计算机的存储容量和计算速度都大大提高。而且为了成功训练网络,要求有海量的数据样本,给它的样本越多,它就学得越好。这些条件在20世纪90年代以前都还不具备。

一直到21世纪才有加拿大计算机科学家杨立昆(Yann LeCun)及其导师欣顿(Geoffrey Hinton)断言,由于计算机技术的飞速发展和网上的大量数字信息,已经到了"深度学习"可以大展宏图的时候。2009年欣顿等人报道,经过"深度学习"训练的神经网络把语音转化为打印出来的文字,达到了很高的准确率;而采用传统的基于规则的方法,长期以来进展甚微。这立刻引起了智能手机业者关注。此后该方法很快被引入识别人脸、识别语音命令、无人驾驶汽车等领域。战胜围棋冠军则不过是最吸引公众眼球的一场表演罢了,就其技术本身而言并无革命性的突破,它靠的是高性能计算机、大数据和深度学习。当然,其宣传效应非同凡响,而且把此技术移植到医学影像识别、语音识别、机器翻译、交通以至城市管理等实际应用领域有很大的价值。

人工智能的第三次崛起正在走向高潮!

1993 年期间,人工智能遇上了第二次"严冬"。

分析人工智能两度遭遇"严冬"之缘由,首先要看到人工智能在发展之初是以符号处理为主流,通过一步接一步的逻辑推理来解决问题。纽厄尔和司马贺曾说过:"物理符号系统中有既充分又必要的工具,来实现人的各种智能。"可是人的智能不完全是符号处理,还包括许多所谓的"亚符号处理",人有非常快速的直觉判断。例如,艺术鉴赏家能够一眼看出赝品,这不是通过一步步逻辑推理得到的。在知觉、模式辨认、导航和学习等许多方面,也都是如此。这些内隐的知识构成了符号处理的背景知识。与此相比,传统人工智能想仅仅用符号处理的方法来解决所有的问题,遇到困难就非常自然了。

跟"符号处理"思潮相呼应,当时对人工智能的许多问题都试图用搜索一切可能解来解决,而实际问题中的可能解数目往往极为庞大,用当时那种性能的计算机根本不可能全部搜索,即便使用当今速度最快、容量最大的计算机都难以做到。因此,要想解决诸如计算机下围棋之类的问题,不能不另辟蹊径。另外,知识表达和知识工程在传统的人工智能研究中起核心作用,但任何常识性规则都有大量例外,计算机若没有足够快的运算速度和足够大的存储容量,就无法胜任海量信息的存储和在合理时间内的提取。

人工智能的第三次崛起

自 20 世纪 90 年代中期以来,得益于计算机技术的飞速发展,芯片的特征尺度从 1971 年的 10 微米下降到 1994 年的 600 纳米。人工智能终于迎来了它的第三次崛起,在数据挖掘、医疗诊断等方面取得了许多应用成果。这次崛起或者说"第三春"的标志性事件,是 1997 年 5 月 11 日计算机国际象棋下棋系统"深蓝"(Deep Blue)战胜世界国际象棋冠军卡斯帕罗夫,宣告了人工智能的"王者归来"。不过这一胜利主要得益于计算机技术的发展,就其思想路线来说,依旧和传统的人工智能一脉相承。与此类似,计算机知识竞赛抢答系统沃森战胜"危险!"抢答人类冠军是另一个标志性事件(详见本系列丛书第一册《脑研究的新大陆》中背景专栏 I K1.1)。

在人工智能领域,除了前述通过程序编制的符号处理方法之外,也有通过

Simon)开发的一个程序能够证明罗素巨著《数学原理》一书头上52个定理中的38个,其中有些证明甚至比当时已知的证明还要简洁和巧妙。到20世纪60年代中期,人们对人工智能充满乐观,而当时的人工智能也确实在应用符号处理方法演示人的高级思维方面取得令世人印象深刻的成绩。司马贺当时乐观地预言:"在今后20年内,机器将能做人所能做的任何工作。"甚至声称他们的"研究解决了古老的心身问题,清楚说明了由物质构成的系统如何会有心智的种种特性"。

然而到了20世纪70年代中期,他们的预言并不见即将实现的征兆,连一些原来认为可以轻而易举实现的课题都未能做到,令人大失所望。美国军方曾对用人工智能方法自动翻译当时苏联的文献资料寄予厚望,结果由于机器不能理解词的歧义,与上下文关联,在译文中犯了许多令人喷饭的常识性错误,比如把"眼不见,心不烦"(Out of sight, out of mind)译成了"盲白痴"(blind idiot)。这是由于当时计算机计算速度和存储容量极其有限,很难取得突破。专家评估认为,比起人工翻译,机器翻译代价既高又不确切。当然,机器翻译到今天已大为改观,在许多场合已可付诸实用,但其中也还有许多问题依然未获解决,在我们的后续通信中还要对此进行讨论。在此后若干年(1974—1980)里,英美政府对人工智能研究的拨款大幅减少,这段时间被称为"人工智能的严冬"(第一次)。

20世纪70年代,由于计算机的存储容量已经相当大,人工智能研究者开始考虑建立知识库的问题。美国计算机科学家费根鲍姆(Edward Feigenbaum)首先提出"专家系统"的思路。20世纪80年代初,"专家系统"在商业上取得成功,人工智能迎来了第二轮的蓬勃发展。所谓专家系统,是一种回答或解决某个特定领域中问题的程序,它运用由专家的知识所建立的逻辑规则来解决问题。1985年,产业界在人工智能方面的投资超过了10亿美元。另一重大事件是1982年日本政府提出"第5代计算机"计划,声称要实现计算机与人对话、翻译、图片解释以及像人一样推理。这也刺激了英美政府重新投资该领域。不过好景不长,到1991年此计划并未实现其主要目标。专家系统为避免"常识问题",只能将应用局限于范围很窄的专门领域;一些原来成功的专家系统由于不能根据新获得的信息进行更新,而使得用户难以承受维护成本。这样在1987—

和弗里曼的话却如此相似!

期待着听到你对这些开放性问题的进一步评论。

凡及

背景专栏ⅢF4.1

人工智能的三次崛起和两次严冬

人工智能的前两次崛起和两次"严冬"

虽然在中外的古代神话和传说中,人们早就提到和生物真假难分的机器,但是对于机器是否也能有智能的第一次科学思考,当属英国数学家图灵提出的"图灵测试"(详见本系列丛书第二册《意识之谜和心智上传的迷思》中背景专栏ⅡF5.1)。不过一般认为,人工智能作为一个科学领域正式诞生于1956年在达特茅斯学院(Dartmouth College)举行的一次学术会议上,会议的倡议书中这样写道:"学习或智能的任何特性在其每个方面都可精确地加以描述,由此就能用机器来模仿。"以后一般把人工智能理解为机器或者软件所表现出来的智能,或指研究如何创造出有智能行为的计算机或软件的学科领域。当然,这里有个没有解决的问题,就是什么是智能。这依然是一个见仁见智的问题。在我们的通信中也讨论了这个问题。

传统人工智能要实现的所谓"智能",包括推理、知识、计划、自然语言处理(通讯)、知觉和物体操控等,其远期目标也包括创造有通用智能的机器。这里的通用智能是指人类所有可能的智能。当时的人工智能专家把认知过程当作某种符号处理过程,以后又发展到用符号表示知识,并把知识作为智能的基础。这就是"人工智能的符号主义思潮"。确实,他们当时编制出了能解代数题、证明逻辑定理和说英语的程序。例如,纽厄尔(Allen Newell)和司马贺(Herbert

量数据进行训练,并消耗大量能量来解决问题,而人脑只要学习少量数据并只消耗 20 瓦。现在即使在弱人工智能的范围里说人工智能击败了人脑,也还为时过早。我不能确定现在的人工智能是否就走错了方向,无论如何它们可以解决许多实际问题,甚至比顶级人类专家做得更好,而且似乎还有广阔的发展空间。然而,正如你所说,如果以为人工智能就应该照现在的样子永远这样发展下去,那可能是完全错误的。而"最大的问题是,要想纠正这些错误代价非常大和耗时"。

你关于启动效应的话让我想起了弗里曼的一段话,他试图解释为什么人们不愿意改变他们的领域,即使他们知道一个新的领域可能更有前途。他在为我和我的同事翻译的他的著作《神经动力学:介观脑动力学的探索》[1]（*Neurodynamics: An Exploration in Mesoscopic Brain Dynamics*）所作的中文版序中写道:

> ……如果这一领域(神经动力学——引用者注)真的前程远大,为什么还很少有人问津?这里要考虑到三条理由。……为什么脑理论研究者如此稀少的第二条更有说服力的理由是,成熟的研究人员在他们受过训练的学科中已作了大量的知识和经济投入。毫无疑问,他们也有强烈而深刻的个人爱好,这对任何科学领域的成功都是必不可少的;他们也满足于继续利用自己熟悉的技术,沿着早已富有成果的研究路线继续取得成果。除非有清楚且非常吸引人的好处,否则几乎不会有动力去改换领域。这些考虑使我们有理由认为,介观脑动力学这一新领域迅速发展成为一个新潮流的最佳希望在于招募那些尚未坚定地致力于现有观点的科学家。仍处于成长期并尚未为大量现有学科知识所累的年轻人,在这方面可能比成熟的科学家更具有实质优势。

虽然你从未读过他上述的话,因为我想你不会去读一本中文书,你的评论

---

[1] 中译本:弗里曼.神经动力学:介观脑动力学的探索[M].顾凡及、梁培基,等译.杭州:浙江大学出版社,2004.——译注

Ⅲ-004　凡及

49

定他们的论点是否会永远正确。无论如何，人脑只是一类虽然有其特殊性的物理系统，这种物理系统确实具有这样的特性，那么为什么其他物理系统原则上就不能具有类似的特性呢？至于我们是否应该开发这样的系统则是另一个问题。

第四个反应是将会有一种新物种，一种将人与智能机器融合起来的混合体，这将是一种超人。他们认为，已经有人植入一些芯片来加强他们的能力。我不明白为什么人们应该冒险在体内植入一些外来的芯片来增强他们的能力，如果他们本身是健康的话，我们已经使用了各种机器来加强我们的能力，这使得我们在祖先的眼中看起来就好像已经成了超人。不管怎么说，在我看来，科学的目的是要使绝大多数人幸福，而不是成为超人。

就像你一样，我也不想贬低人工智能在近几年里取得的进展。就在十年前，当对计算机和人脑进行比较时，人们往往会说，尽管计算机自发明以来，在进行计算和逻辑运算方面，计算机比人脑算得更快，也更准确；但在物体识别和动作方面，一个 3 岁大的孩子的脑可以轻松打败任何计算机，例如从一群人中识别出母亲的脸或穿过繁忙的街道。现在人们再也不能说这样的话了。如你所知，我不熟悉技术和人工智能，在这方面我是一个门外汉，我不能确定我的印象是否正确。我认为，现在取得这样的突破并非偶然，十年前还不可能发生，因为当时没有像现在这样强大的计算机，也没有互联网共享大量数据，没有这些条件，深度学习仍然只是某种理论成就。有人说，深度学习是模仿脑的胜利，然而，正如我们以前讨论过很多次的那样，这最多只是一个受脑启发而取得的成就。CNN 的组织与脑完全不同！深度学习为人工神经网络提供了通过训练数以亿计的数据解决实际问题的能力，而不像它的祖先，如感知器，只能解决玩具世界中的许多简化问题。深度学习的神经网络具有一定的提高技能的能力，根据王培的定义，他们是有智能的，但是这种改进有一定的局限性，迟早会达到一定的饱和水平。脑没有这种限制，他们不仅可以改进一些特殊技能，而且还可以学会新技能。所以深度学习仍然是一种弱的人工智能技术。它不属于通用人工智能，更不用说是有内心世界的人工智能！

你的话绝对是对的，即使在弱人工智能的范围内，人工智能机器需要用大一些人,比如王培博士正在研究通用人工智能(抱歉的是,我还没有读过他有关这个问题的原创性论文),虽然我不知道现在应用方面是否已经有了任何成功的通用人工智能。看来现在人工智能的所有成功应用都属于弱人工智能,它只能解决一些非常特殊的问题,比如下围棋、自动驾驶等。尽管许多人现在都会谈论起人工意识,甚至有一个刊名为《人工意识》(*Artificial Consciousness*)的国际期刊,不过当人们甚至还不能给意识下一个可操作的定义时,我非常怀疑现在是否已经到了研究这个问题的时机。在我看来,对你根本不了解的东西进行逆向工程是荒谬的。另外,我不明白为什么现在人们要想创造出一种有自己心智和意志的人工智能机器人,即使目前这事还不可行。开发人工智能来帮助人类去做一些人自己做不到的事情是一回事,而开发出有自己心智和意志的人工智能机器人则是另一回事,是否应该像禁止化学武器或生物武器一样也禁止这样的研究?

在中国,对阿尔法狗胜利的反响与你在信中所说的很相似。很多人相信"人类的末日将近""机器和机器人将接管控制权",你在信中清楚地解释说,情况并非如此。然而,这给人们造成了这样的印象:"谷歌和深心开发了一种非常强大而对人类构成威胁的 AI 系统",这使得谷歌在竞争对手面前占据了非常有利的地位。

第二个反应是人们担心在阿尔法狗之后还会有阿尔法医师、阿尔法律师、阿尔法会计师、阿尔法出租车司机等,许多人会失去工作。事实上,自第一次工业革命以来,类似的担忧曾一再以不同的形式反复重演。当然,很多老职位会丢失,但也会出现新的工作。在上海,由于阿里巴巴或京东的网购,许多小商店已经关闭,但是客户从网站订购商品所需要的送货人员数量急剧增大,尤其在节日期间更是如此。当然,我不能确定新增职位的数目是否比丢失的旧职位还要多。然而,在历史上,情况就是如此,并且给了人们更多的空闲时间去做创造性工作或享受生活。这次的情形是不是也是如此,我不确定,但政府和社区应该提前关注这个潜在的社会问题。

第三个反应是来自一些生物学家,他们说机器永远不会有创造力、洞察力、心智和意识。虽然就目前和可预见的将来来说,情况确实都是如此,但我不确里得知的那样,弗里曼几乎工作到了他的最后一天!据说由他和其他人一起编辑的一本新书很快就会出版。多么好的一个人走了!他会永远活在朋友们的心中。

正如你在信中所提到的那样,IBM的深蓝计算机在1997年击败了国际象棋世界冠军卡斯帕罗夫,这可能是人工智能第三次崛起的第一声春雷。你的话让我想起了深蓝国际象棋弈棋系统设计师许峰雄(F. H. Hsu)博士在他的书《"深蓝"揭秘:追寻人工智能圣杯之旅》[1] (*Behind Deep Blue: Building the Computer That Defeated the World Chess Champion*)中所说的话。他宣称这场比赛是在两种扮演不同角色的人之间进行的:一位是棋手,另外一位是这个工具的制造者。卡斯帕罗夫两次对阵深蓝,结果不同。在1996年的比赛中,作为棋手的人赢了;而在1997年的比赛中,作为工具制造者的人赢了。至于深蓝是否有智能的问题,他的回答是,深蓝根本就没有智能,它只是一种制作精良的工具,只能表现出某种行为,就好像是在某些有限的领域中有智能似的。你指出这种领域就是"具有有限规则集的公理化弈棋"。卡斯帕罗夫是一名真正有智能的棋手,虽然他在比赛中是输家。举个例子,卡斯帕罗夫多次抱怨说在比赛中有一些欺骗行为,许峰雄说深蓝决不会像卡斯帕罗夫那样凭空指责别人。我认为许峰雄和你已经把这件事解释得很清楚了(我不想在这里再引用你的话了,因为你懂得的比我多☺),为什么还有人要宣称阿尔法狗AI系统击败了人类呢?!尽管打败围棋冠军比战胜国际象棋冠军要困难得多,但我认为这两者在原则上并没有本质上的区别。所以我对阿尔法狗的胜利也不会感到惊讶。唯一让我感到意外的是,这一天比包括我在内的许多人预料的要早得多!

顺便说一句,在你的信中,你说过你"相信从长远来说人工智能会取得成功,甚至强人工智能也会成功"。你在这里所说的"强人工智能"指的是什么意思啊?正如我们所知道的那样,对"强人工智能"一词有两种不同的理解,即通用人工智能和有意识的人工智能。一般来说,我同意你对两者的期望。现在有

---

[1] 有中译本: Hsu F H. "深蓝"揭秘:追寻人工智能圣杯之旅[M].黄军英,等译.上海:上海科技教育出版社,2005.——译注

Ⅲ-004　凡及

（深蓝国际象棋弈棋系统并没有智能；强人工智能和弱人工智能；通用人工智能和人工意识；不应该开发有自己心智和意志的人工智能机器人；人工智能的社会反响；需要大数据训练和高耗能是当前人工智能发展的瓶颈之一）

2016－04－28

亲爱的卡尔：

　　我很遗憾地告诉你一个不幸的消息。我刚刚从弗里曼教授的博士生、长期合作者莱斯利·凯（Leslie Kay）[1]教授那儿得知，弗里曼于本月24日在伯克利家中平静地去世了。我很遗憾不能去那里参加他的葬礼，只好请她代我向他的家人致以最深切的慰问。神经科学界失去了一位伟大的科学家和思想家，我失去了一位像导师一样的朋友。

　　他随时准备帮助别人。正如我以前告诉过你的那样，我多次向他求助，他总是及时给我建议，即使是非常微不足道的事情。例如，有一次我问他"neural"和"neuronal"这两个术语有什么区别，这让我在阅读文献时感到困惑。他告诉我，这两个词几乎是一样的，只有当你想特别强调某事和神经元密切相关时，那么你就只能用"neuronal"，而不是"neural"。尽管这是一个非常简单的问题，但我很难在我的国家找到合适的人请教。不是以英语为母语的神经科学家可能会感觉到这种微妙的差异，但可能并不能肯定地给我一个明确的答案；而英语语言和文学教授则恐怕完全不了解神经科学。正如我从莱斯利和其他同事那

――――――――――

[1] 美国计算神经科学家。——译注

似乎并不是神经科学中的每个人都认识到了这一点。

我也很喜欢你在这方面引用了王培，以及他在讨论中提出的补充意见。适应性方面确实对此增加了一个相关维度，有助于区分不同类型的智能行为。你所引用的法布尔的著名泥蜂实验给出了一个极好的例子。我也和你一样特别喜欢王培把技能和智能区分开来的想法。但是当你想要测量昆虫或人类的纯粹的"不带技能"的智能时，也会碰到问题。当你测试一个能够学习的系统时，不管它是生物系统还是基于硅芯片的人造系统，时间和经验都要发挥作用。随着时间的推移，智能可以提高能够学习的系统的技能。对泥蜂来说，进化起着重要作用，每只现代泥蜂都享有数百万只泥蜂通过进化造就的本领。我们人类的情况要好得多，因为在语言的帮助下（当然还有祖父母☺），我们可以积累、转移、应用（或理清）我们的祖先和我们的伙伴先前获得的知识。

时间不仅是这个过程中的一个主要因素，而且也是我写作时的一个主要因素，因为我再次用尽了时间和空间，尽管我还没有回答你的所有问题，甚至还没有告诉过你有关艾伦的事。☹

我很抱歉成为这样一个不守纪律的笔友，希望你能原谅我。

祝好，早安/晚安。

卡尔

师的头脑中抢先灌输非常重要,谷歌投入了大量资金以引领潮流并在先入为主方面独居鳌首。从我周围年轻的 AI 朋友来看,谷歌的做法确实起了作用,因为许多人都愿意赶上 TensorFlow 的潮流。从工程师的角度来看,当然这是可以理解的,人们总希望站在胜利者一边。他们希望知道什么会成为新标准,而不想把时间浪费在学习可能很快就会过时的框架、工具和语言上。但是在 IBM 部门中快乐地当上一名 COBOL 程序员的 40 年光景已经一去不复返了。要是我说得对的话,现代神经网络架构的基础不牢,真正的创新还在我们前面,还没有公认的实际标准,因为长江后浪推前浪来势汹涌。

要是站在中国人工智能制造商的立场上,我不会只依靠拥有接近 10 亿用户和相关的大量数据的优势。我宁愿寻找可以用较少数量的数据集,尤其是用较少能量就能解决问题的方案,就像一个 2 岁小孩所能做到的那样。现在我们已经看到机器可以做到图像识别、语言理解和翻译之后,如果能看到有解决方案不仅只是从线性的尺度上来说要更好,而且要好过好几个数量级,我也不会感到惊讶。而未来的这种飞跃并不一定非来自拥有大量资金和资源的工业重量级的大型研究实验室不可。在这方面,也可以是基于正确的思想提出完全不同的方法或算法的某个小集体。这种情况在过去就经常发生,当工业主流都把希望寄托在那些显然不合适的原则上时,这种情况尤其可能再次发生。无论如何,前景令人兴奋,我很好奇想看到接下来会有什么让我们吃惊的事发生。

我不得不承认,我的观点感觉上不那么好,我也提不出更好的解决方案。但我很高兴你、弗里曼和冯·诺伊曼似乎都同意我的观点,也就是将计算机与人脑相类比有不妥的地方。

我很高兴你选择布赖滕贝格小车作为出发点来讨论智能的要素。浏览一下文献就会给人留下下列印象,有多少作者写有关智能的著作,那么对智能这一术语的定义也就有同样那么多。要想寻求"正确"的智能定义是徒劳的,因为在开放系统中,关于现象的所有术语都是可变的,而不是像在数学或逻辑那样的封闭系统中的公理。只有当涉及对某个智能行为的具体理论的实证检验时,智能定义才可能多少有点帮助。实现某个变量的操作化过程往往很艰难,也很烦琐,必须始终根据相应的行为进行调整。这是一个几乎显而易见的道理,但的教育和培训,从感觉上和表现上都就像是一个属于某强大教会的神职人员。他们所用的拉丁语就是 COBOL,而其教义问答就是描述著名的 IBM -370 -大型机架构的手册。

有一阵子微软看起来像是 IBM 的继任者而成为主宰者,但他们没跟上互联网与相关的搜索引擎、社交网络和手机革命的形势,很快就失去了对操作系统和编程语言的控制权。现在是谷歌和苹果在定义操作系统和相关语言。

在所有这些创新和革命中,语言、平台和相关工具始终是关键因素。人脑所学习的第一种计算机语言,似乎与学习母语一样具有定义性(defining)并持续产生重大影响(formative)。从一种编程语言改换成另一种有不同逻辑的编程语言,就像从母语改换为外语一样困难。

大多数人总是通过在以前语言的基础上进行模仿来使用他们的新语言。当"面向对象编程"在社会上流行起来后,我看到许多优秀的程序员为一些新概念伤透脑筋,而且始终未能真正熟悉它,因为他们总是试图在他们习以为常的旧有基础上模仿这种新逻辑。

至少就学习语言而言,在我们的心智架构中有些部分的行为,似乎就像只能写入一次的只读存储器芯片一样。同样的机制似乎是造成我们大多数人从未摆脱过我们母语特有的方言和腔调的原因。你可以立刻听出一位用英语演讲的人是否是德国人或法国人。德国人通常学不会以正确的方式发英语中的"V"音,而以"W"音代替;典型的法国人也不能正确地读出德语中的"H"音。我想甚至当你听到有人努力用纯正的普通话讲话时,你也可以听出他是中国哪个地区的人。

虽然要想纠正发音错误和改进口头表达能力还是有可能的,但这要做出很大的努力,因此大多数人不愿意这样做,而宁愿保持他们在心智和语言方面早先养成的习惯。要想改变我们最先学习到的语言的基本原则是极端困难的,这使我非常怀疑乔姆斯基(Chomsky)所说的存在一种天生的、普适语法的想法,他认为这种语法植根于人脑最低层次的某处。

无论如何,谷歌的经理和 IT 行业中的每个人都知道这种启动效应(priming effect)。现在一个人工智能的新时代刚刚开始,成为领导者和在年轻一代工程是因为他们有更多的用户，因此可以使用大量的数据来训练他们的神经网络。这可能是对的，至少对于真需要数亿张图片或文件来教会神经网络区分猫和狗这样的情形来说是如此。这种教学是通过计算这些图片的元素之间差异的统计量来进行的，并且你拥有的例子越多，得到的统计结果也越好。它真的有效，有时甚至非常好，而在其他一些情况下，这种办法也会失败。但无论如何，它需要大量的硬件和能量。

当我想到一个 2 岁大的孩子可能会在看了 10 或 20 只猫和狗后也能解决同样的任务时，这让我觉得这里有什么地方不对劲。也许这是超级计算机所犯的另一个同类错误，超级计算机即使从理论上说起来也必须要消耗千兆瓦级的功率才得以匹敌靠 20 瓦运行的脑的能力。对我来说，这些算法网络看起来甚至有根本性的错误，这些网络被冠以"神经"的荣称，就好像它们真的模仿了我们的脑似的。无论是谁，只要观察得更仔细一点的话，就都会发现事情并非如此。之所以觉得有这样的工具非常好，只是因为它们比我们以前有过的工具要好得多，所以我们必须认识到，实际上这只是刚刚迈出的一小步。

问题是这一步也可能会引向错误的方向，就像我以前提到过的蒸汽机那样。大问题是纠正这种错误代价很高也很费时。当你开始按照某种不适当的原理来建立一个行业时，在很长一段时间内要想纠正这样的错误几乎是不可能的。

对我来说，谷歌大张旗鼓上演的围棋胜利主要是一场营销活动，其目的是将全世界的 AI 程序员吸引到他们的 AI 平台 TensorFlow[1] 上来。在这里，他们与脸书、苹果、百度、微软、IBM、亚马逊等竞争，这确实是一场非常重要的竞赛。它可以与 20 世纪 80 年代的 PC 操作系统之战，以及目前最流行的编程语言的竞争相当。IT 巨头 IBM 在计算机界垄断了半个世纪，但在操作系统之战中败给了微软，而且从未从中恢复过来。在过去，一家公司的 IT 中心就被称为 IBM 部门，运行这些机器的人员都接受了 IBM 操作系统和 IBM 编程语言方面

---

[1] TensorFlow 是一个开源软件库，用于各种感知和自然语言处理任务的机器学习。目前被 50 个团队用于研究和生产许多谷歌商业产品，如语音识别、Gmail、谷歌相册和搜索等。——译注

愿意发出 AI 的可怕后果的警告。

当然谷歌在应用神经网络方面正在取得一些进展。但是,在 IBM 的象棋演示(以及在这 20 多年中摩尔定律依旧生效)之后,战胜围棋冠军的胜利并不能说明机器就优于最好的数学家。这就像是计算机在不到一秒钟就能算出 π 的一百万位小数,凭此就说计算机优于最好的数学家一样的荒谬。公理化的逻辑游戏只是真实世界极小的一部分,这一领域正是递归神经网络和专用硬件能充分发挥其长处的地方。这是库兹韦尔所倡导的应用容量更大、速度更快的硬件方法真正可行的一个领域。谷歌的公关人员这次取得了胜利,但最近神经网络在应用方面还取得了更多的重要进展,却几乎没有人注意到。

多年来人工智能面临的最大挑战之一就是人脸识别。即使识别不同种类的动物,对于一台机器来说也是一项非常困难的任务。多年来,我一直用计算机和算法难以区分猫和狗为例,说明人工智能几乎没有取得什么进展。最近在脸书开始分析人们在那里存储的无数图片之后,这一点已经发生了变化。现在,在他们用数亿张照片训练了自己的网络后,它们不仅能很好地区分猫和狗,而且还能很好识别出具体到是哪个人。顺便说一句,他们并不像谷歌那样使用递归神经网络(RNN),而是由脸书的人工智能研究组(FAIR)负责人杨立昆(Yann LeCun)发明的卷积神经网络(CNN)。CNN 是早期感知器的一种更有智能的变体,感知器是专门为图像识别设计出来的第一种人工神经网络。

脸书做出的进步在我看来更为重要,而且非常令人惊叹,因为有可能从视频中识别人是一种真正的突破,而计算机赢棋却不是。有趣的是,FAIR 也在尝试使用 CNN 进行自动翻译,谷歌也用 RNN 来实现自动翻译。

在改进机器翻译方面确实还大有可为,并且在我们面前还有大量应用领域前景光明。迄今为止,谷歌和微软提供的文本翻译器对于简单的文字和事情,例如了解互联网网站大概在说什么,有时是相当有帮助的。但用于复杂的文本甚至文学,结果往往相当差、误导和无用,甚至完全荒谬。

然而,这将是 AI 领域的竞技场,因为谁要是能够发明可以阅读和听取书面或口头语言并真正理解的机器,他们就将会改变比赛。

有人说中国的百度、腾讯这样的人工智能玩家在这场比赛中占有优势,这就远远超过了 20 年前的深蓝。要是有一个存储了所有曾下过的棋局的数据库,那么整个事件空间就全知道了,机器也就可以访问到所有的棋局。对于这种由简单规则规定好了,按部就班运作、然而可能性极多的应用,计算机的蛮力确实很有帮助,但这与你和王培所谈论的智能无关。

如果阿尔法狗建议他的程序员们,不要为这样毫无意义的演示而浪费数亿美元、耗费数千兆瓦·年的电,而且提议只要围棋冠军输了,就给他一千万美元。那么阿尔法狗就真走出了确有智能、出人意料(也非常幽默)的一步。这也让程序员有机会将阿尔法狗引入下一个智能层次:道德规则。这两种做法从公关效果上来说是一样的,而这笔钱本可以花在更有意义的问题上。☺

所以,尽管我相信从长远来说人工智能会取得成功,甚至强人工智能也会成功,但你知道,我对这场表演并不满意。然而,社会反响表明,这在市场营销和公关方面都是一次巨大的成功,因为所有媒体人都热衷报道。不可避免地有许多警告者站出来,一起声称人类的末日即将来临,这次是因为机器和机器人将会控制世界,这种感觉被煽动了起来。当然从一场相当愚蠢的弈棋胜利中就得出这样的结论纯属胡说八道,但它让每个人都相信谷歌和深心已经开发出了一种强大的 AI 系统,它对人类构成了威胁。精明的炼金术士都知道,当专家警告说炼金会带来灾难性的后果时,他们以实现其许诺的奇迹而募集资金的机会就增大了。

如果专家们声称在任何情形下我们都不要炼金,否则的话,经济就会崩溃,这对推销炼金术就太好了。这类警告提高了炼金术士的知名度,并把讨论的注意力都吸引到了这种坏结果上去。至于说到可以炼出金子来的许诺,几乎无人能加以判断,也就变得无关紧要了。对于公众和潜在的投资者来说,这种警告中的情绪性成分似乎成了理性警告。人类是容易逃跑的动物,当人群中有人发出警报时,人们往往不假思索地恐慌逃跑。这种反射一再起作用,其可靠程度一如你的泥蜂例子。因此,每当有人发出"末日将临"的警告时,人们都会去读启示录,没有人会再去问炼金(或是随便什么许诺过的奇迹)是否可能的问题。所以,一个聪明的炼金术士总是和警告者为伴。人工智能的大师也是如此,特别是当他们知道可以依靠硅谷商界人士的帮助时,而后者总是乐于提供帮助并

Ⅲ-003 卡尔

（阿尔法狗;人脸识别;自动翻译;人工智能需要
的下一个突破点:降低能耗和小样本学习;语言学习;
启动效应;技能不等于智能）

2016-03-20

亲爱的凡及:

　　谢谢你的亲切来信、节日问候和允许我引用你极妙的冰山隐喻!

　　我一直在思考你的许多论点和你给出的有启发性的见解,但有鉴于最新消息,我想从你在附言中提到的话题开始,也就是谷歌阿尔法狗人工智能系统和围棋冠军李世石之间的比赛。现在结果已经出来了,阿尔法狗在五局中赢了四局。这是在 1997 年 IBM 的深蓝计算机战胜国际象棋世界冠军卡斯帕罗夫,以及稍后 2011 年沃森(Watson)战胜知识抢答"危险!"冠军之后,机器再一次击败了人类。

　　好吧,计算机和人工智能界很少有人对此感到惊讶,大多数人都猜到了这一结果,其中也包括我在内。事实上,我倒想知道李世石怎么能够赢了一场,我觉得这整个运作更像是一场巨大的市场营销,几乎没有多大证据表明人工智能在质上有了突破。或者正如我的一位拥有信息学博士学位的年轻朋友在他有关人工智能的论文中所说:"对于计算机来说,还没能在家庭游戏之外大显神威。"他是对的,因为下棋只有一组有限的规则,并基于公理之上,这是计算机在数百万个并行内核上运行算法的天堂,也是模糊的单个人脑的噩梦。而且如果下棋中可能状态的事件空间越大,计算机的这种优势就越大。如果棋盘上有100×100 格,而不是普通国际象棋中的 8×8 格,那么计算机的优势就会更大。由于围棋比起国际象棋来说有更大的活动空间,所以阿尔法狗所具有的优势也附：你一定也注意到了 1 月 27 日的一条消息，深心公司开发的围棋弈棋系统"阿尔法狗（AlphaGo）"[1] 与前世界冠军李世石（Lee Sedol）下个月将进行比赛。人们正在热议 AI 机器是否能够击败全球顶尖的围棋棋手，尽管它在去年 10 月击败了欧洲冠军樊麾先生。我对结果很好奇，你怎么看？

_____

[1] AlphaGo 可译作"阿尔法围棋"。本书采用了报纸期刊广泛采用的、大众较熟悉的"阿尔法狗"译法，更显生动一些。

层面向上的线性因果关系的方向发展,然而,他们似乎很少关注"全局状态向下组织单个神经元的活动"。这仅仅是我的印象,我不确定是否真的如此,你可以告诉我,我的印象是否有道理。

你关于技术是否应该总是通过数值计算才能发展的论点非常有启发性。高电压和低电压之间翻转所消耗的能量使得这种机器非常耗能。你的许多极好的例子表明,要实现与数字技术相同的功能也可能有更简单和节能的方法。当思想小球落入吸引子的某个吸引域时,往往难以跳出来。⊗

你的例子让我想起了拉马钱德兰博士用以治愈幻肢痛的镜盒。一些截肢者可能会感觉到失去的手指紧紧地抠入手掌,这让他们感到无法忍受的痛苦。拉马钱德兰博士知道,视觉往往在多模态的竞争中胜出,他认为如果他能找到一种方法让病人看到失去的拳头松开,疼痛就可能会消失。当然,现在我们有各种各样的虚拟现实技术,但是这在 20 世纪 90 年代非常昂贵,可能他所有的经费都不足以购买这样的仪器。因此,他只是买了一面大镜子,把它放在一个大盒子的中间,将盒子分成两半,并在镜子两侧的盒子前壁上开了两个孔,以便让病人能够把他的胳膊放进洞里。然后病人可以在镜子中看到健康手的像,就好像是他失去的手一样。当病人松开健康的拳头时,病人注意到他的幻影拳头也像是镜子里的像一样松开了,疼痛消失了!

当然,你可能知道这个故事,但是当我第一次在他的经典著作《脑中魅影》中读到这个故事时,我很兴奋,我也把它翻译成了中文。拉马钱德兰是一位利用低科技手段解决难题的天才!顺便说一下,因为这本书中有 12 句话对我来说太难理解了,所以我写信给弗里曼请他帮忙。他立即给了我一个详细的解答,虽然他的视力不太好,健康状况也不太理想。他是一位如此慷慨和善良的绅士!

我期待着听到你对语言、艾伦和其他故事的进一步评论。然而,天已经放亮了,我只好等你的下一封信了。

早上好!

凡及

外。……以因果学说的名义否认我们作为人类对我们自己的未来有作出选择和决定的能力是荒谬的。当我们在作这种选择或决定时,我们行使了我们的因果力量,并体验为自由意志。"[1]

这样,弗里曼给出了关于自由意志的合理解释,这个问题让布莱克莫尔和几乎所有接受她采访的科学家都感到困惑:"坦率地说,在我开始采访之前,我已经预料到,几乎每个人都会在理智上拒绝自由意志的想法,但是如果没有任何这种信念却很难过日常生活……正如约翰逊(Samuel Johnson)所说的令人难忘的话:'所有的理论都反对有自由意志;而所有的经验都支持有自由意志。'"[2]

弗里曼指出了这个矛盾的来源。正如你曾经说过的那样,布莱克莫尔在写《意识对话》之前没有采访弗里曼是一件非常可惜的事情。顺便说一句,当我与凌瀚思[3](Hans Liljentström)博士讨论因果关系问题时,他告诉我说,我们的共同朋友哈肯(Herman Haken)是第一个提出循环因果关系问题的人,大多数科学家都忽视了这一点,他们仍然坚持线性因果原理和古典决定论的世界观。

还有一个重要观点虽然我也曾感觉到过,但是不能像弗里曼在该文中那样清楚地表达出来,那就是:"有两件事将人类与其他所有生物区分开来。一个是包括脑在内的人体的形状和功能,它是由三十亿年的生物进化给予我们的。另一个是 200 万年的文化进化给我们的遗产。"[4]后者让我们发展技术,并使我们的生活改变如此之快。这是我们在前几封信中讨论过的主要话题之一。

我完全同意你的评论:"弗里曼的想法可以为 HBP 和 BRAIN 倡议将来的研究提供一种非常有前途的理论框架。"然而,我不确定这两个巨型项目是否考虑了循环因果关系的重要性。在我看来,他们的方法可能主要还是沿着从微观

---

[1] Freeman W J. Consciousness, Intentionality and Causality[J]. Journal of Consciousness Studies, 1999, 6(11-12): 143-172.

[2] Blackmore S. Conversations on Consciousness[M]. Oxford: Oxford University Press, 2005.

[3] 瑞典计算神经科学家,他的这个中文名是笔者给他起的,并告诉其含义是"飞越浩瀚的思想",他很高兴以此作为自己的中文名。——译注

[4] Freeman W J. Consciousness, Intentionality and Causality[J]. Journal of Consciousness Studies, 1999, 6(11-12): 143-172.

当有一位喜欢恶作剧的科学家将昆虫和卵拿走后,这是一种它从未遇到过的情况,它无法预测自己行动的结果!

虽然法布尔的泥蜂提供了一个很好的例子,说明一个没有智能的生物也可以表现出复杂的行为,但我们仍然不知道它的脑里发生了些什么。布赖滕伯格的小车则给出了一个更好的例子,因为我们可以毫无疑问地彻底了解其内部的简单机制!

我很高兴你赞同我对皮尤悖论的看法。事实上,我听说类似的悖论已经有很长时间了,有人说我们无法理解我们自己的脑,就像我们无法抓着自己的头发往上拉,而把自己拉离地球一样。当然这种说法甚至更不靠谱,因为原则上说起来,只要换个旁人就可以把你提起来。你有关心智的社会网络的话让我想起了对这些悖论的解释。因此,如果我的解释能够在某一天发表的话,那么我们两人应该共同分享对这一解释的发现权。

去年 1 月 30 日是弗里曼 89 岁的生日,在上海,我们通常把 89 岁生日当成大事,所以我给他发了一封带有电子生日卡的电子邮件。尽管去年他的妻子去世后,他的健康状况并不是很好,但他仍然在努力工作,并与他的学生和同事一起出版了几本书和论文。

是的,他的论文《意识、意向性和因果性》非常浓缩且不易阅读。我同你的感觉几乎一样,"我仍然不能确定是否已经掌握了弗里曼所要说的一切",我也同意你对这篇论文所说的其他的话。

这篇论文中给我印象最深的一点是,他强调了循环因果关系在意向性中所起的作用。

"意向性不能用线性因果关系来解释,因为按照这个概念说起来,行为必须归因于环境和遗传决定因素,而不会为自我决定留下任何余地。""觉知和神经活动并不是两个没有关系的并行过程。在时间顺序上也没有哪个作为主体而产生或推动另一个。循环因果关系可以用于在几个层次上的解释,而不用去找某个主体。""普适确定性的假设非常重要,并具有广泛的影响。按照这一假设,人类行为的原因仅限于环境和遗传因素,自决的因果力量被排除在科学考虑之分开来。他认为,智能是在缺乏足够知识和资源的情况下进行适应的能力,是通过经验提升技能的能力。布赖滕贝格小车可能非常善于跟踪光线,但它不能以其经验来提高这种能力。因此,根据王培的标准,布赖滕贝格小车虽然有技能,但是没有智能。

布赖滕贝格小车让我想起了法布尔(J. Henri Fabre)的经典著作《昆虫记》[1](Souvenirs Entomologiques)中的一个故事,其中描述了泥蜂的行为。他发现泥蜂可以精巧地用泥土造巢作为育儿室。这个巢就像一个罐子,用来保存它的卵和储存捕获的昆虫作为幼虫的食物。它杀死一只昆虫并将其带入巢中,并在这只昆虫上产卵,然后飞走捕获另一只昆虫并将它放在第一只昆虫的上面。如此继续下去,直到在巢中有了 10 只或更多昆虫足够喂养幼虫发育成泥蜂为止,这时它就会封闭洞口以保证安全。这样,幼虫会先吃最早放进去的昆虫,最后吃最新鲜的昆虫。你可能会很佩服它为后代想得多么周到。然而,法布尔对泥蜂玩了个恶作剧,看看泥蜂是否真的很聪明。每次当泥蜂把一只昆虫放入巢中飞走后,他就把昆虫和上面的卵(如果有卵的话)拿走。但是,泥蜂从来也没有注意到它的巢空了,而只是继续捉昆虫进去,直到最终封闭了洞口,即使这时巢里已空无一物!因此,虽然泥蜂看起来聪明而且有很高的建筑技巧,但其复杂的行为只是天生的一种固定程序,不能适应情况的变化。它没有智能!

我特别欣赏王培将智能与技能区分开来的想法。无论技巧多么复杂,如果不能改进,只是固定在同一水平上,那就不能称之为智能。霍金斯强调,"预测而不是行为才是智能之所在。""智能是以记忆和预测世界模式的能力来衡量的……"[2]我常常想智能是主体在遇到一个从未见过的问题时仍然能够解决的能力。我的想法可能与上述诸定义是一致的。当然,如果某个主体面临一个从未遇到过的问题,即其解决问题的知识和资源都不够,而主体仍然可以预测其行动的结果,那么他就是有智能的!法布尔的可怜的泥蜂并没有智能,因为

---

[1] 有中译本:法布尔.昆虫记[M].陈筱卿,译.杭州:浙江文艺出版社,2005.——译注

[2] Hawkins J & Blakeslee S. On Intelligence[M]. New York:Levine Greenberg Literary Agency, Inc, 2004.

Ⅲ-003　凡及

（智能；自由意志；循环因果关系和线性因果关系；要实现与数字技术相同的功能，也可能有更简单和节能的方法）

2016－02－24

亲爱的卡尔：

非常感谢你的节日问候。也祝你和你的家人农历新年快乐！

如果你在写作中引用我的冰山隐喻，这将是我很大的荣幸。当然，这不是中国的一句老话，而是我自己的一句"新话"。☺

你将科学研究的描述性部分与解释性部分（"什么""何处"与"如何""为什么"）区分开来是绝对正确的。在很长一段时间内，生物学研究主要限于前一部分，即使在今天的某些生物学研究领域，情况也依然如此。在一开始时，知道"什么"和"何处"是重要的，但知道"什么"和"何处"并不意味着完全知道"如何"和"为什么"。原因很明显，但是，很多人似乎忘记了这样一种显而易见的真理，并对前者感到满意。你关于布赖滕贝格小车的故事给人启迪，我以前不知道此事。如果我们不知道它是如何工作的，我们可能会以为它"知道"什么对自己有好处，我们可能会认为它有一个指导其行为的价值体系。因此，仅仅基于其行为来判断主体的内心世界是非常值得怀疑的。不幸的是，媒体上关于人工智能的许多报道都是这样做的。

你有关布赖滕贝格小车的故事也引发了这样一个问题："什么是智能？"这个故事表明你不能仅仅通过行为来判断智能。你的故事让我想起了王培在中文网络杂志《赛先生》上的一篇文章[1]，在文中他强调了必须把技能和智能区

---

[1] 在《赛先生》上，王培发表了以人工智能为主题的系列文章，在他的文章《人工智能迷途：计算机的高技能等于高智能吗？》（2016－02－17）中讲了智能与技能的根本区别；而在《人工智能：何为"智"？》（2015－08－31）中则提出了他有关智能的定义。——译注

目标是用非侵入性成像技术采集健康年轻成年人的脑回路和连接信息，并对此进行分析和共享。

这一招标获得了全世界神经科学家的热烈响应。2010 年 9 月 15 日，NIH 宣布最后有两个组中标。一个是华盛顿大学、明尼苏达大学和牛津大学的联合研究组（WU－Minn－Ox），他们以神经生物学、神经信息学和神经成像分析见长，获得了 3 千万美元的资助。另一个中标的是由麻省总医院和加利福尼亚大学洛杉矶分校组成的联合组（MGH－UCLA），他们的目标是建立用于扩散成像的特殊设备，得到了 1 千万美元的资助。此计划于 2010 年正式启动，到 2016 年 6 月完成了其第一阶段。

关于此计划的后续进展和评论，我们还要在以后的通信中介绍。

图Ⅲ K2.1 中的左右两图是两种很简单的情况。左面小车的光敏元件都联结到同侧马达,而且当光强增大时,马达的转速也增大。在这个时候,如果在现场右前方只有一个光源,那么由于右后轮的转速大于左后轮,小车就会偏开光源行驶,好像是要躲开。右面小车的两个光敏元件则都联结到对侧马达上,因此离光源较远一侧的马达转速较快,小车就会驶向光源,表现出"趋光性"。

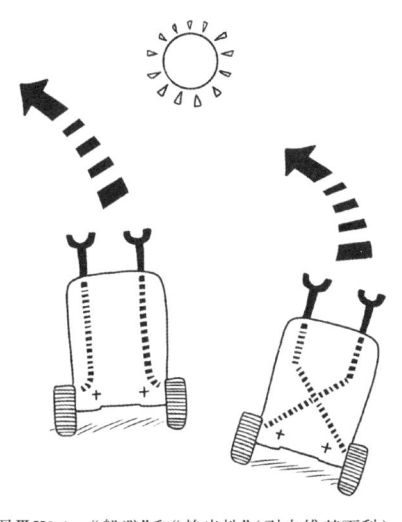

图Ⅲ K2.1 "躲避"和"趋光性"(引自维基百科)

如果现场有多个光源,那么这样的小车还能表现出更复杂的行为。这些行为看上去似乎是有目的的,表现得相当灵活,甚至好像有智能。一位不知就里的旁观者还可能以为其内部机制一定非常复杂,其实却如此简单!

背景专栏Ⅲ K2.2

人类连接组计划

为了研究正常年轻人的宏观连接组,2009 年美国国立卫生研究院(NIH)宣布对人类连接组计划(Human Connectome Project,简称"HCP")的公开招标,其我本想回过头来讨论你关于机器翻译的可能性的问题,这是 AI 现有弱点的一个典型案例。为什么我们在这个领域仍然做得不好,其原因与语言要涉及内容和意义有关。它是浓缩的文化,在一个裸脑的硬件中是检测不到这样的东西的。在我们的脑中如何会产生语言,这个问题属于你所能想到的最主观的部分。实际上语言就是我所谈论的随时间积累知识的一个绝佳例子,因此它是这个难题的核心,可能是我们能够解决的最后一批问题之一。

但这是一个全新的领域,尽管这封信又写得太长了,我甚至还没有来得及谈到艾伦,但我还是希望在未来的信中开始这场讨论。也许在开启讨论大门之前,你可以先告诉我,我对弗里曼的文章充满热情是否有充分根据还是有些夸大其词了。当然,我希望是前者,并向中央之国这一领域的一位值得称道的大师致以最良好的祝愿。

晚安!

卡尔

背景专栏 Ⅲ K2.1

布赖滕贝格小车

控制论学家布赖滕贝格提出了一种思想实验,使一个非常简单的机械装置的行为看起来好像具有智慧。他想象有一辆玩具小车,这个小车的两个后轮分别为一个独立的马达所驱动。车身前部的左右两侧各有一个光敏元件,按其接收到的光的亮度而加速或减慢其连接到的驱动马达的转速。因此,按照当光强增大时光敏元件使其连接到的马达转速是增大还是减小,以及光敏元件是连接到同侧马达还是对侧马达,就分成 4 种情形。每种情形下小车表现出不同的行为,例如趋向光源,或是躲开亮光等,这些行为和某些昆虫的行为类似,因此一位不明就里的人还可能以为这个小车有自己的意向,甚至智能。

然而,滤波并不一定要通过数字计算完成。模拟式无线电和电视系统中充满了滤波器,通过组合无源元件(如电容器、电磁线圈和电阻器)分离各种信号。这种滤波进行得非常快,几乎不需要成本,这就像用机械筛将大小不一的砾石分开一样。你也可以使用相机、计算机和机器人并花费大量的能量进行相同的分离。我们已经习惯于数字计算和算法的概念,因此我们经常会忘记,早就有了容易得多、便宜得多、有时还快得多的解决方案。镜子就是一个很好的例子。经典的涂银玻璃镜以光速反射入射光,无需额外的能源。今天人们常把智能手机当镜子用。在这里,带有 CCD 传感器芯片的相机被用于图像的数字化处理,将其发送到 CPU 后,应用一些算法,经过一些延迟,并耗费了大量能量之后才投放到显示器上。另一个例子是将太阳光束分解成彩虹中各种颜色的玻璃棱镜,它以光速工作并且不需要任何成本。你也可以使用传感器芯片阵列、CPU、巧妙的算法、彩色电视,再花费很多能量之后做到类似效果。

把脑说成是一种从噪声中过滤出意义的模拟过滤系统,我不能确定这种想法一定就能解决问题。但是当我将脑的能耗(20 瓦)与 IBM 的莫德哈所称的仿神经结构芯片进行比较时,后者在将来某一天也许可能需要花费千兆瓦的能耗才能完成同样的工作,这让我思考这种数字计算方法中是否有什么地方根本错了。

我的一位朋友乌尔曼(Bernd Ulmann)是模拟计算机的顶尖专家(并且是世界上收集模拟计算机最多的人之一[1]),至少他没有反对我的想法。☺

很高兴知道我们、弗里曼和霍金斯并非仅有的怀疑脑是计算机这种说法的人。感谢你引用了埃德尔曼和托诺尼在《意识的宇宙》一书中很多有力的论点,这些论点在我看来都是成立的。而你所说的关于松本元的话也和弗里曼的立场非常接近。

啊,连埃德尔曼和托诺尼的书你也翻译过了!哇,你透露的消息再一次让我印象深刻。在这个领域中有没有你没有翻译过的重要著作或者你不知道的作者?☺

---

[1] http://www.analogmuseum.org/english/

上述一切你当然都知道，人脑中有一个关于外部世界的内部模型（也就是外部世界在内心中的表现）的想法也并不新鲜。但我发现不同凡响的是，弗里曼尝试着用这个非常基本的原则作为人类行为和有意向动作的完整概念的出发点，不仅仅局限于解释肌肉的运动，而是一直延伸到了意识层面。

有关这个领域，你比我熟悉得多，因此我想听听你对这个概念的看法，以及我的下列想法是否有错：弗里曼的想法可以为 HBP 和 BRAIN 倡议将来的研究提供一种非常有前途的理论框架。我很想知道 C.科赫对此会怎么想。

在这个问题上最吸引我的一点是，这种想法对我来说很熟悉，这让我想起了控制论中的类似想法。我这里所说的控制论并非冯·诺伊曼数字计算机意义下的控制论，而是由维纳在他的闭环控制理论和以后的自动机理论中所提出的那些原则。问题在于这些发展几乎全部都是沿着数字化的思路展开的，而神经元和脑则是模拟系统。说到这里，值得注意的是，当初维纳是围绕控制论和高炮模拟计算机闭环控制这样的问题而发展起自己的想法的。

无论如何，维纳的下列想法在技术领域非常成功：系统有一个它所希望的通过在物理世界中行动而达到的状态，在行动之后将感觉输入与内部期望进行比较，并且在以后的循环中使差异最小化。这是我们在机器人学和 AI 中所看到的对所有机器进行控制的主要思想。

关键的一点是这种适应是否必须通过数值计算来完成（使用今天的数字计算机进行这样的计算是非常耗能的），换言之，是否有更好的方法。

剑桥的沃尔珀特（Daniel Wolpert）是计算生物学和感觉运动控制方面的专家，他指出，再传入原则也可以用卡尔曼滤波器的功能来描述。[1]

这是一种很有趣的想法，因为卡尔曼滤波器是现代数字信号处理中使用最广泛的算法之一。最初它是为了改善低功率卫星发射机和地球之间的通信而发明的，而当涉及区分各种控制系统中的信号和噪声问题时，它也是工程师们的首选。

---

[1] Daniel M, Wolpert et. al. An Internal Model for Sensorimotor Integration[J]. Science, 1995, (269S.): 1880 – 1882.

态构成意识,意识可以变得越来越复杂,最终包括自我觉知在内。有意向性的动作不需要觉知,而随意动作(voluntary acts)需要自我觉知。对动作/知觉周期的觉知就成了主体(agency)线性因果关系的认知隐喻,人们将这个隐喻应用到世界中的物体和事件上以预测和控制它们,并分配社会责任,因此线性因果关系是社会契约和技术的基石。"[1]

弗里曼在这里所表达的见解定义了一个巨大的研究领域,足以成为好几篇博士论文的题目。这篇文章非常浓缩,不容易阅读,因为你不仅要熟悉神经生物学的一些概念,还要熟悉物理学、逻辑学、混沌理论、控制论、哲学等。因此,这 36 页让我读得很辛苦,而且我仍然不能确定是否已经掌握了弗里曼所要说的一切。对我来说,这似乎是他终身研究的精选,以及他留给后代研究人员的遗产。这也是他试图为肉体、脑和心智中的自组织的一般理论做出贡献。

对弗里曼来说,正如你在他的所有作品中看到的那样,肉体与物质世界的相互作用是他讨论问题的起点。在这里他非常强调再传入原则。这一原则基于这样的想法,即对于每个想要做的动作,都有一个期望要与动作产生的感觉输入进行比较。如果发生差异,那么在后续动作周期中的动作之后,脑-身体系统就要努力减小期望值与动作之后感觉输入之间的差异,直到达到最小值。你在坎德尔的"圣经"中找不到再传入原则这个术语,而被称为"传出拷贝(efference copy)",它被用来解释肌肉运动系统如何控制和稳定运动(例如抓握、移动物体或在运动时改善注视稳定性),以及脑如何从输入的感觉信号中去除由于自身运动所产生的信号。关于这一原则,大家最熟悉的一个例子就是为什么我们不能给自己搔痒,而别人却可以,这是因为传出的运动信号的内部拷贝与传入的感觉信号相同。我们的感受正是我们所期望的,因而不能让自己感到惊讶。你不仅不能搔痒自己,你也不能讲笑话给自己听而使自己发笑。[2] ☺

---

[1] Freeman W J. Consciousness, Intentionality and Causality[J]. Journal of Consciousness Studies, 199, 6(11-12):143-172. 弗里曼赞扬了 Erich von Holst 和 Horst Mittelstaedt,这是因为他们早在 1950 年就介绍了再传入原则。

[2] https://en.wikipedia.org/wiki/Efference_copy

的回答当然更加微妙也更好。

你是对的,解决办法是聚合许多脑和心智的能力,这可以成指数级地提高单个脑和心智的力量。这样做并不限于在一个特定的时间点上并行地进行,而且还可以串行地在多个世代(祖父母在这里扮演了主要的角色)汇集起许多代智慧和知识。这给了皮尤悖论一个完美的答案。许多脑和他们积累起来的智慧将能够认识单个的脑。凡及,在这个问题上你做得很棒!

现在当我再想这个问题时,我发现这个悖论的主要缺陷在于,皮尤忽视了这样一个事实,即一个与世隔绝的脑几乎什么都做不了,而拥有恰当知识的脑则几乎什么都能做。关键的一点是心智可以分享他们的内容,并不断积累、聚焦和磨炼他们的知识。这正是在技术中所发生的事情,也是为什么物质的自组织在经过几十亿年的积累达到目前水平之后,我们的工程能力能在数百年内就取得了爆炸性发展的原因。

无论如何,是你解开了皮尤的悖论。你可能在考虑发表这一见解,如果你能提及是"祖父母重要性"理论启发了你,使你找到了解决的方案,那么我将倍感荣光。由此我们两个也许会名扬天下,也许会富甲一方。☺

你的朋友弗里曼殚精竭虑地研究了这种似乎深深扎根于我们脑中的因果机制,也研究了因果关系和意向性在建立意识中所起的作用。他为此工作了数十年,这一巨大努力的结果发表在他的《意识、意向性和因果性》(*Consciousness, Intentionality, and Causality*)一文中。我只是在最近才发现这一点,感谢你激发了我对他的工作的兴趣。与此同时,在我对他的观点更为熟悉之后,我发现他在这里所阐述的是他理论的核心,他所说的相当重要,也很恰当。

在这篇文章的摘要中,他说道:

"根据源于实用主义、格式塔心理学、存在主义和生态心理学的行为学理论,关于世界的知识是通过有意向的动作继之以学习获得的。用这里所描述的神经动力学的话来讲,如果要做某个动作的意向通过再传入(reafference)而被觉知到,它就被认为是一个原因。如果一个动作的结果通过本体感觉和外部感觉而被觉知到,那么它们就被认为是一种后果。一系列这种觉知(awareness)状部主体的极其复杂的理论,尽管实际上根本就没有这样的主体,而且解决这个难题的方法其实相当简单。[1]

我们的头脑似乎偏爱由因果关系主宰的故事,在其中有某些主体及其意图对发生的事情负责。但布赖滕贝格告诉我们,为了产生复杂的行为并不一定要有什么不断在做计算的主体。他的小车在没有使用超级计算机和大量数值计算的情况下,也能令人惊讶地表现出像人一样的行为。他的小车实质上只是一种用技术实现的思想实验,这有助于我们认识到,可能有相当聪明的机器或动物能够在复杂的世界中成功地生存下来,而无需做大量的计算,也不需要中央控制机构。昆虫,特别是蚂蚁,是现实世界中这种可能性的很好例子。它们的脑非常小,但是却成功地建立起了很复杂的社会系统。

布赖滕贝格是一位非常独立的思想家,也是一位对神经学与 AI 进行跨学科研究的专家,已经对 AI 工程师产生了深远的影响,对布鲁克斯(Rodney Brooks)及其昆虫机器人的影响尤其大。对我来说,他在这方面是最聪明的人之一,尽管他并没有由此而赢得什么声誉。这可能是因为他大部分时间里都在用纸和铅笔帮助思考,而从未申请过数十亿美元。☺

我很感激你没有把我的"祖父母重要性"理论视作纯粹的废话。你引用了皮尤悖论"如果人脑简单到我们可以清楚地认识它,那么我们就会太简单了而无法认识脑"作为补充说明,这是我之前从没想到过的。

我以前也听说过这个悖论,但从来没有像你那样深入地考虑过。我没有多想,认为这是一个伪命题。我的解决办法是问下列问题:"一个只能举起 100 千克的人能解决如何举起 10 000 吨重的难题吗?"答案是肯定的,因为在几乎所有我们体力能做的事情上,我们都能造出性能优于人类的机器。对于我们的脑可以执行的大部分任务来说也是如此。计算机在记忆方面更好,计算速度更快,可靠性也更强,技术传感器系统的性能就其适用的物理和化学的范围方面来说也远远超过了人类。实际上,我们的心智能力中只有很小一部分是机器不能完成的。那么为什么机器就不可能做到其余的方面呢?但是你对这个悖论

---

[1] http://en.wikipedia.org/wiki/Braitenberg_vehicles

在各门学科中这两部分的比重是非常不同的。通常情况下,我们越不了解某个系统是如何工作的,我们就越强调描述性部分。部分原因是人们希望如果对某种现象知道得越多,越是能够更准确地描述这种现象,那么我们也就越接近认识这个谜题。

有时候这确实有帮助,但这种学究气的努力常常只是使我们忙忙碌碌,给我们一种以为认识清楚了的幻觉,因此甚而可能妨碍我们进一步破解现象背后的谜团。

我和你的看法一样,这个魔术师的故事指出"许多研究工作的缺陷,例如许多脑成像的结果"。正如你所说,这种描述性研究不是没有意义,但还不够,让我们希望 BRAIN 倡议不要沦为一个只是为了产生更多的描述性数据的计划吧。当然,如果我们真的像很久之前林奈(Carl Linnaeus)[1]为比较植物学所做的那样,也为脑的分区和活动事件建立起一个分类系统,那也很好。

但即使生物学家能想办法描述世界上所有的植物,完美地刻画这些植物并把它们分类到林奈的系统中去,它也还是不能揭示所有植物这种神奇的多样性背后的谜团、DNA 的功能或进化的作用。即使对于你正确地呼之为"魔术师团队"的表演来说也是如此,这一团队在我们的脑中起奇妙的作用并给了我们意识。

其实我们甚至不知道这种作用是否真的很复杂,或者我们是否忽略了很简单的原则,而解决方案其实可能就在我们眼前。

布赖滕贝格(Valentino Braitenberg)[2]发明了一种奇妙的人造生物,也就是著名的"布赖滕贝格小车"。他以此说明有些看似复杂而难以理解的行为其实可能只是由非常简单甚至原始的内部机制引起的。外部观察者可能会觉得很难甚至不能理解这些机器的内部编码,尽管其内部的"智能"非常原始,仅基于几个电话继电器、简单的传感器和电动机。当一位研究者想了解其内部逻辑时,他可能会偏离正道而创建起一种理论,即关于控制这种机器的强有力的内

---

[1] Carl Linnaeus(1707—1778),瑞典植物学家、医生和动物学家,被认为是现代分类学之父。——译注
[2] Valentino Braitenberg(1926—2011),意大利神经科学家和控制论学家。——译注

Ⅲ-002 卡尔

（科学研究中的描述性部分和解释性部分；为了产生复杂的行为，并不一定要有什么不断在做计算的主体；脑网络和积累起来的智慧将能够认识单个的脑；再传入原则与闭环控制）

2016－01－09

亲爱的凡及：

首先祝你和家人新年快乐！

感谢你对可能存在祖母细胞一事微妙而又机智的评论。我非常喜欢你那个吃了七块饼的伙计的笑话，此人相信是最后一块饼让他吃饱了，因此认为要早知道的话就可以节省大部分的钱。我更加喜欢你用下面这句简洁而明快的比喻来归结这个意思："撞沉'泰坦尼克号'的并不只是冰山之巅"。这句话我（几乎）可以确定不是中国的老话。☺如果这是你想出来的话，我请你允许我在将来的写作中使用它，因为人类推理的因果关系链不管用的情形太多了，在这种情况下就完全适用这句话。当然，如果你允许我使用的话，我肯定会说明是引用了你的话，表明你才是首创者。

你的这个比喻和你所引用的格拉齐亚诺的话"是魔术师让它变成这样的"，触及了研究方法、我们心智推理的内在机制，以及我们内在有一种用因果关系来解释的渴望等这一切中的一个关键问题。

我相信这种对因果解释的渴望是我们的心智最喜欢的取舍标准，它也决定了我们如何做科学研究。科学研究的描述性部分是对物体和现象进行命名、分类、系统化和测量，这与解释部分完全不同。在前一部分中，我们涉及的是"什么"和"何处"的问题，而后一部分则是关于"如何"和"为什么"的问题。

正如你在上一封信中所说的那样，起初我也误解了 HCP 的目标，只有在阅读了相关材料后，我才明白 HCP 的目标是绘制脑区之间的连线图，而不是神经元之间的线路图。这项任务要简单得多，也更加实用。即便如此，他们把任务设定为在头五年只限于绘制健康的年轻人脑区间连线的任务，并将发育中和老龄化的健康脑以及患有各种脑疾患病人脑区间的连线问题留待以后再研究。该项目的一项重要任务是要在更为坚实的基础上将每个大脑半球分割成许多脑区，以前布罗德曼（Brodmann）只是根据细胞构筑把皮层划分为 52 个区域。这将为他们进一步的工作打下坚实的基础。他们还使用了四种不同的 MRI 方法来绘制这些不同区域之间的解剖和功能连接。[1]

艾伦脑科学研究所由艾伦夫妇于 2003 年创立，现任所长是 C.科赫。2011 年 11 月，他们开始了一项艾伦鼠脑连接图谱（Allen Mouse Brain Connectivity Atlas）研究，其目的是绘制鼠脑中的三维高分辨率连接组，并仅限于在视觉系统的某些区域，以此作为研究的第一步。他们还启动了另一个项目——艾伦细胞类型数据库（the Allen Cell Types Database），其目的是对鼠脑中的神经元进行分类。

与马克拉姆的 HBP 和奥巴马总统最初对 BRAIN 倡议的说法相比，上述项目的目标都要有限得多，也更切合实际。他们专注于采用相对成熟的方法收集重要的基础数据。虽然他们的工作量很大，需要发展新技术以使工作更有效、更完善，还需要把不同背景的专家组织在一个团队中，但是正如承现峻所说，如果有足够的资金支持，这些任务最终会取得成功。这些项目可能是大科学、团队科学和开放科学的良好尝试。然而，我认为这可能并不是组织所有科学研究的一种普遍方式，特别是那些需要创造性和洞察力的科学研究。

随着"白天来临"，我必须在这里停下来。

祝好！

凡及

————————————

[1] http://humanconnectome.org/ccf

响突触强度的变化。通常在人造系统中碰不到这些极为重要的特性……

最后……高等脊椎动物脑的最令人吃惊的特性,是有一种我们称之为复馈(reentry)的过程。……在脑内相互联结的区域之间,并行信号不断地进行着循环的相互交换,这些相互交换持续协调着这些区域在时空两个方面彼此间的映射活动。……

当然,我不认为他们已经列出了脑和计算机之间的所有差异。已故的松本元(Jen Matsumoto)列出过另外两个区别:脑编制自己的程序,自行搜索数据。此外,脑通过主动地与环境相互作用来"训练"自己,也许你可以列出更多的差异。无论如何,埃德尔曼和托诺尼提出了一些值得思考的东西,特别是他们有关脑是一种选择性系统,而不是一种计算系统的想法。正如你所知,霍金斯也强调,脑不是一种计算系统,而是一种记忆和预测系统。无论如何,脑的计算机隐喻似乎相当令人怀疑。

我很高兴你喜欢王培博士的想法。我也不认识他,我只是偶然在一份中文网站杂志《赛先生》上看到他的几篇文章,非常喜欢它们。这些文章是用中文写就的,所以对我这样一个外行来说,阅读和理解起来就容易许多。我已经下载了他的所有出版物,包括这些从他的主页(https://cis.temple.edu/~pwang/)[1]上下载的中文文章,我想在稍后再阅读。我同意你对他的所有评论,并且很高兴有一天能与你讨论他的想法。

哇!你认识艾伦本人!他也对库兹韦尔提出批评!艾伦在他的论文《奇点不近》中对库兹韦尔的批评让后者非常恼火,以至于库兹韦尔在其著作《如何创造心智》的"第11章 异议"中花了近五分之四的篇幅针对艾伦的批评进行抗辩。

好吧!现在让我们回过头来讨论有关人类连接组计划(HCP)和艾伦脑科学研究所的事吧。

---

[1] 在这个网站上,你可以找到许多有趣、富有教育意义的材料,尤其是"面向未来 AGI 研究者的教育建议"部分:https://cis.temple.edu/~pwang/AGI-Curriculum.html 这是我们所能找到的有关这一话题的最佳列表。它根据读者的知识基础进行了很好的分类,一步一步地引导读者认识最新科学的最高水平。我们强烈推荐对人工智能感兴趣的读者登录阅读。

是,有时我根本就猜不出译者为什么会这样翻译!因此,在我看来,对于日常生活翻译而言,由于翻译实例数量众多,且其语法结构并不复杂,一般说来,机器翻译并没有太大问题。然而,对于很少被提及的内容,具有复杂的语法结构并且需要真正理解,我怀疑机器翻译是否可以击败合格的人类翻译。由于我是 AI 的外行,你是专家,我想听听你对机器翻译的看法。

我喜欢你关于人脑与电"脑"之间差异的评论,其中的一些差异计算机之父冯·诺伊曼早在 60 多年前就已提到过!你的评论还让我想起了诺贝尔奖获得者埃德尔曼和托诺尼在其著作《意识的宇宙》[1]一书中的下列话语,他们甚至在书中专门写了一节"脑不是计算机"。

……脑的组织和功能特性看起来并不像是脑在执行一系列指令或计算。我们知道,在相互联结的方式方面没有任何一种人造装置与脑一样。首先,脑中几十亿个联结都不是精确的……从最精细的尺度上来看,没有任何两个脑会是一样的,即使同卵双胞胎的脑也不完全一样。……这些观察给基于指令和计算之上的脑模型提出了根本性的挑战。……

从我们正在勾画的图景中得出的另一个组织原则是,发育史和本身经历的结果都独一无二地印记于每个脑中。……随之而来的个体差异并不只是噪声或误差,它们能影响到我们记忆事情的方式。……现在还没有哪一种人造的机器在设计时把这种个体的多样性作为一条主要原则来加以考虑……

如果我们比较脑所接收的信号和计算机接收的信号,我们会发现脑有许多其他的特有性质。首先,呈现在脑面前的世界当然不是一串确切无误的信号,这和计算机是不一样的。脑使得动物能感觉到它的环境,从各种各样变化多端的信号中归结出一些模式来,并引起运动。……

我们也曾经指出过,脑中有一些特殊的有弥散性投射的核团——价值系统。这个系统会向整个神经系统发出信号,通知遇到了突出的事件,它也会影

---

[1] 有中译本:埃德尔曼,托诺尼.意识的宇宙[M].顾凡及,译.上海:上海科学技术出版社,2002.很不幸,这一旧译本中犯了笔者在信中批评的许多错误,2019 年重新出版修订本,实际上几乎是重译本。——译注

观点并不同意,但我非常喜欢这个故事。事实上,这个故事指出了许多研究工作的缺陷,例如许多脑成像的结果。当然,我并不是说这样的结果毫无意义,但这还不够。虽然知道是谁造成的也很重要,但更重要的是要说清楚是如何做到这一点的!至于祖母细胞,我很抱歉地说,它甚至还不是魔术师,而只是某个魔术师团队中的一员。

你关于祖父母在进化中发挥作用的理论非常有趣。我完全同意除了自然选择之外,将知识和经验传授给他人,特别是年轻一代,对于人类来说非常重要,祖辈可能在其中扮演重要角色,这使得人类得以如此迅速地增强自己的能力,并最终成为万物之灵。祖父母在这个过程中有很大的贡献。[1] 社交互动也起着重要作用。这让我想起了皮尤(Emerson M. Pugh)的一句话:"如果人脑简单到我们可以清楚地认识它,那么我们就会太简单了而无法认识脑。"[2]他的悖论让我困惑了很久,然后我认识到,一方面,虽然他的话似乎也有合理之处,因为人脑是如此复杂以至于单个脑很难认识自己。然而,另一方面,并不只是单个脑认识自己,而正如你所说的那样,试图认识单个脑的是不断演化着的脑的社会网络。脑的社会网络要比单个脑复杂得多!因此,虽然我不指望在某一天我们可以宣称"啊哈!现在,所有关于人脑的奥秘都被发现了!",我们仍然可以逐渐接近这个目标,即使我们永远也无法完全做到这一点。

正如我在上封信中告诉你的,当我检查我的翻译草稿时,如果发现有可疑之处,我就会去查阅原文,在多数情况下我会发现这是由于在翻译时没有看懂而造成错误。当我阅读翻译的科普书时,我经常会发现一些难以理解之处,查阅原书,大多数时候,我发现原因都是一样的——译者不明白作者在说什么,他可能只是逐字地直译或者参照其他类似的翻译,就像机器翻译一样。更糟糕的

---

[1] 在卡尔发表这一看法之后,最近译者看到有报道说,有科学家发现人类社会当成员可以活到祖孙同堂的时间之后,得到了加速发展。这似乎是对卡尔祖父母理论的支持。当然由于这一发现还是相关性,究竟是由于人能活到做上祖父母而加速了社会发展,还是因为由于社会加速发展人才能活得更长?或者互为因果?这一发现并没有解决。——译注

[2] 这段话首先出现在美国核物理学家乔治·皮尤(George Edgin Pugh)在1977年出版的书《人类价值的生物学起源》(*The Biological Origin of Human Values*)一书中,不过作者说明这句话是其父亲老皮尤(Emerson M. Pugh)所说,而他也是一位物理学家。——译注

Ⅲ-002　凡及

（解决"如何"的问题才是科学中的重要问题；个体脑和社会脑；机器翻译；计算机和脑；人类连接组计划；艾伦脑科学研究所）

2015－11－22

亲爱的卡尔：

　　你真是一位真正的学者，我不得不表示佩服。即使是升级成为外祖父这样一件事，也会引起你的科学思考。☺我真的很荣幸成为第一个听到你的"祖父母重要性"理论的人。至于"祖母细胞"，我并不怀疑实验事实，问题在于要认识一个人的祖母不仅仅取决于一个"祖母细胞"，相反，必须有一个回路甚至一个负责识别的系统，然而，我们对这种回路或系统几乎一无所知。撞沉"泰坦尼克号"的并不只是冰山之巅，海平面下的冰山山体才是撞沉邮轮的主因！或者如一个古老的中国笑话所说：一个人吃了一块饼，仍然饥肠辘辘，所以他吃了第二块，一直等到他吃完第七块饼，这才饱了。他说："要是我早点知道就好了，我只要吃最后一块饼就行了，这样我会省下大部分钱！"连接组学研究可能有一天会解决这个问题。

　　宣称是祖母细胞使人认出其祖母的说法，让我想起了美国神经科学家格拉齐亚诺（Michael S. A. Graziano）在他的新书《意识与社会脑》（*Consciousness and the Social Brain*）中讲的一个故事。当一位父亲和他的儿子看魔术表演时，魔术师的手法非常高明，父亲问儿子："吉米，你觉得他们是怎么做到的？"儿子回答说："很明显呀，爸爸。"父亲很惊讶，说："真的吗？你看出来了吗？有什么诀窍？"儿子说："是魔术师让它变成这样的。"祖母细胞就是魔术师！但它是如何做到的？尽管我不得不承认我对格拉齐亚诺博士在他的书中所表达的主要如果你还没有读过的话,你可能会发现它很有意思。[1]

我不能确定王培开发的纳思系统(Non-Axiomatic Reasoning System,简称"NARS",非公理化推理系统)是否就是 AI 的最终解决方案,但至少他们对主流 AI 的批评看起来很重要且对我有帮助。当然,我喜欢王培的主要观点,即计算机和脑在本质上是不同的系统。所以这很有帮助——谢谢! 我期待与你讨论这些问题。

好吧,当你告诉我你发现了关于 C.科赫在艾伦脑科学研究所和人类连接组计划(HCP)中的一些有趣的事,并且要到下封信中才告诉我,你就成了现代的莎赫札德。☺

当然,我很想知道更多信息,急着等你告诉我究竟是些什么新闻。无论如何,因为你翻译过《意识探秘》,你一定认识 C.科赫,这很好。因此,如果真有什么令人激动的新闻,也许我们可以问他更多细节。你知道吗? 我碰巧也认识艾伦[2](Paul Allen),因为我曾与他旗下的一家公司进行过一些联合技术开发。这与替代的非语言编程方法(alternative,non-verbal programming methods)和数据库有关。但是,由于我的信早已写得太长了,尽管我还没有来得及讨论你提出的所有那些有趣的问题,但是如果你愿意,我还是可以在下一封信中告诉你更多。☺

最好的祝愿。

卡尔

---

[1] http://www.iiim.is/2010/05/questions-about-artificial-intelligence/

[2] Paul Gardner Allen(1953—2018),美国发明家、投资者、考古学家和慈善家,微软的两位创始人之一,他投资成立了艾伦脑科学研究所、人工智能研究所、细胞科学研究所等。2018 年 10 月 15 日去世。——译注

也会不知所措。在阅读一篇文章时,你无法立即在头脑中建立起整个库,就像你无法在20分钟内学会如何弹钢琴一样。你可以在20分钟内了解钢琴上的琴键以及如何敲击琴键,但是如果你要想学会就得每天都弹,并练习20年。这同样适用于各种运动、数学、物理和其他一切。要是头脑里没有某种程度的预存知识,那么就不可能真正读懂一本书。你可以借助纯粹的死记硬背通过下一次考试(如果你的教授不是费恩曼的话),但那不是理解。

这也正是你所抱怨的问题的核心所在。但是并没有简单的解决办法,因为"没有免费的午餐"这句老话也适用于精神食粮。科学本身对其内容标准化可以有所帮助,事实上自然科学的历史也就是规范术语、方法和工具以及标准化的历史。在这种标准化的领土上找到方向比在开放的混乱环境中要容易得多。通常,你可以通过其标准化程度来判断科学发展的阶段。有些人甚至说,在一门学科没有这样的坐标系统之前,就不能称之为科学。他们说,物理学只是在牛顿提供了必要的理论框架之后才成为一门科学。在这种观点之下,你很难把神经生物学称为科学。因此,美国BRAIN计划中较为温和的路线似乎是使神经生物学成为一门科学的一个更合理的步骤,而在我看来,最初的HBP路线更像是人类自大狂妄的一种和善版本。

谢谢你告诉我关于王培博士的信息和他认为图灵测试不够恰当的想法。他是一位非常敏锐的思想家,我喜欢他的主页上刊登的信息。我以前并不知道他的名字,但当我读到他的《通用智能理论》(*General Theory of Intelligence*)时,不知为什么对我有所触动。他使用的论点让我想起了戈尔策尔(Ben Goertzel)早些时候发表过的一些内容。再稍微搜索一下就可知道,王培和戈尔策尔是天普大学(Temple University)的同事,并共同出版过一本书——《通用人工智能的理论基础》(*Theoretical Foundations of Artificial General Intelligence*)。在戈尔策尔批评霍金斯的《智能论》后,我在2005年读到过这种想法。王培和他说的观点虽然不属于主流,但是非常有趣和有启发性。

我不知道你读过哪些和王培有关的文章,也不知道《有关人工智能的三个基本误解》(*Three Fundamental Misconceptions of Artificial Intelligence*)一文是否在其中。

Ⅲ-001 卡尔 15

有染的消息一样。

塞尔的"中文屋"在程度上要差一点,但它之所以广为流传也得益于同样的机制。对于西方人来说,中文是一种陌生而且难懂的语言。中文文本上的象形文字看起来很神秘,并且与其丰富而复杂的文化联系在一起,因此中文看起来充满了异国情调和神秘,鲜有西方人能真正懂中文。这至少是深深植根于西方大多数人心中的神话。所以当塞尔选择中文作为他在这个自动化翻译室中的人所必须翻译的语言时,他就把工作难度提到了最高,从而调动了读者的情绪。我敢打赌,如果他选择把 Java 程序翻译成 Python 编程语言作为例子,并将他的思想实验称为"翻译室",那么今天没有人会谈论它。如果薛定谔选择在放射性衰变发生后翻一枚硬币而不是把猫杀死,那么除了专家之外,没有人会谈到它。但是薛定谔和塞尔都很聪明(或很幸运),他们的这些术语不是通过狭窄的理性之门,而是通过广阔的情感之门进入数百万人的头脑,结果荣登引用指数的榜首,并且可能会永远留在那里。☺

当然,这样的隐喻对于译者来说有两个不同的问题,因为这种隐喻既高度依赖于各种语言的文化背景,又依赖于主要读者的知识水平。对读者所具有的知识水平的预期,决定了还必须提供一些相应的附加信息。另一个难点是,在源语言中有些和其文化密切相关的东西在目标语言世界中可能并不存在,而译者也要为此找出适当的隐喻。内容越是专门化,越容易找到确切的表达和标准化的术语。当一位法国作者写了一篇关于弦论中的具体数学问题的文章时,翻译成中文的难度可能会比翻译法国哲学家所写的有关"自由意志"的作品的难度要小。至少当中国译者懂弦论时是如此,其关键在于,著者和译者双方都通过漫长而艰苦的工作学习过这门学科的主要内容。在有关爱因斯坦的许多电影中可以引证一个很好的例子说明这个问题。一位社会人士拜访爱因斯坦,并问他:"爱因斯坦教授,你能用简单的语言向我解释相对性原理吗?"爱因斯坦说:"不!"毋庸置疑,爱因斯坦的回答并非势利,这确实是不可能的。其实,不仅仅在这样复杂的科学问题上是不可能的,要想让一个从未见过棒球(或足球)比赛的人明白棒球比赛的魅力,这同样也是不可能的。

如果受众的头脑里对某件事的核心内容毫无所知,那么即使是最好的译者列举了许多我们行为不理性的非常有趣的例子,这些例子成了其他科学中经常提到的科普心理学知识的一部分。

为了认识我们的心智究竟是如何工作的,也许要加以区分的系统还不只有这两种,但不管怎么说,卡纳曼至少使我们懂得为什么我们常常会不假思索地谈论一个我们不太理解的概念,而不觉得有什么问题。他也解释了为什么我们会把注意力投放到事情的一个实际上并无关系的方面上去。著名的"薛定谔的猫"就是一个很好的例子。有时候,这种难以理解和神秘的概念似乎特别有吸引力,而这只可怜的猫似乎就是其原型。"薛定谔的猫"成了哲学家的宠儿,也成了他们最喜欢用的隐喻之一,这已成了现代科学文献中的必用语。在引用它的许多人中通常是为了证明量子力学的哥本哈根解释的有效性,我不知道他们中究竟有多少人认识到薛定谔发明这个思想实验的唯一目的是为了证明哥本哈根解释的荒谬性:按照这一假设,在你看猫之前,猫可以既死又活。

塞尔著名的"中文屋"也是一样,这可能仅次于"薛定谔的猫",是科学中引用得最多的隐喻之一。正如你可以从库兹韦尔的解释中看到的那样,你正确地指出他的解释绝对是错误的,如果不说是愚蠢的话,每个人都可以引用它来说明他们想要说明的任何东西。我相信这两个隐喻的巨大成功与他们的内在品质和原本意义无关,但与它们在情绪上所产生的影响有很大关系。正是这种把神秘的东西与一些非常熟悉和众所周知的东西放在一起说,使得这些术语令人印象深刻并被牢牢记住。虽然没有人真正了解海森堡的不确定性和玻尔的互补性意味着什么,或者量子叠加和量子纠缠究竟是什么意思,但每个人都知道猫、玻璃管和锤子是什么。尽管不是每个人都知道氰化物究竟为什么如此危险,但大家都知道这是一种致命的毒药。有关可怜猫的命运的故事牵动着人心,我们的脑喜欢这种故事。

我们的脑不是设计来做数学和抽象物理学的。脑进化成一种做对象识别、意义检测、故事记录和编造故事的器官。你能想出一个比这种精心谋杀一只猫更令人激动的故事吗?所以我们的脑很容易接受这种故事,实际上这正是美国人所说"no-brainer(不动脑筋)"的一个完美例子。对此我们不需要学习;我们一听到就记住了,其容易的程度和我们听到一位著名演员与另一位演员的妻子是我们的头脑通过逻辑推导得出的,而是通过内心情绪、冲动和欲望制定的,这些又通过观察别人做什么以及哪些行为得到了奖励、哪些行为则受到了处罚而加以控制。

像这样的行为可能看起来很古怪,前后不一致、自相矛盾,而且当我们重新思考我们做过的事情时,我们常常发现我们不应该一时冲动、鲁莽行事。但是我们有意识的思维相对缓慢,无法在大多数互动场合做出理性决策。当你处于行动过程中时,你没有足够的时间来充分详细地考虑和理解一切。在我们信息不足、或缺乏对因果关系或行为后果的全面理解的情况下,我们的脑非常擅长"不加思索"地行动。而这正是脑的正常运作模式。理性和科学的推理是非常罕见的,也是一种人为的思维方式,你必须努力学习才行,就好像是在学弹钢琴。

当我在卡尔斯鲁厄大学做博士后时,我就理性原则问题研究了多年。当时理性原则是许多有关人类行为和决策理论的基石,但在我看来却完全不是这么回事。所以1980年我就挑选这个问题做我授课资格论文答辩的主题。论文标题是"*Zur Frage der Angemessenheit des Rationalitätskalküls in den Handlungs- und Entscheidungstheorien*"("关于理性演算在行动和决策理论中的适用性问题")。我以年轻科学家的典型激情和热忱进行攻击,彻底摧毁了把理性作为人类行为和决策的核心原则的想法。[1]

然而,当时我没有注意到,在以色列有两位聪明的科学家特韦尔斯基(Amos Tversky)和卡纳曼(Daniel Kahneman)早已认识到这种范式的弊端,于是致力于研究另一种更为实际的理论,后来卡纳曼因此而在2002年获得诺贝尔经济学奖。与此同时,卡纳曼把人类思维分成两个层面(系统1:快速、本能和情绪性;系统2:更慢,更审慎,更合乎逻辑),这成为经济决策理论中的新范式。在他的畅销书《快想和慢想》[2](*Thinking，Fast and Slow*)(2011年)中,他

---

[1] Schlagenhauf, Karl (1984). Zur Frage der Angemessenheit des Rationalitätskalküls in den Handlungs-und Entscheidungstheorien, in: Lenk, Hans (Hrsg.) (1984) Handlungstheorien-interdisziplinär Bd. III, Zweiter Halbband, Wilhelm Fink Verlag, München, 680 - 695.

[2] 有中译本:丹尼尔·卡纳曼.思考,快与慢[M].胡晓姣,李爱民,何梦莹,译.北京:中信出版社,2012.——译注

我要指出的是,我们的知觉和意向都不是基于逻辑推理或前后一致性的,而且我们往往预见不到我们行动的后果。我们头脑中的理性监控实际上是一个编讲故事的过程,它让我们相信我们按理性行事,并使我们有了理由相信我们行事一贯的幻觉。

虽然科学家们受到的训练就是去发现逻辑缺陷和前后矛盾之处,但这些事情在日常生活中却并不重要。当你注意听一些在别的地方都表现得非常聪明的人的讲话时,你可能在几秒钟内就发现他们的话中会有从逻辑上来说自相矛盾之处。但这似乎根本就算不上什么问题,他们自己和听众都毫不在意。只要故事讲得生动并符合听众之所想,逻辑缺陷就会像噪声一样被过滤掉。这有一条很好的理由。逻辑规则只适用于我们每天所说的各种各样话中的一小部分。基本上它们只能用于适用命题演算的公理系统,在这种系统里可以证明一个命题为真或为假,就像数学一样。正如哥德尔(Kurt Gödel)告诉过我们的那样,即使在这样的系统里要适用逻辑规则也还有一定的限制。然而,我们大多数的陈述和论点属于一类不能用"对""错"或"自相矛盾"这样的话来加以判定的情形,因为它们包含主观偏好。在这种情况下,即使是最基本的形式逻辑规则,如传递性也不再适用。

你会发现有人在巧克力冰激凌和香草冰激凌中会挑选前者,在香草冰激凌和草莓冰激凌中又会挑选香草冰激凌。但是当你让他们在草莓冰激凌和巧克力冰激凌之间挑选时,他们却可能会选择草莓冰激凌。关于这一行为是否愚蠢的争论和关于蓝色是否比红色更美丽的争论一样不会有任何结果。

而当问题涉及处理与义务和允许相关的规范性陈述(normative statements)时,情况就更糟了。多年来,哲学家和逻辑学家试图建立一种叫做"道义逻辑"(deontic logic)的规则,这可能有助于我们基于逻辑基础之上,根据一些高级规则对规范性陈述进行逻辑运算,从而得以告诉其他人什么是允许的和什么是不允许的。但是这一切都是徒劳的,对它的探索就像试图发明一种永动机或寻找上帝存在的证据一样,没有任何结果。尽管如此,人们却一再尝试。形式逻辑仅适用于我们所面对的所有陈述和论据中非常小的一部分。我们的脑知道这一点,因此对这种罕见事件几乎不在意。行为准则(normative orientation)并不猫"[1]（Schrödinger's cat）或是"中文屋"时，他是否确实掌握了他所说的话的精神实质也很难说。这些确实是非常复杂的事情，不容易理解。因此，人们如何不假思索地使用这些术语令人感到惊奇。量子跃迁几乎成了家喻户晓的词汇，当记者想表达一个巨大的革命性进步时，他们经常使用这个词。看来并不是所有人都认识到，量子跃迁是物理世界中可能的最小迁移。所以实际上这与他们想告诉我们的情况正好相反。

但是我们的脑并不因此而不知所措，因为它总能设法"嗅到"说话人的意思。有时它也可能把本来有意思的事当成了胡说八道，但这对生活并不造成多大问题并且可以应付自如。当你试图用某个变量除以另一个有零值的其他变量时，数字计算机就会崩溃。程序员必须非常小心，以避免有前后矛盾和逻辑错误，因为冯·诺伊曼机器不允许有错误。我们的脑并非如此，它对犯错是非常宽容的。事实上，它会轻松容易地跳过一处处错误，并且只是猜测了事。我们的知觉模糊不清，而我们的心智根据这些模糊的知觉却得出相当可靠的解释，这简直难以置信。[2] 但同时正是这一点使我们得以可靠而前后一致地行事。与冯·诺伊曼机器不同的是，我们的心智面对缺失和基础不牢都没有问题，并且一直生活在前后矛盾和使人困惑的输入之中，或者至少自以为可以做到这一点。冯·诺伊曼早已完全认识到了这一点，可能比他的许多后继者都要认识得更为清楚。

---

[1] 这是薛定谔提出的一个思想实验，他自己对这个实验是这样描述的："把一只猫关在一个密闭的铁质容器里面，并且装置以下仪器（注意必须确保这套仪器不被容器中的猫直接干扰）：在一台盖革计数器内置入极少量放射性物质，在一小时内，这些放射性物质至少有一个原子衰变的概率为50%，而没有任何原子衰变的概率也同样为50%；假若衰变事件发生了，则盖革计数管会放电，通过继电器启动一个锤子，锤子会打破装有氰化物的烧瓶。经过一小时以后，假若没有发生衰变事件，则猫仍旧存活；否则发生衰变，这套机构被触发，氰化物挥发，导致猫随即死亡。用以描述整个事件的波函数竟然表达出了活猫与死猫各半纠合在一起的状态。"——译注

[2] 这里可以举一个例子。下文是这段话的充满打印错误的英文原文：

Not so for our brian, whcih is very forrgiving when it comes to mistkes. Actualy, it hops from misttake to mistake with ease and just gusses arround. It is the vagueness and fuziness of our percerption and the relativly solid interprtation which our mind dirrives from it which is so treachrous. But at the same time this is what allows us to act in a very robbust and persitent way.

但是只要读者有相当的英语水平，就能很容易地读懂上面这段文字，而不理会其中的种种错误。——译注

数字是基本的"线",复杂的"线"如认识到每个圆的圆周长与其直径之间有某种特定的关系,很久以前有人把这种关系称为 π。有时候我们知道这种"线"的发明者,并以其名字命名以示尊重。但通常我们并不知道那些非常有用的"线"的发明者是谁。所以我们既不知道勺子、梳子、轮子或船的发明者,我们也不知道是谁发明了冶金、酿酒或造纸[1]。一个人也许可以发明这一切,但人的生命不够长。这就像一个好的程序员可以编制库中他使用的所有部分一样,但不可能什么都自己编。

如果从这个角度来看成人的脑,那么它的大部分内容并不是由其所有者产生的,而是从其他人那里继承或借用来的。不过事情在现实生活中可能还要更复杂一点,但是我希望这些例子已经清楚地说明了不可能只通过分析单个的脑就认识网络系统。

但这是一个全新的领域,在今天的信中,我还是回过头来再谈谈你有关两个人的主观感受的相似性以及翻译所遇到的困难上的见解,这些想法在我眼中都是正确的。我们可能在主观体验特性是否可以归结到产生它的物理、化学和生物学机制的问题上有不同见解,但我完全同意你所说的把机器也说成有意识或精神活动的问题,以及在这种情况下常常不经意地使用不合适的拟人化术语的问题。把自然语言处理称为"自然语言理解",并且说机器人"思考"和"感知"不只是语言草率的问题,而是一种不可接受的错误。

而且,我也同意你说的在成功地翻译之前首先需要理解内容。我并不像你那样是这方面的专家,但常常会有这样的印象:报道科学发现或技术成就的译者和记者并不完全了解他们正在谈的问题。

但我也相信我们所有人在生活中经常会遇到类似的情况。你所说的"理解"问题并不仅仅限于机器人或人工智能系统,它也是人际交流中的一个问题。例如,很难判断你的邻居所看到的蓝色是否与你看到的完全一样。但是,当一位作家谈到"量子跃迁"[2](quantum leaps)"薛定谔的

---

[1] 当然,在中国历史上是蔡伦发明了纸,但是西方人并不知道。——译注

[2] 在中文中"量子跃迁"一词并没有作者在这里所讲的这种转义,但是在英语里,"quantum leap"却可以转义成"跃进""巨大突破"的意思。例如 The scale of migration took a quantum leap in the early 1970s.(20 世纪 70 年代初,移民的规模骤然扩大。)——译注

Ⅲ-001　卡尔　　9

并且受到了实质性的、有道理的批评。这就是为什么我不喜欢祖母（和祖父）暗地里受到这样一种错误假设的嘲笑，并且认为他们应该受到更合理的理论的尊重。因此，我提出了一种祖父母在人类进化过程中所起作用的理论。你是第一个听到它的人！许多关于进化的生物学教科书都是以基因为中心的，一切都是关于 DNA 和突变。这种狭隘的观点认为在你将一半的 DNA 传给你的后代后，你已经完成了你的生物责任，仅此而已。虽然这可能对果蝇和其他不关心其后代的物种来说是对的，但对人类来说却并非如此。

DNA 不能把本体所获得的经验和物种的集体知识传给后代。所有的处世之道都必须一次又一次地从头开始被"安装"到一个年轻的脑中。虽然父母必须承担起养活、战斗、保护和谋生的责任，但祖父母在系统化、汇总和筛选知识以及教授相关知识和技能方面做了大部分工作。

一个如此依赖学习的物种需要超过一代的长辈才能恰当地完成这项工作，这就是为什么当人类群体中有成员的寿命能超过一个繁殖周期并帮助完成这项工作时，它为进化提供了优势。在家族中拥有经验丰富的智慧老人为团体的社会凝聚力做出贡献，也提供了巨大的选择优势。这就是为什么祖父母如此重要！我奇怪我为什么花了那么长的时间才认识到这样一件明显的事。

我不确定我是否是第一个有此想法的人。但我确信，如果以前有人也有过同样的想法，那么这个人肯定是位祖父母。☺

如果你愿意的话，我想更详细一点地讨论这个和社会有关的问题。我们前些时候在谈及模因和规章时，也曾经简短地谈到过这些问题。我的印象是，大多数神经生物学家和心理学家过于关注单个脑。他们倾向于忽视人脑只有作为脑网络的一部分才能正常工作的事实，而且事先需要有大量的知识才行。

我们都有这样的感觉，即我们的知识和技能是我们自己不可分割的一部分，我们主观地拥有它们。这种观点有其合理的一面，但并非完全如此。

我们体验到的我们自己（personality）实际上是由许多"线"编织而成的织物。其中有一些"线"是我们自己纺成的，比如行走、游泳或骑自行车的能力。但其他"线"则是其他人纺成的，并由我们的父母、朋友、老师或书籍作者教给我们。我们使用它们就像程序员调用库中的现成函数一样。语言中包括一大堆这样的"线"。字母和

Ⅲ-001　卡尔

（社会的脑；翻译和理解；计算机和脑是本质上不同的两种系统；逻辑规则只适用于我们每天所说的各种各样话中的一小部分，基本上只能用于适用命题演算的公理系统；理性和情绪）

2015－10－07

亲爱的凡及：

感谢你的来信，和你对我升级为外祖父的祝贺。这么说吧，我不得不承认，我以前并不特别渴望成为外祖父。但在阿迪和我身上发生了在所有祖父母身上都会发生的事。亚历克斯刚一出生，我们就立刻爱上了这个男孩，并且非常喜欢他。但是，这种经历改变了我的观点，因为这让我对我的新角色进行思考，以及思考从一般意义上来说的祖父母的作用问题。

当然，你很熟悉"祖母细胞"的概念，"祖母细胞"概念假设认为像祖母的脸这样的事物可以被大脑中的单个神经元所识别。[1] 这个理论存在很多缺陷，

---

[1] "祖母细胞"这一术语来自美国神经科学家莱特文（Jarome Lettvin）根据一本小说里的人物波特诺伊（Alexander Portnoy）编的一个故事：波特诺伊的妈妈非常专横，这使他一想到他的母亲就非常痛苦。这时正好有一位神经外科医生阿卡希维奇（Akakhi Akakhievitch）在脑中发现了有 18 000 个神经元只对自己的母亲起反应。波特诺伊于是就请医生把他脑中所有这些细胞都一一清除掉。手术以后，医生为了检验效果就问了他两个问题：
① "你还记得你每周四晚上都爱吃的薄饼吗？"
"当然记得，太好吃了。"
② "那么是谁给你做这种饼的呢？"
波特诺伊茫然地看着医生不知所对。
莱特文把这个故事讲给学生听以后，还煞有介事地告诉他们，阿卡希维奇医生接下来正在研究祖母细胞呢。他的故事大获成功，不胫而走成了科学家们热议的话题，而"祖母细胞"也就成了一个术语。——译注

样的问题,并考虑得更为深入。王培批评说,以通过图灵测试作为智能的定义,太过以人为中心。他认为,通过图灵测试只是智能的充分条件,但不是必要的。他还认为,在人工智能的历史上,主流是让计算机做一些以前只有人脑才能做的事情,而不是让机器与人类无法区分。他说,即使图灵自己也明白,如果一台机器要想通过图灵测试,那么它就不得不学会如何假装愚蠢或撒谎,否则它总会因为某些原因而不像人类,例如它的计算速度太快。不幸的是,在大多数谈论图灵测试的论文中,很少提到这一点。

是的,我和 C.科赫的博士生侯晓迪一起翻译了他的书《意识探秘》[1]（*The Quest for Consciousness*）。当我们不明白某些句子的含义时,我们问了他 70 多个问题,他拨冗回答了我们所有的问题。他在为中译本所写的序中说道:

"翻译任何文字都是一件极耗心力的工作,它需要译者首先理解纸面上文字背后的含义,然后才能将其组织润色成另一种语言。在一份成功的译著里,你应该感觉不到有译者介身其中——原作者与读者就像在直接进行交流一样。"

这让我又想起了上面提到的问题,机器翻译如何能"理解纸面上文字背后的含义"。他和格林菲尔德（Susan Greenfield）几年前就意识究竟是全局性性质还是局域性性质有过一次辩论。非常有趣！我还发现了他领导的艾伦脑科学研究所以及今年刚刚通过第一阶段的人类连接组计划（HCP）的一些非常有趣的内容。但是,由于我的信已经太长了,不得不在稍后再来讨论这些问题。

最好的祝愿。

凡及

---

[1] 中译本:克里斯托夫·科赫.意识探秘:意识的神经生物学研究[M].顾凡及,侯晓迪,译.上海:上海科学技术出版社,2010.——译注

就像你一样：

"我感到惭愧，直到最近，甚至也在我们的通信过程中，不假思索地用了这种信息处理的观点，就好像它是理所当然的一样。但事实并非如此，我也和大多数人一样连想都没有想，只是在我们讨论的过程中，我才意识到这是一个非常具有误导性的想象出来的视角，这种观点是逻辑学家和数学家将其引入到他们自己对此不甚清楚的生物学问题中去的。"

当然，我知道《计算机与脑》这本书。事实上，有一本已经在我的书架上躺了半个多世纪，纸张都已经变黄和发脆了。但是，我很惭愧地告诉你，我从来都没有通读过此书。我于1966年6月购买了它，当时正是一个特殊年代，在后来的十年里，我都没有机会阅读它。那段时间之后，我忙于其他事情，也没有想到要读它。看完你的信后，我把书从书架上小心地拿出来，浏览了一遍。是的，即使从一开始，冯·诺伊曼就发现了计算机和脑之间的巨大差异，如果他没有过早去世，他很可能会深入钻研这个问题。在他那个时代，把计算机称为"电脑"就已经在媒体上很流行了，今天在中国这种说法仍然很普遍。但是，他早已注意到脑并非按数字化方式工作，在脉冲串中插入或删除一个锋电位并不会太大改变它所携带的信息，但是，如果脉冲序列所携带的信息是用二进制来表示的话，那它们的值会完全不同！

我见过马尔斯伯格，曾在鲁尔大学他的办公室里拜访过他，但这也是我唯一一次和他的个人接触。然而，我也很羞愧地承认，当时我并不知道他对广为流行的脑信息处理范式持负面态度。我只知道他是首先提出同步振荡可能在解决"绑定问题"中起关键作用的人之一。也许你可以告诉我更多关于他的消息，特别是他关于脑的模拟特性而不是数字特性的思考。

至于说到图灵测试，尽管我对在最近一封电子邮件中所提出的问题已经考虑了很长一段时间了，但在读了王培博士的一篇论文[1]之后，发现他提出了同

---

[1] https://cis.temple.edu/~pwang/

顺便说一句，我刚浏览了一下库兹韦尔的书《如何创造心智》[1]（*How to Create a Mind*）。在最后一章"异议"中，他批评了塞尔的"中文屋"思想实验。他认为，尽管房间里的人不懂中文，但整个系统——屋里的人加上规则手册一起懂中文，否则系统如何能用中文正确地回答问题？因此，库兹韦尔认为，沃森明白他说的是什么。他的论点是错误的，是"中文屋"规则手册的编写人员懂中文，而不是规则手册本身或"中文屋"懂中文。

总之，我不能否认这样的可能性，即有一天我们可能有一台有意识的机器。无论如何，脑本身也是一种物理系统，它确实有它的内心世界，没有任何逻辑上的理由排除其他物理系统也会产生内心世界的可能性，如果这种系统足够复杂到满足涌现意识的充分必要条件。然而，由于我们对涌现这种内心世界的充分必要条件知之甚少，所以宣称人工智能的发展将创造超级智能来支配我们的星球还为时过早。这就像说万物都有结束的一天一样，地球也是如此。当然，这种预言理论上讲并没有错，那么我们现在是否就应该担心地球的毁灭，就像一些邪教信徒一样？总之，虽然我们应该知道存在这样的可能性或风险并且应该关注它，但我们不必太担心这种可能性，至少目前是这样。

我非常喜欢你关于"讨论"的讨论，这是绝对正确的。人们习惯于沿着惯常的思路去思考，每个习惯想法似乎都很自然，很少从另一种角度去怀疑和思考。在讨论过程中，对手可能会挑战这种习惯性思维，并促使人们从另一个角度来考虑。当然，前提是这个人应该有一个开放的头脑，就像你在上一封信中强调的那样。否则，就对应上了中国的一句谚语："对牛弹琴"。

习惯性思维非常固执，正如你在上一封信中分析的那样：

"有时候我们被误导了，并不是因为我们是某个特定学派的忠实信徒，而是因为我们像其他人一样不假思索地使用了某些术语、概念或观点……把脑看成为信息处理装置则是现代神经科学中的一个例子。"

――――――――――――

[1] 有中译本：雷·库兹韦尔.人工智能的未来[M].盛杨燕，译.杭州：浙江人民出版社.——译注

人类：我的男朋友让我来这里。

伊莉莎：是你的男朋友让你来这里的？

人类：他说我大部分时间都很郁闷。

伊莉莎：听到你很郁闷，我很抱歉。

人类：这是真的，我不开心。

伊莉莎：你认为来这里会帮助你不要不高兴吗？

魏岑鲍姆发现，即使患者完全意识到他们正在和一个简单的计算机程序对话，人们仍然会把它看作是一个真实的、能思考的、关心他们问题的人。这种现象现在被称为"伊莉莎效应"。魏岑鲍姆后来写道："我当时没有料到……和相当简单的计算机程序只谈了短短一会儿，就可能会引起正常人的强烈妄想。"今天，我们也可以在许多地方看到伊莉莎效应。[1] 因此，在我看来，主观性问题不仅对意识研究非常重要，而且对人工智能研究也很重要。拟人化对于向公众推广人工智能非常有用，但我们必须清楚，这也容易引起误解。

"奇点临近"，机器人有解放自己的意志并摧毁人类，类似的警示语广为流传。仿佛危险就近在眼前。然而，正如我们前面所讨论过的那样，即使对我们自身而言，我们仍然不明白这些内在属性是如何涌现出来的，或者如我所争辩的那样，它们涌现的充分必要条件是什么。现在我们知道人类意识涌现的一些必要条件，比如德阿纳的"印记"，或者更为普遍的一些必要条件，比如埃德尔曼和托诺尼的整合性与分化性，或托诺尼的整合信息。但是，我们仍然不知道充分必要条件。托诺尼和 C.科赫认为，非零的整合信息对于有意识是充分和必要的。但是，这将导致泛灵主义[2]（panpsychism）的结论。我们还没有任何机器拥有这样的内心世界。看起来开发这种机器［也就是塞尔所说"强人工智能（strong AI）"］还有很长的路要走，而且应不应该开发这样的机器也还是一个值得研究的问题。

---

[1] https：//en.wikipedia.org/wiki/ELIZA_effect

[2] 泛灵主义认为万物都有灵魂。——译注

的。但是,对于一台机器,即使它表现得像一个正常人,我仍然不确定它是否有意识。当然,如果一台机器做得如此精巧,没有人能够将它与人区分得清,那么人们会以为它也是一个人,因此认为它是有意识的。

然而,至少在现在或在可预见的将来,都没有这样的机器,并且也没有任何机器有任何心理活动,虽然人们已经使用了许多拟人化的术语来描述其行为。例如,有时人们称自然语言处理为"自然语言理解",尽管机器根本就不"理解"语言。正如我之前告诉过你的,由于翻译质量不好,我在读一些科学书籍的中译本时很恼火。我发现所有的错误都是因为译者不理解作者想要表达的意思。他们只是一字一句地进行直译,就像基于语法规则和字典的旧机器翻译一样,或是参照了错误的翻译。如果一个机器翻译系统学习这样的翻译样本,不管使用怎样高级的深度学习,翻译怎么可能会正确?我认为任何深度学习系统都不能"理解"真正的意义,它只能从大数据样本中学习。

这让我想起了一件事,如果我向许多用中式英语说话的朋友学习讲英语,那么我的口语必然也是中式英语!有时候,人们宣称他们开发了一种情感机器人,实际上它只有对应情绪的表情,但没有任何感受。人们也会用"思考""感知"等来描述机器人,这会导致混乱。记者喜欢用这种方式表达,这我可以理解,这种讲法能吸引读者的眼球。然而,一些科学家也喜欢用这种方式来表达,最终连他们自己也搞糊涂了。有时候,拟人化的影响是如此强大,人们会相信一个无意识的机器也有它自己的头脑。事实上,即使在人工智能的早期,麻省理工学院计算机科学家魏岑鲍姆(Joseph Weizenbaum)于1966年开发了一种聊天程序[1](chatterbot)伊莉莎(ELIZA)。伊莉莎模仿了一位罗杰斯式[2]心理治疗师,主要是将病人的话改写为问题:

---

[1] 许多人把它译为"聊天机器人",因为这个词的英文词和机器人的英文词有相同的词尾 bot。不过在美国工作的人工智能专家王培认为,最好把机器人这个词留给实体的硬件,此处译为聊天程序更好些。——译注
[2] 卡尔·罗杰斯(1902—1987),是美国心理学家,人本主义心理学的主要代表人物之一。从事心理咨询和治疗的实践与研究,并因"以当事人为中心"的心理治疗方法而驰名。1947年当选为美国心理学会主席,1956年获美国心理学会颁发的杰出科学贡献奖。——译注

Ⅲ-001　凡及

（"主观性"是意识研究和人工智能研究中都碰到
的最棘手的问题;翻译和机器翻译;拟人化和伊莉莎
效应;计算机和脑）

2015－08－24

亲爱的卡尔:

请接受我最衷心的祝贺! 祝贺你荣升为外祖父。☺也请祝贺阿迪、杰尔卡和于尔根。当然,也向亚历克斯[1]本人致以祝贺,如果他听得懂我的祝贺是什么意思的话。也感谢你允诺让我一同分享你对年轻大脑发育的新奇发现,这确实非常有趣!

我完全同意你的话:"虽然我们永远不会确切地知道其他人心中究竟发生了什么,但我的假设是,这种主观愉悦在你我心中必定非常相似。☺"我认为其原因是我们作为人,都有着相似的脑结构,而且我们都毕业于科学和技术专业,几乎在我们的一生中都在跨学科领域工作,所以我猜测同样的刺激会引起类似的脑活动。但是它们决不会完全一样,这是由于我们的基因组不同,连接组不同,经历不同,社会和文化环境也不同。

你的话让我又想起了"主观性"这个意识研究和人工智能研究中都碰到的最棘手的问题。我们已经讨论过很多关于意识主观性的问题:这种主观性如何从一个物理的脑中涌现出来的问题是否有一天可以用物理定律来解释,或者只是承认它是某种类型的人脑活动的不可还原的涌现特性。对这个问题,我们还不能达成共识。如果一个人的行为像正常人,我们相信他或她必定是有意识

---

[1] 全名为亚历山大·塞茨(Alexander Seitz),亚历克斯是昵称。

Ⅲ-010 卡尔 ... 142

事实胜于雄辩;中国人工智能的崛起;人工智能的未来

跋 ... 153

译后记 ... 168

致谢 ... 170

Ⅲ-005 卡尔 ...66

格伦的循环过程;社会规章和行为;智能的定义

Ⅲ-006 凡及 ...76

智能与学习能力;人类连接组计划完成第一阶段工作;HBP 进入正式实施阶段;各国脑计划

Ⅲ-006 卡尔 ...83

大科学;帕金森定律;智能和技能;莫拉韦茨悖论

Ⅲ-007 凡及 ...92

智能的定义;他人的心智;仿神经结构芯片

背景专栏 Ⅲ F7.1 ...97
米德和仿神经结构工程

Ⅲ-007 卡尔 ...99

深入元层次;科学写作文化;仿神经结构芯片;新工业革命

Ⅲ-008 凡及 ...107

机器翻译;AI 在可预见的未来都只是一种工具

Ⅲ-008 卡尔 ...112

机器翻译;人际交流;深度学习

Ⅲ-009 凡及 ...123

机器翻译;自由意志

Ⅲ-009 卡尔 ...127

自由意志;线性因果关系;循环因果关系;机器翻译;中国《新一代人工智能发展规划》

Ⅲ-010 凡及 ...137

阿尔法狗;新一代人工智能发展规划

Ⅲ-002　卡尔　...22

科学研究中的描述性部分和解释性部分;为了产生复杂的行为,并不一定要有什么不断在做计算的主体;脑网络和积累起来的智慧将能够认识单个的脑;再传入原则与闭环控制

背景专栏ⅢK2.1　...29
布赖滕贝格小车

背景专栏ⅢK2.2　...30
人类连接组计划

Ⅲ-003　凡及　...32

智能;自由意志;循环因果关系和线性因果关系;要实现与数字技术相同的功能,也可能有更简单和节能的方法

Ⅲ-003　卡尔　...38

阿尔法狗;人脸识别;自动翻译;人工智能需要的下一个突破点:降低能耗和小样本学习;语言学习;启动效应;技能不等于智能

Ⅲ-004　凡及　...45

深蓝国际象棋弈棋系统并没有智能;强人工智能和弱人工智能;通用人工智能和人工意识;不应该开发有自己心智和意志的人工智能机器人;人工智能的社会反响;需要大数据训练和高耗能是当前人工智能发展的瓶颈之一

背景专栏ⅢF4.1　...50
人工智能的三次崛起和两次严冬

Ⅲ-004　卡尔　...54

人工智能领域正在发生范式转换;从深蓝到深度学习;并行计算

Ⅲ-005　凡及　...63

"智能"不是"解决具体问题的能力",而是"获得解决具体问题的能力的能力";人工智能的分类;艾伦脑观察站;大科学

目录[1]

专家推荐 　... i

导读 1 　... iv

导读 2 　... vi

导读 3 　... viii

自序 　... ix

丛书内容概览 　... xiv

Ⅲ-001　凡及 　... 1

"主观性"是意识研究和人工智能研究中都碰到的最棘手的问题;翻译和机器翻译;拟人化和伊莉莎效应;计算机和脑

Ⅲ-001　卡尔 　... 7

社会的脑;翻译和理解;计算机和脑是本质上不同的两种系统;逻辑规则只适用于我们每天所说的各种各样话中的一小部分,基本上只能用于适用命题演算的公理系统;理性和情绪

Ⅲ-002　凡及 　... 17

解决"如何"的问题才是科学中的重要问题;个体脑和社会脑;机器翻译;计算机和脑;人类连接组计划;艾伦脑科学研究所

---

[1] 本目录给出了书信列表和专家点评信息,其中包含了每封信所讲到的主题和关键概念,并列举了一些与主题相关的专栏资料供读者参考。

计算机与脑(Ⅱ-010卡尔,Ⅲ-001凡及,Ⅲ-001卡尔,Ⅲ-002凡及)

人工智能的下一步(Ⅲ-003卡尔,Ⅲ-004凡及,Ⅲ-008凡及)

新工业革命(Ⅲ-007卡尔,Ⅲ-010卡尔)

中国《新一代人工智能发展规划》(Ⅲ-009卡尔,Ⅲ-010凡及,Ⅲ-010卡尔)

5. 意识之谜

困难问题和简单问题(Ⅱ-001凡及)

意识的神经相关机制(Ⅱ-007卡尔)

意识的神经全局工作空间假设(Ⅱ-008凡及)

主观性(Ⅱ-001凡及,Ⅱ-005凡及,Ⅱ-006凡及,Ⅱ-007凡及,Ⅱ-007卡尔,Ⅱ-008凡及,Ⅱ-008卡尔,Ⅲ-001凡及)

私密性(Ⅱ-001凡及,Ⅱ-002凡及,Ⅱ-005凡及,Ⅱ-007凡及,Ⅱ-008凡及,Ⅱ-009凡及,Ⅱ-009卡尔)

心智上传(Ⅱ-005凡及,Ⅱ-005卡尔,Ⅱ-006凡及,Ⅱ-006卡尔,Ⅱ-007凡及,Ⅱ-007卡尔,Ⅱ-008凡及,Ⅱ-009卡尔)

意识研究的前景(Ⅱ-005凡及,Ⅱ-005卡尔,Ⅱ-006凡及,Ⅱ-006卡尔,Ⅱ-007卡尔,Ⅱ-008凡及)

自由意志(Ⅲ-003凡及,Ⅲ-009凡及,Ⅲ-009卡尔)

6. 大科学计划

大科学(Ⅰ-005卡尔,Ⅰ-006凡及,Ⅲ-005凡及,Ⅲ-006凡及,Ⅲ-006卡尔)

欧盟人脑计划(Ⅰ-001凡及,Ⅰ-001卡尔,Ⅰ-005卡尔,Ⅰ-007卡尔,Ⅰ-010卡尔,Ⅰ-011凡及,011卡尔,Ⅱ-006凡及,Ⅱ-009卡尔,Ⅱ-010卡尔,Ⅲ-006凡及)

美国脑计划(Ⅰ-007卡尔,Ⅰ-008凡及,Ⅱ-006凡及,Ⅱ-010凡及,Ⅱ-010卡尔)

人类连接组计划(Ⅲ-002凡及,Ⅲ-006凡及)

艾伦脑科学研究所(Ⅲ-002凡及,Ⅲ-005凡及)

理论、模型和仿真（Ⅰ-001卡尔，Ⅰ-002凡及，Ⅰ-004卡尔，Ⅰ-006卡尔，Ⅰ-010凡及，Ⅰ-011凡及，Ⅱ-003卡尔，Ⅱ-008卡尔）

循环因果关系和线性因果关系（Ⅲ-002卡尔，Ⅲ-003凡及，Ⅲ-005卡尔，Ⅲ-009卡尔）

3. 脑研究中的一些未解问题

脑研究是万里长征（Ⅰ-001凡及，Ⅰ-005凡及，Ⅰ-005卡尔，Ⅰ-006凡及）

神经元（Ⅰ-002凡及，Ⅰ-009卡尔，Ⅰ-010凡及）

功能柱可能并非皮层的标准模块（Ⅰ-002凡及）

记忆（Ⅰ-007凡及，Ⅰ-008卡尔，Ⅰ-009凡及）

情绪（Ⅰ-007凡及，Ⅲ-001卡尔）

连接组（Ⅱ-009卡尔）

脑是一种信息处理系统还是创造意义的机器？（Ⅱ-009凡及）

社会的脑（Ⅲ-001卡尔，Ⅲ-002凡及，Ⅲ-002卡尔）

4. 人工智能中的争论问题

智能（Ⅰ-001卡尔，Ⅰ-002卡尔，Ⅱ-005凡及，Ⅲ-003凡及，Ⅲ-003卡尔，Ⅲ-004凡及，Ⅲ-005凡及，Ⅲ-005卡尔，Ⅲ-006凡及，Ⅲ-006卡尔，Ⅲ-007凡及）

"图灵测试"和"中文屋"思想实验（Ⅱ-005凡及，Ⅱ-010凡及，Ⅱ-010卡尔）

神经网络和深度学习（Ⅱ-001卡尔，Ⅲ-004卡尔，Ⅲ-008卡尔）

机器翻译（Ⅲ-001凡及，Ⅲ-001卡尔，Ⅲ-002凡及，Ⅲ-003卡尔，Ⅲ-008凡及，Ⅲ-008卡尔，Ⅲ-009凡及）

阿尔法狗（Ⅲ-003卡尔，Ⅲ-010凡及）

人工智能领域正在发生范式转换（Ⅲ-004卡尔）

奇点（Ⅱ-002卡尔，Ⅱ-003凡及，Ⅱ-004凡及）

摩尔定律（Ⅱ-004凡及，Ⅱ-004卡尔，Ⅱ-005卡尔）

脑启发计算和脑样计算（Ⅰ-010凡及）

仿神经结构系统（Ⅰ-010卡尔，Ⅰ-011凡及，Ⅲ-007凡及，Ⅲ-007卡尔）

逆向工程（Ⅱ-003卡尔，Ⅱ-004卡尔）

弱人工智能和强人工智能（Ⅲ-004凡及）

丛书内容概览

XV

丛书内容概览[1]

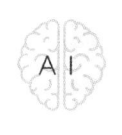

本套丛书共包含31对书信,讨论了有关脑和人工智能的一系列开放性问题。这些讨论和争辩贯穿整套丛书,但是在各个分册中侧重面又有所区别。其中第一册《脑研究的新大陆》包含11对书信(编号Ⅰ001－011),重点是讨论脑研究中的开放性问题;第二册《意识之谜和心智上传的迷思》包含10对书信(编号Ⅱ001－010),重点是讨论对意识研究的不同见解和心智上传的可能性问题;第三册《人工智能的第三个春天》包含10对书信(Ⅲ001－010),重点讨论人工智能的潜能和前景。在所有三册中也讨论到科学方法论和科学组织的问题。三册图书的书信内容是按照时间进行排序的。为了便于读者从整体上了解整套丛书的内容,我们提供了如下列表:

1. 引言(Ⅰ－001凡及)

2. 科学方法论

兴趣派与规矩派(Ⅰ－002卡尔,Ⅰ－003凡及,Ⅰ－003卡尔,Ⅰ－008卡尔,Ⅱ－003凡及)

自然和工程采用不同的方法(Ⅰ－002凡及,Ⅰ－003凡及,Ⅰ－005凡及,Ⅱ－004卡尔,Ⅱ－009卡尔)

不同学科的不同思想习惯(Ⅰ－003凡及)

科学家之间的竞争和合作(Ⅰ－003卡尔)

学术争论(Ⅰ－004凡及,Ⅰ－004卡尔,Ⅰ－005凡及,Ⅰ－006凡及,Ⅱ－002卡尔)

---

[1] 这是一个按信件内容进行分类的目录,也可以说是索引。括号内的数字表示信件的编号,表示在此信件内有这方面的内容,但并非只有这方面的内容。

对于那些对我们的推理总结以及我们所得到的结论感兴趣的人，我们增加了一个比通常要长得多的跋，总结了我们最重要的发现。对于好奇的读者，就像在读侦探小说时一样，急不可耐地想及早知道凶手是谁，那么可以先阅读这个跋。

当然，我们并不声称已对所有问题都给出了答案。事实上，我们的探索甚至还没有完成，许多问题仍然没有得到回答，而新的问题又产生出来了。我们也并不声称我们比其他人懂得更多。在某些部分中我们表现出来的自信和坚定的语气不应该被误解为对自己立场的绝对把握。这只是老式辩论文化中惯用的方法。这样做只是要把某种立场尽可能清楚地表达出来，这并非是为了捍卫这种立场，而是为了请对方对它进行反驳。

我们的许多假设、结论和评论可能都是错误的或不完整的，需要尽快予以纠正。问题在于我们不知道哪个是错的。无论如何，在我们经常大肆批评别人之后，我们也准备好接受读者的批评。

如果我们的看法最后被证明是错了，我们可能并不会因此感到高兴，但如果我们想要取得进展，这是不可避免的。我们都认为，在不断通过实证研究挑战理论的过程中，理性思考是增加我们知识的最佳方式。但是理性思考本身并不能代替在现实世界中的实践活动，而在这个过程中，理性思考也不能代替研究人员和工程师的好奇心、勇气和雄心壮志，尤其对年轻一代来说更是如此。

对于一些人来说，看到我们对自己心智之谜所知之少，以及我们在认识心智问题上进步速度之慢，可能会感到失望。而引人注目的是，在该领域的技术方面，则进展要快得多。还有些人可能会把这当成是进入一个非常有前途的工作领域的机会。在这个领域中，对于那些准备打破传统的人来说，可望收获令人难以置信的有价值的发现。

我们都确信我们并没有做出任何重要的发现。但是我们都希望能激发某些人才来试试运气，并找到魔法城堡的新入口。并非所有人都会成功，但我们希望许多读者能够发现我们的见解是有帮助的，并且会像我们在过去6年里所做的那样，享受在这魔法城堡中的漫步。

顾凡及，卡尔·施拉根霍夫

自序

美国先进创新神经技术脑研究（Brain Research through Advancing Innovative Neurotechnologies，简称"BRAIN"）倡议的启动，阿尔法狗（AlphaGo）击败前世界围棋冠军和自动驾驶汽车上路等。我们跟踪了这些令人印象深刻的事件，并讨论了如何评价它们的重要性。一些新进展支持了我们的推测，并鼓励我们进一步讨论。一些进展甚至超出了我们最好的期望，我们不得不重新考虑我们的观点并从错误中吸取教训。所有这些都激发了我们的讨论，重新聚焦要讨论的问题，引发新的争论，有时这会使我们改变想法。在某些问题上，我们达成了共识，对于另一些问题我们仍然存在分歧，还有某些问题我们从未找到过任何答案。

我们并不指望所有读者都会阅读所有的信件。一些人可能是神经生理学方面的专家，他们只想知道可以指望从"仿神经结构"芯片中得到些什么，而另一些读者则可能是熟练的人工神经网络编程人员，他们希望更好地理解把生物神经元和人工神经元相提并论有什么问题。还有些人可能对脑的生理细节或计算机架构都不太感兴趣，而是对一些我们两人都仔细考虑过的问题感兴趣，这些问题包括意识的涌现、自由意志、模因的意义、自组织、循环因果关系或是在生命科学和工程学科中的研究组织问题。我们曾请关心科学组织的一些朋友审阅过本书的草稿，他们觉得我们有关官僚主义大科学弊病的讨论对他们很重要。还有些人可能会忽略技术细节，并喜欢像看旅行纪录片那样，观看两位作者在生物学和计算机技术这一困难的交叉领域中摸索前行的故事。

我们提供了两种不同的目录，那些只对特定问题感兴趣的读者，可以由此找到包含这些主题的信件。

把我们之间的通信出版成书是后来才想到的。开始想到的读者群是那些对科学和技术感兴趣的雄心勃勃的外行人，他们希望更好地了解在这两个热门的领域中的事实和迷思。但是当我们的讨论在一些地方深入到许多细节之后，如果没有适当的背景知识就很难理解，本书的中文部分用专栏、脚注和插图进行解释，给予读者辅助的材料，从而知道我们所讨论的问题的有关基础知识。增加这些材料需要花费大量时间，但丛书内容也因此而丰富起来。为了方便阅读，我们把本丛书设计为三卷。

清醒分析。

这是我们在这次热潮开始之前就已经在尝试做的事情，如果也有人想了解这些领域中正在发生的事，并将事实、流行观念、现实希望、梦想和营销噱头区别开来，那么我们愿意和他们分享我们的见解。

你现在读的既不是一本科学教科书，也不是典型的科普书，更不是对这两个学科的系统或完整的介绍。我们所做的更像是随意漫游，从一个领域转悠到另一个领域，随着我们的意愿不时停下来深入探究。我们只是受到好奇心的驱使，当我们想要更准确地理解事物或者当我们觉得需要填补我们的知识空白时，我们就会加倍努力。通常，我们喜欢对知识追根究底，也包括我们不同文化的历史回顾。但是，尽管我们的探索看似无序，我们觉得，通过我们持续的、有时甚至是有争议的辩论，我们得到了如果选用了更系统的方法得不到的见解。

我们都喜欢从孩子的视角来看问题，他们会提出简单的问题，以了解真相。有时孩子可以看到皇帝的新衣并不像所说的那样华丽。但是我们也不想过于夸大，因为说我们就是著名童话故事《皇帝的新衣》中那个勇敢说出看不到别人"看到"的东西的孩子，就未免太自以为是了。

然而就 HBP 而言，卡尔坚持认为，从很早开始，当其他一些人还在赞扬它的时候，凡及就认识到这个令人印象深刻的计划存在缺陷。

我们在早期的信件中花费了大量的精力来说明并使自己确信在 HBP 的概念中有多处错误，我们不应该对此计划期待过高。由此开始了我们的通信，它成为探索脑和心智及其与人工智能和计算机技术的可能联系的许多基本方面的良好试验田。

今天，在这个项目的名声在公众面前已严重受损之后，这种批评很常见，而我们过去的批评在一些人看来似乎有点像在打"落水狗"。也许现在一般性的批评甚至过多了，因为在我们看来，HBP 概念中也确实有一些有趣的部分值得再作尝试。

除了讨论有关脑和人工智能的各种迷思之外，我们还讨论了理性思维和意识问题。在此期间，在脑科学和人工智能研究中都发生了若干重要事件，例如

自序　　　　　　　　　　　　　　　　　　　　　xi

其次,凡及有许多对此深感兴趣和挑剔的读者,卡尔有很多人(年轻的科学家,工程师,企业家以及工业和政府部门的管理成员)在这个问题上征求他的意见。所有这些人都有理由要求我们所说既非信口道来也不肤浅。

我们的讨论是从一个问题开始的。凡及在考察了后来名满天下的欧盟人脑计划(Human Brain Project,简称"HBP")的技术概念之后,向汉斯·布劳恩问了一个和特定类型神经元有关的问题。汉斯把这个问题转给卡尔,我们小小的旅程就从2013年1月正式开始了。这样就有了一系列电子邮件,我们的讨论从神经元开始,延伸到人工智能的最新发展以及某些人所谓的中美之间的技术和贸易战。

本书就是将我们的通信经过重新组织以后的结集。其中的信件都是按照昔日的辩论文化传统写成的,按照这种传统,科学家们在精心思考的信件中交换看法并进行有争议的讨论。当然这并非我们的发明。事实上,这是一百年前科学的黄金时代科学家们进行交流和完善他们的想法的常用方式。在推特和短信服务大行其道的当今,这看起来有些过时,现在所有内容都必须以标题表达,几秒钟内即可读完,讲得快也忘得快。

对于更习惯于达成共识的年轻科学家来说,我们信件中的对抗性语气可能读起来有些奇怪。然而,应该提到的是,对抗方法是目前在最先进的人工神经网络应用中引入的一种非常有前途的技术。以老式的对抗方式进行交流可能非常耗时且要求很高,但对于那些喜欢深入探究以便彻底了解真相的人来说,它也可以非常高效和有益。

如今人们已不太习惯写长信了,但信件比普通出版物有一个很大的优势。它们不那么正式,为创造性甚至猜测留下了更多空间;它们使说话的人更容易改变立场,从而向对方学习。你还可以用更平易的语气提出更为尖锐的问题,并直抒某个想法。我们发现这种方法对于我们感兴趣的、内容迅速变动的领域非常有用,在这里没有什么东西已有定论,而且还流传着种种迷思和概念滥用。

特别是在中国宣布将在2030年成为世界主要人工智能创新中心的雄心之后,许多人对我们已经关注了很长一段时间的那些问题感兴趣起来了。

为了清楚起见,需要对相关的科技现状以及有可能实现的前景和极限进行

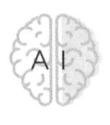

自序

本书即使不说是绝无仅有，也是很独特的。两位作者是在不同的文化氛围中成长起来的，从未谋面。卡尔是一名德国的工程师和企业家，而凡及则是一名中国的脑科学家和科普作家。我们是在6年前通过一位共同的朋友，神经生理学家汉斯·布劳恩（Hans Braun）教授的介绍结识的，此后一直就脑研究和人工智能（Artificial Intelligence，简称"AI"）方面的问题进行通信。

我们成了好朋友，通信频繁，甚至超过了与一些多年老友的通信。我们对脑、心智、意识和人工智能之谜的共同兴趣维系着我们的友谊，我们以极大的热情共同关注着这些领域中的迅速发展。

我们都喜欢理性思考的方法，而且我们总是渴望追究事物的原因和理由，而不是随大流或囿于学究式的思维。由于我们经历的不同，我们的观点也有明显差异，卡尔在产业界工作，而凡及则在学术界工作。凡及的工作主要是创造知识和传授知识，而卡尔则致力于如何通过技术应用来利用知识。

在跨学科领域和多种技术行业中工作几十年后，我们都到了法定的退休年龄。然而，科学家和企业家是永不"退休"的，因此我们以更大的热情去利用自己的时间和经验。我们享受由此得到的自由，我们不用再为前程操心，也不用考虑要给同行留下深刻印象。不再受到这些约束而只凭自己的兴趣行事真是妙不可言，我们都非常享受这一点。

就像卡尔经常说的那样：我们就像两只自由而快乐的鸟儿，可以待在喜欢的任何一棵树上，讨论感兴趣的东西。但这并不意味着我们就漫无目标或要求不高。

首先，我们总是要求自己尽可能好地了解我们感兴趣的复杂领域中所发生的事情，并评估它们将如何发展。

导读 3

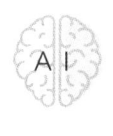

　　在 2018—2019 岁末年初的假日期间，我正巧同时读了阿加莎·克里斯蒂的犯罪小说《尼罗河上的惨案》和卡尔·施拉根霍夫与顾凡及《脑与人工智能》的通信手稿。两者均以其丰富的内容和写作风格，同样让我感到兴奋。

　　科学书《脑与人工智能》的主题对社会既及时又重要！作者们以脑研究和人工智能的大量事实非常坦率地讨论了这些问题，这足以证明他们在这些领域的扎实知识。我虽然并非是这些领域的专家，但是，作为固态物理学家和材料科学家，我在前几年的研究活动中多次涉及这些主题。当时我研究过仿神经结构系统的材料和过程，并把机器学习应用于高分辨率层析成像和光谱数据处理的三维成像高级数据分析算法。我确信人工智能方法对于更好更快地获取信息非常有用，但是，正确的策略应该是将基于数学、自然科学和工程学的坚实知识与（大）数据驱动的信息学这样两个方面结合起来。

　　"往日科学家之间辩论文化"的风格，即通过信件相互尊重地交换论点，以逻辑方式（循因求果的理性思考，我们物理学家就是这样训练出来的）步步相扣，基于可靠信息之上发展知识，这对我来说都有耳目一新之感。我们不应该忘记这些技能，这一点在当今这个推特和短信的时代显得尤为重要。

　　我希望广大读者都会喜欢阅读这本书，从中受到启发，并像本书的写作风格那样，即既热情又慎重地参与讨论作者所提出的那些当今议题。

埃伦费里德·切希

教授，博士

德国弗劳恩霍夫陶瓷技术与系统研究所系主任

欧洲材料研究学会理事

直到几年以前还不以人工智能研究者自况，而现在却在对自己没有真正研究过的问题发表"权威意见"了。凡此种种，也就难怪学界在这里开始和商界、政界，甚至娱乐界难分彼此了。

我正是因为对目前人工智能界的一派"繁荣景象"并不乐观，所以才更觉得二位前辈垂范之可贵。具体观点都可以商榷，但如果不能以科学的态度讨论科学问题，目前的人工智能热只能以再次坠入冰河收场。我希望读者（尤其是有志于科学的青年）从这本书中不仅仅学到科学知识，而且学到探讨科学问题时的正确心态和姿态。我确信在近年出版的大量关于人脑和人工智能的书籍中，本书应该算是最深刻的之一。读者们一定可以从中受益，并被激发出更深入的思考与研究。

王培

美国天普大学副教授

通用人工智能学会副主席

《通用人工智能》杂志执行主编

导读 2

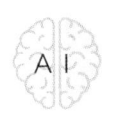

　　顾凡及和卡尔·施拉根霍夫二位前辈邀我为他们的对话体著作写些评论，我就恭敬不如从命了。

　　去年读到二位先生的来往邮件，令我有耳目一新之感。其中最触动我的是二位在讨论中所表现出的科学素养和君子之风。二位先生都已经是学有所成的人物，但仍然对科学怀有赤子之心，对相关领域的进展敏感、好奇，愿意以开放的心态去学习自己不熟悉的内容，同时又对没有充分证据支持的结论抱有清醒的警惕。二位在讨论中互相尊重，但又不因此回避思想交锋；既不轻易放弃自己的观点，也不觉得被对方说服是丢脸；希望对科学研究有所贡献，却并不以近期的名利为导向。这不正是进行科学讨论时所应该有的态度吗？

　　前面提到的这些品质可以说是既广为人知又难得一见，尤其是在和人工智能相关的讨论之中。近年来这个领域一下子从清锅冷灶变成烈火烹油，以至于各路人物都加入了这场大合唱。在众生喧哗之中，自说自话的多，偶有交流也往往不是互捧就是互贬，认真考虑对方意见后坦诚回应的反而少见。至于所表达的观点，也常常不是"跟着感觉走"就是"随着大流走"，唯缺在充分了解前人工作之后的独立思考。

　　这种情形自然和学术界的"大气候"直接相关，但种种怪相在有关人工智能的讨论中表现尤甚，也是由于这个领域的特殊性。一方面是因为思维之谜公认是科学研究中最复杂的问题之一，在这个领域中还没有经过历史考验的研究规范作为共识基础，而另一方面出于对思维现象的直接感受，人人似乎都有见解。尤其显著的现象是各路名人的"跨界发言"，好像聪明人自然会知道"聪明"是怎么回事。在人工智能界内部，由于"人工智能"观念的多面性和历史变迁，很多"人工智能专家"实际上只熟悉他们自己赖以成名的某个局部领域，现在产业界已经有了许多方法可以适应这些需求，我立刻就联想到了开源软件和硬件方法，我非常期待在本书的下一版中会大大增加这方面的内容！

作者们观察到"大量的研究还没有积累到（能创造出）一种理论，来帮助我们解释和认识心智的功能……"，以及过去五十年左右时间里在该领域一直缺乏重大突破，这些都表明我们迫切需要某种可靠的模型或方法。期望与实际应用之间的这种脱节为媒体上的许多夸大之词提供了空间，甚至将随意一个进展都描绘成了"终结者"式的飞跃。正如作者们所指出的那样，现在有一个机会可以纠正这个问题：为了加速进展而在人脑计划（HBP）上的大量投入也许属于花销不当，因为我们对脑实际上是如何工作的还知之甚少，不足以对其进行仿真。当然，它也从侧面证明，通过大量的资本投入，我们确实有能力取得突破性的进展并获取大量的知识。

在欧洲，即使是不那么严格地按照自由市场或新自由主义方法提出解决方案，也会面临过于乐观或忽略消极面的指责风险。不过在这种情形下，我们也许可以通过加快技术多样化借鉴这一模型的成功之处，同时采用更加平衡的方法，从而减少一些对 AI 的过多、夸大的宣传，而加以消化吸收。

来自硅谷及其他全球技术中心的例子有很多，虽然其中既有正面的，也有负面的（如数据安全性及滥用、假新闻、回音室效应、不择手段的营销和做广告等）。不可否认的是，从信息通信技术促进发展的角度来看，信息技术对社会和政治都有影响，对许多社会大众也产生了积极的作用。现在，从技术在商业和公共领域中的实际应用和滥用情况来看，瓶中的"精灵"已被释放出来。把学术兴趣、商业利益和实际经验结合起来，尤其是在机器学习、专家系统和各种 AI 技术领域中加速创造和应用"智能系统"，也许可以为其中一些负面影响提供"更好"的解决方案。

我真诚地希望凡及和卡尔的书能够激发科学、政治和产业界之间新的合作方式，以创造下一波"真正的"通用人工智能，从而造福所有人。

拉斐尔·拉古纳
Open-Xchange 首席执行官
德国颠覆性创新开发署主任

导读 1

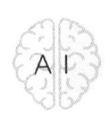

　　这是一场令人耳目一新的对话,一位是来自中国的脑科学家,另一位是多才多艺的德国工程师。如果把这场对话的标题改写成《人工智能,科学和万物》(*AI, Science and Everything*),也将会非常合适。顾凡及和卡尔·施拉根霍夫的话题涵盖了很多方面,并对于在 AI、科学、政策制定,产业界和政治领域工作的支持者和从业者,结合他们所面临的一些关键问题,提出了一些全新的、原创性的见解和想法。这些想法也是有抱负的年轻一代迫切需要知道的,他们想要了解的不只是这枚"冰球"在什么地方,更想知道球将打到哪里。

　　在回顾一些颇具争议性的讨论时,这本引人入胜的读物毫无保留地提出了自己的见解,是对那些炒作 AI 的大众传媒及相关舆论炮制者开出的一帖绝佳的清醒剂。

　　作者们带领我们进行了一次漫长的旅程:首先回顾了科学方法论的问题;随即通过讨论脑研究中的问题和变化,为探讨人工智能做好铺垫;在很快地对"意识"这一还不太清楚的概念进行辩论之后,最终以研讨棘手的"效率"问题收尾。在这里"效率"一词的意思是指怎样才能最好地加速学术界的进步,并创设一种良好的环境——不把资源浪费在错误的方向上,也不把资源浪费在范围过广而在哲学上误入歧途的点点滴滴的进步上,而是应该指导实际活动,从而使德国和欧洲在研究上取得能被有效地应用到产业环境中去的突破,并加强自身在美国和中国等国家都大力发展的人工智能领域上的竞争力。

　　尽管作者们并没有非常详细地讨论这个问题,但他们明确表示需要在价值链上注意将 AI 解决方案推向市场,而产业界和学术界之间的知识转移和合作是做到这一点的基础之一,并将有助于消除风险投资(VC)和媒体夸张宣传所造成的某些影响。

版的大量关于人脑和人工智能的图书中,本书应该算是最深刻的之一。

——王培(美国天普大学副教授,通用人工智能学会副主席,

《通用人工智能》杂志执行主编,

《哥德尔、艾舍尔、巴赫——集异璧之大成》译者之一)

一本扣人心弦犹如阿加莎·克里斯蒂侦探小说的科学书。

"传统的科学辩论文化"风格,即在书信中严肃地交换意见并基于坚实的基础推进认识,令人耳目一新。我们不应该忘记这些技能,这在一个推特和短信的时代更显得尤为重要。

——埃伦费里德·切希(德国弗劳恩霍夫陶瓷技术与系统研究所系主任,

教授,欧洲材料研究学会理事)

专家推荐 iii

中国历史悠久的智慧加上神经生理学的专业知识与西方社会科学和信息技术间的碰撞,"我们需要这样一本书吗?"我的回答绝对是肯定的。

——格特·豪斯克(德国慕尼黑理工大学退休教授,
《生物控制论》杂志前主编)

我真诚地希望由凡及和卡尔撰写的充满原创思考和见地的图书,能够激发科学、政治和产业界之间新的合作方式,以创造下一波"真正的"通用人工智能,从而造福所有人。

——拉斐尔·拉古纳(Open-Xchange 首席执行官,
德国颠覆性创新开发署主任)

作者以理性的视角和客观的思维,带着读者一同探讨智能时代背后的飞腾与迷思。

——梁培基(上海交通大学生物医学工程系教授,
中国神经科学学会理事)

无论你是对人工智能与大脑充满懵懂的好奇,或者是已经在某一个具体科学领域有所涉猎,又或是想给自己的业余生活增加些情趣的读者,相信这本真诚、有趣而深刻的对话录都能带给你丰富独特的体验。

——宋蔓(美国加利福尼亚大学圣迭戈分校博士生)

这是一本发人深思的书,所有对脑科学和人工智能感兴趣的读者都应读一读这本书。

——唐孝威(中国科学院院士,浙江大学教授)

一本如何进行科学讨论的范本,所有对脑科学、认知科学、人工智能、科学哲学等领域有兴趣的读者都会喜欢这种观点和思想的碰撞。我确信在近年出

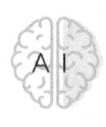

专家推荐

这是一本奇书。对人类洞察力的本质以及我们如何能在对此的认识上取得进展感到好奇的每个人,阅读此书都可以有所得。

——马修·贝特格(德国图宾根大学伯恩斯坦计算神经科学中心主任,
教授,deepart.io 联合创始人)

这是一本不同寻常的书,很明显地从许多其他神经科学出版物中脱颖而出。应该向所有希望挣脱传统的主流研究窠臼的 IT 专家和神经科学家推荐此书。

——汉斯·阿尔贝特·布劳恩(德国马尔堡菲利普大学生理学教授,
生理学与病理生理学研究所神经动力学研究组负责人)

一位德国工程师和一位中国生物物理学家之间的精彩对话,我们应该感谢他们写了一本有关人工智能、脑活动和意识的好书。

——陈宜张(中国科学院院士,第二军医大学教授)

我如痴似醉地读完了全书,激动得想在有一天会把书拍成电影。

——塞尔达尔·多甘(德国导演兼电影制作人)

我以前只有在很少一些情况下才享受过这样的读书乐趣。它是思想和见解的宝库,时而有趣,时而非常严肃,但总是非常乐观。

——迪特马尔·哈霍夫(德国马克斯·普朗克创新与竞争研究所所长)

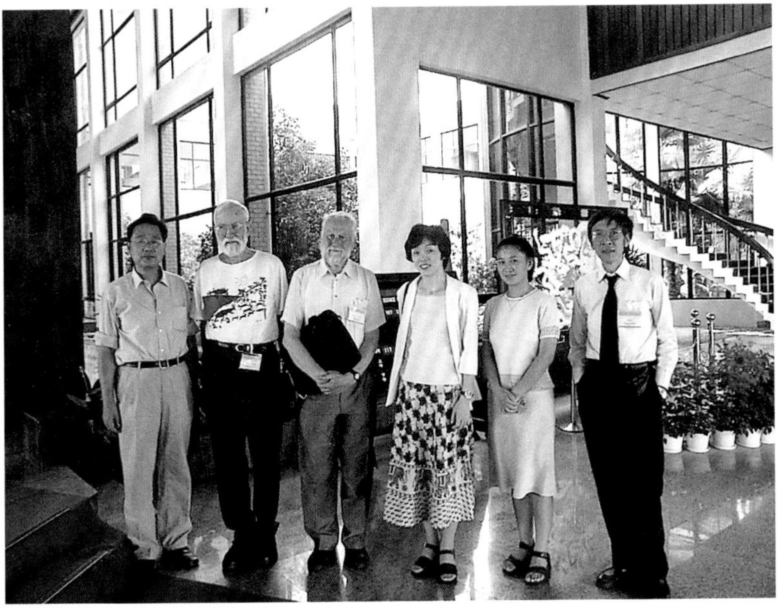

自左至右：童勤业、沃尔特·弗里曼、赫尔曼·哈肯、山口洋子、孟欣、顾凡及

自左至右：沃尔特·弗里曼（Walter J. Freeman）、刘岭、布劳恩（Hans A. Braun）、徐得名、范兴（Petra Mayer）、郝光曼、哈肯（Herman Haken）

The Authors

Fanji Gu

Prof. Fanji Gu is an emeritus professor of computational neuroscience at the School of Life Sciences, Fudan University, Shanghai, China.

He graduated in mathematics from Fudan University in 1961, worked in Department of Biophysics, Science & Technology University of China from 1961 to 1979, then worked in School of Life Sciences, Fudan University since 1979. He worked as a visiting research associate in the Department of Physiology & Biophysics, University of Illinois at Urbana-Champaign from 1983 to 1985.

His major is computational neuroscience. He published 3 monographs and about 100 papers. After his retirement in 2004, he was the managing editor of the journal *Cognitive Neurodynamics* from 2006 to 2011. He edited three proceedings of international conferences.

After that he became a popular science writer on brain science. He wrote 6 popular science books on brain science and translated seven books, including Walter Freeman's *Neurodynamics*, Christof Koch's *Quest for Consciousness*, Gerald Edelman's *A Universe of Consciousness* and V. S. Ramachandran's *Phantoms in the Brain*. His popular science books won seven awards, including the 2017 best seller in China, the 2016 Shanghai Municipal Award of Science and Technology (science popular book, third grade), the 2015 Shanghai Science Popularization Education Innovation Award (science popular book, second grade) etc. He himself also won a Merit Award conferred by the 2013 Fourth International Conference on Cognitive Neurodynamics held in Sweden, the 2017 Shanghai Science Popularization Education Innovation Award (individual contribution, second grade), and the 2018 Shanghai Award for Science & Technology (third grade).

He is now the honorary president of Shanghai Chapter of University of Illinois Alumni Association and a member of the editorial board of WeChat's subscription channel "Fanpu".

The Authors

Karl Schlagenhauf

Karl Schlagenhauf is a serial entrepreneur in the field of new technologies and has founded numerous start-ups in the high-tech sector in Europe and the USA. He was CEO of IFAO (Institute for applied organizational research) and ADI Software, a company with a focus on RDBMS, Multimedia and Internet-Applications in industry and banking, both of which he spun off from the University of Karlsruhe in the 1980s.

He stepped down as CEO in 2003, and is now chairman of the board of ADI Innovation and runs a Family Office. He also serves on the board of companies where he is invested and has served on the boards of several technology companies such as AP Automation + Productivity (now Asseco Solutions), Brandmaker, CAS Software, JPK Instruments and Web.de.

As an inventor he holds patents in the field of secure remote control via the internet and Nano-robots for protein analysis based on atomic force spectroscopy.

He is an advisor to private equity firms and governmental bodies and also a coach to young entrepreneurs.

While his focus has always been on leading edge technologies and their impact on social systems and the delicate orchestration of human, intellectual und financial resources, he has extended his interest from software, electronics and manufacturing to the life sciences and artificial intelligence.

He holds a Master's degree in Economics/Industrial Engineering, a PhD in Philosophy and a "Venia Legendi" for Sociology and Theory of Science from University of Karlsruhe (now KIT).